Dinesh C. S. Bisht and Mangey Ram (Eds.)
Computational Intelligence

De Gruyter Series on the Applications of Mathematics in Engineering and Information Sciences

Edited by
Mangey Ram

Volume 3

Computational Intelligence

Theoretical Advances and Advanced Applications

Edited by
Dinesh C. S. Bisht and Mangey Ram

DE GRUYTER

Editors
Dinesh C. S. Bisht
Department of Mathematics
Jaypee Institute of Information Technology
Noida
India
drbisht.math@gmail.com

Mangey Ram
Graphic Era Deemed to be University
Department of Mathematics; Computer Science and Engineering
566/6 Bell Road
248002 Clement Town, Dehradun
Uttarakhand
India
drmrswami@yahoo.com

ISBN 978-3-11-065524-7
e-ISBN (PDF) 978-3-11-067135-3
e-ISBN (EPUB) 978-3-11-066833-9
ISSN 2626-5427

Library of Congress Control Number: 2020936753

Bibliographic information published by the Deutsche Nationalbibliothek
The Deutsche Nationalbibliothek lists this publication in the Deutsche Nationalbibliografie; detailed bibliographic data are available on the Internet at http://dnb.dnb.de.

Cover image: MF3d/E+/Getty Images
Typesetting: VTeX UAB, Lithuania
Printing and binding: CPI books GmbH, Leck

MIX
Papier aus verantwortungsvollen Quellen
FSC www.fsc.org FSC® C083411

www.degruyter.com

Acknowledgments

The Editors acknowledge De Gruyter for this opportunity and professional support. Also, we would like to thank all the chapter authors and reviewers for their contribution.

https://doi.org/10.1515/9783110671353-201

Preface

In this modern era, we need instruments that can be applied to the problems arising in interdisciplinary fields. In this sense, computational intelligence (CI) is a developing computing approach for the forthcoming several years. CI gives the litheness to model the problem according to given requirements. It helps to find swift solutions to the problems arising in numerous disciplines. These methods mimic human behavior. The origins of CI are found in fuzzy logic, data analysis and intelligent systems. CI constitutes of following three main parts:

A. Fuzzy systems
B. Neurocomputing
C. Nature-inspired algorithms

Above stated techniques are widely used in the field of engineering, management and health. These techniques can be clubbed together to give better performances. Every technique has its own strength. Fuzzy logic has the capacity to cope up with imprecise data, linguistic terminology or vague form of the information. Neural networks can learn with examples and can produce extraordinary results. Nature-inspired algorithms are inspired by nature's way of adaptation. This book aims to cover recent advancements in the field of CI. It will cover the theoretical as well as the application part of the computational intelligence. The projected audience for this book will be scientists, researchers and postgraduate students who can use these techniques directly as black-box or can modify according to their research problems.

Dinesh C. S. Bisht
Mangey Ram

https://doi.org/10.1515/9783110671353-202

About the Editors

Dr. Dinesh C. S. Bisht received his Ph.D. with major in Mathematics and minor in Electronics and Communication Engineering from G. B. Pant University of Agriculture & Technology, Uttarakhand. Before joining Jaypee Institute of Information Technology he worked as an assistant professor at ITM University, Gurgaon, India. The major research interests of him include Soft Computing and Nature Inspired optimization. He has published many research papers in reputed national and international journals. He has published four book chapters in reputed book series. He has been awarded for outstanding contribution in reviewing from the editors of Applied Soft Computing journal, Elsevier.

Dr. Mangey Ram received the Ph.D. degree major in Mathematics and minor in Computer Science from G. B. Pant University of Agriculture and Technology, Pantnagar, India. He has been a Faculty Member for around 12 years and has taught several core courses in pure and applied mathematics at undergraduate, postgraduate and doctorate levels. He is currently a Professor at Graphic Era (Deemed to be University), Dehradun, India. Before joining the Graphic Era, he was a Deputy Manager (Probationary Officer) with Syndicate Bank for a short period. He is Editor-in-Chief of *International Journal of Mathematical, Engineering and Management Sciences* and the Guest Editor and Member of the editorial board of various journals. He is a regular Reviewer for international journals, including IEEE, Elsevier, Springer, Emerald, John Wiley, Taylor and Francis and many other publishers. He has published 200 plus research publications in IEEE, Taylor & Francis, Springer, Elsevier, Emerald, World Scientific and many other national and international journals of repute and also presented his works at national and international conferences. His fields of research are reliability theory and applied mathematics. Dr. Ram is a Senior Member of the IEEE, life member of Operational Research Society of India, Society for Reliability Engineering, Quality and Operations Management in India, Indian Society of Industrial and Applied Mathematics, member of International Association of Engineers in Hong Kong, and Emerald Literati Network in the U.K. He has been a member of the organizing committee of a number of international and national conferences, seminars and workshops. He has been conferred with "*Young Scientist Award*" by the Uttarakhand State Council for Science and Technology, Dehradun, in 2009. He has been awarded the "*Best Faculty Award*" in 2011; "*Research Excellence Award*" in 2015; and recently "*Outstanding Researcher Award*" in 2018 for his significant contribution in academics and research at Graphic Era Deemed to be University, Dehradun, India.

Contents

Acknowledgments —— V

Preface —— VII

About the Editors —— IX

List of contributing authors —— XIII

Part I: **Theoretical advances**

Aneesh Wunnava, Manoj Kumar Naik, Bibekananda Jena and Rutuparna Panda
1 Nature-inspired optimization algorithm and benchmark functions: a literature survey —— 3

Gunjan Goyal, Pankaj Kumar Srivastava and Dinesh C. S. Bisht
2 Genetic algorithm: a metaheuristic approach of optimization —— 27

Adam Price, Thomas Joyce and J. Michael Herrmann
3 Ant colony optimization and reinforcement learning —— 45

Tanmoy Som, Tuli Bakshi and Arindam Sinharay
4 Fuzzy informative evidence theory and application in the project selection problem —— 63

Mališa R. Žižović and Dragan Pamučar
5 The fuzzy model for determining criteria weights based on level construction —— 77

Divya Chhibber, Dinesh C. S. Bisht and Pankaj Kumar Srivastava
6 Fuzzy transportation problems and variants —— 91

B. Farhadinia
7 Hesitant fuzzy sets: distance, similarity and entropy measures —— 111

Manal Zettam, Jalal Laassiri and Nourddine Enneya
8 A prediction MapReduce-based TLBO-FLANN model for angiographic disease status value —— 119

Part II: **Advanced applications**

Tanmoy Som, Pankhuri Jain and Anoop Kumar Tiwari
9 Analysis of credit card fraud detection using fuzzy rough and intuitionistic fuzzy rough feature selection techniques — 135

Shruti Kaushik, Abhinav Choudhury, Nataraj Dasgupta, Sayee Natarajan, Larry A. Pickett and Varun Dutt
10 Evaluating single- and multi-headed neural architectures for time-series forecasting of healthcare expenditures — 159

Ankita Mazumder, Dwaipayan Sen and Chiranjib Bhattacharjee
11 Optimization of oily wastewater treatment process using a neuro-fuzzy adaptive model — 177

M. Mary Shanthi Rani and P. Shanmugavadivu
12 Deep learning based food image classification — 197

Vandana Khanna, B. K. Das and Dinesh Bisht
13 Particle swarm optimization and differential evolution algorithms: application to solar photovoltaic cells — 209

S. N. Kumar, A. Lenin Fred, Parasuraman Padmanabhan, Balazs Gulyas and Ajay H. Kumar
14 Multilevel thresholding using crow search optimization for medical images — 231

Index — 259

List of contributing authors

Aneesh Wunnava
Department of Electronics and Communication Engineering
Institute of Technical Education and Research ITER
Siksha O Anusandhan
Odisha, India
aneeshwunnava@gmail.com

Manoj Kumar Naik
Department of Electronics and Communication Engineering
Institute of Technical Education and Research ITER
Siksha O Anusandhan
Odisha, India
naik.manoj.kumar@gmail.com

Bibekananda Jena
Department of Electronics & Communication Engineering
Anil Neerukonda Institute of Technology and Sciences ANITS
Andhra Pradesh, India
bibekananda.jena@gmail.com

Rutuparna Panda
Department of Electronics & Telecommunication
VSS University of Technology VSSUT
Odisha, India
r_ppanda@yahoo.co.in

Gunjan Goyal
Department of Mathematics
Jaypee Institute of Information Technology
Noida, India
gunjangoyal.04@gmail.com

Pankaj Kumar Srivastava
Department of Mathematics
Jaypee Institute of Information Technology
Noida, India
panx.16j@gmail.com

Dinesh C.S. Bisht
Department of Mathematics
Jaypee Institute of Information Technology
Noida, India
drbisht.math@gmail.com

Adam Price
University of Edinburgh
Institute for Perception, Action and Behaviour
Informatics Forum
Edinburgh, United Kingdom
A.Price-7@sms.ed.ac.uk

Thomas Joyce
University of Edinburgh
Institute for Perception, Action and Behaviour
Informatics Forum
Edinburgh, United Kingdom
joyce@biomed.ee.ethz.ch

J. Michael Herrmann
University of Edinburgh
Institute for Perception, Action and Behaviour
Informatics Forum
Edinburgh, United Kingdom
michael.herrmann@ed.ac.uk

Tanmoy Som
Department of Mathematical Sciences
Indian Institute of Technology (BHU)
Varanasi, India
tsom.apm@iitbhu.ac.in

Tuli Bakshi
Department of Computer Science
Calcutta Institute of Technology
Howrah, India
tuli.bakshi@gmail.com

Arindam Sinharay
Department of Information Technology
Future Institute of Engineering and Management
Kolkata, India
arindam.sinharay@gmail.com

Mališa R. Žižović
Faculty of Technical Sciences in Čačak
University of Kragujevac
Kragujevac, Serbia
zizovic@gmail.com

Dragan Pamučar
Department of Logistics
Military Academy
University of Defence
Belgrade, Serbia
dpamucar@gmail.com

Divya Chhibber
Department of Mathematics
Jaypee Institute of Information Technology
Noida, India
divyachhibber@gmail.com

Bahram Farhadinia
Department of Mathematics
Quchan University of Technology
Quchan, Iran
bfarhadinia@qiet.ac.ir

Manal Zettam
Informatics, Systems and Optimization Laboratory
Department of Computer Science, Faculty of Science
Ibn Tofail University
Kenitra, Morocco
zettammanal@gmail.com

Jalal Laassiri
Informatics, Systems and Optimization Laboratory
Department of Computer Science, Faculty of Science
Ibn Tofail University
Kenitra, Morocco
laassiri@uit.ac.ma

Nourddine Enneya
Informatics, Systems and Optimization Laboratory
Department of Computer Science, Faculty of Science
Ibn Tofail University
Kenitra, Morocco
enneya.nourddine@uit.ac.ma

Pankhuri Jain
Department of Mathematical Sciences
Indian Institute of Technology
Banaras Hindu University
Varanasi, India
pankhurijain20@gmail.com

Anoop Kumar Tiwari
Department of Computer Science
Banaras Hindu University
Varanasi, India
anoop.phd2014@gmail.com

Shruti Kaushik
Applied Cognitive Science Laboratory
Indian Institute of Technology Mandi
Himachal Pradesh, India
shruti_kaushik@students.iitmandi.ac.in

Abhinav Choudhury
Applied Cognitive Science Laboratory
Indian Institute of Technology Mandi
Himachal Pradesh, India
abhinav_choudhury@students.iitmandi.ac.in

Nataraj Dasgupta
RxDataScience Inc.
Durham, USA
nd@rxdatascience.com

Sayee Natarajan
RxDataScience Inc.
Durham, USA
sayee@rxdatascience.com

Larry A. Pickett
RxDataScience Inc.
Durham, USA
larry@rxdatascience.com

Varun Dutt
Applied Cognitive Science Laboratory
Indian Institute of Technology Mandi
Himachal Pradesh, India
varun@iitmandi.ac.in

Ankita Mazumder
Department of Chemical Engineering
Jadavpur University
Kolkata, India
chemical.ankita91@gmail.com

Dwaipayan Sen
Department of Chemical Engineering
Heritage Institute of Technology
Kolkata, India
dwaipayanju@gmail.com

Chiranjib Bhattacharjee
Department of Chemical Engineering
Jadavpur University
Kolkata, India
c.bhatta@gmail.com

M. Mary Shanthi Rani
Department of Computer Science & Applications
Gandhigram Rural Institute (Deemed to be University)
Gandhigram, Tamil Nadu, India
drmaryshanthi@gmail.com

P. Shanmugavadivu
Department of Computer Science & Applications
Gandhigram Rural Institute (Deemed to be University)
Gandhigram, Tamil Nadu, India
psvadivu67@gmail.com

Vandana Khanna
Department of EECE
The NorthCap University
Gurugram, India
vandanakhanna@ncuindia.edu

B. K. Das
Department of EECE
The NorthCap University
Gurugram, India
bkdas45@gmail.com

S. N. Kumar
Amal Jyothi College of Engineering
Kanjirappally, Kerala, India
appu123kumar@gmail.com

A. Lenin Fred
Mar Ephraem College of Engineering and Technology
Marthandam, Tamilnadu, India
leninfredip@gmail.com

Parasuraman Padmanabhan
Nanyang Technological University
Singapore
ppadmanabhan@ntu.edu.sg

Balazs Gulyas
Nanyang Technological University
Singapore
balazs.gulyas@ntu.edu.sg

Ajay H. Kumar
Mar Ephraem College of Engineering and Technology
Marthandam, Tamilnadu, India
ajayhakkumar@gmail.com

Abdul Rheem
Department of Applied Science and Humanities
Faculty of Engineering and Technology
Jamia Millia Islamia
New Delhi, India
abdulraheemsre84@gmail.com

Iftikhar Ahmad
Department of Applied Science and Humanities
Faculty of Engineering and Technology
Jamia Millia Islamia
New Delhi, India
iftiphd@gmail.com

Musheer Ahmad
Department of Applied Science and Humanities
Faculty of Engineering and Technology
Jamia Millia Islamia
New Delhi, India
mahmad@jmi.ac.in

Part I: **Theoretical advances**

Aneesh Wunnava, Manoj Kumar Naik, Bibekananda Jena, and Rutuparna Panda

1 Nature-inspired optimization algorithm and benchmark functions: a literature survey

Abstract: In the evolution process of humans, nature has played an important role in the transformation of mathematical investigation knowledge to modern human technological advancement. Nature optimizes the work based on biological, physical and chemical principles integrated with numerous mathematical models. People have learned the different principle and structure to design optimization algorithm that has been widely used in theoretical analysis and practical applications. In this chapter, we try to review several common nature-inspired algorithms with analysis systematically. Also, we try to incorporate the standard benchmark functions that are needed to evaluate the optimization algorithm. Through the chapter, we will discuss the original principle and the diversity of the algorithm in different practical applications.

Keywords: Nature-inspired optimization algorithm, benchmark functions, dragonfly algorithm, crow search algorithm, salp swarm algorithm, artificial butterfly algorithm, sooty tern optimization algorithm, seagull optimization algorithm, Harris hawks optimizer, squirrel search algorithm

1.1 Introduction

From ancient civilization, it can be observed that nature is the best source of knowledge for advancement of humankind. In the whole life of any biological species, they try to find the most favorable environment for nutrients so that they can produce their offspring. To get the most favorable environment, the biological species must search for nutrients in an environment, that may be optimal. Based on the behavior, searching pattern and foraging behavior of the different biological species, many researchers

* **Manoj Kumar Naik,** Dept. of Electronics and Communication Engineering, Faculty of Engineering and Technology, Siksha O Anusandhan, Bhubaneswar, Odisha – 751030, India, e-mail: naik.manoj.kumar@gmail.com
Aneesh Wunnava, Dept. of Electronics and Communication Engineering, Faculty of Engineering and Technology, Siksha O Anusandhan, Bhubaneswar, Odisha – 751030, India, e-mail: aneeshwunnava@gmail.com
Bibekananda Jena, Dept. of Electronics and Communication Engineering, Anil Neerukonda Institute of Technology and Science, Sangivalasa, Visakhapatnam, Andhra Pradesh – 531162, India, e-mail: bibekananda.jena@gmail.com
Rutuparna Panda, Dept. of Electronics and Telecommunication Engineering, Veer Surendra Sai University of Technology, Burla, Sambalpur, Odisha – 768018, India, e-mail: r_ppanda@yahoo.co.in

https://doi.org/10.1515/9783110671353-001

developed mathematical optimization model to investigate and applied in technological advancement.

The design variable, fitness function or optimization function, search space, and solution space are four important parts of the mathematical optimization model. The optimization process can be described as "The solution space (optimal solution) obtained from the search space with the help of fitness function based on design variable." The basic principle of any optimization algorithm is based on exploration and exploitation. The exploration enlarges the search space to give more exposer to searching agents within their lifetime. The exploitation is finding the optimal value close to the best solution. So, any optimization algorithm tries to trade-off between exploration and exploitation. The evaluation of optimization algorithm is very much essential to justify the effectiveness in terms of faster convergence, less complexity and to get optimum value with the help of benchmark functions.

In the year 1975, Holland et al. [1] proposed a genetic algorithm (GA) based on the Darwinian theory of evolution, survival of fittest based on the mathematical operator: crossover, mutation, fitness and selection. The GA is one of the most popular population-based optimizations ever known.

Then in the year 1995, Kennedy et al. [2] proposed particle swarm optimization (PSO) based on the social behavior of movement of a bird flock or fish school. The PSO algorithm coined with the help of the following parameters: current position, velocity, fitness function, personal best position and global best position of a particle or agent. Through the years, PSO became a popular technique in optimization due to simplicity, with many researchers investigating the PSO to enhance its capability and application to various problems [3].

Derigo et al. in the year 1996 came up with ant colony optimization (ACO), the first variant ant system [4] that is modeled using the knowledge of foraging behavior of some ant species. The operating principle of ACO can be observed by *stigmergy* type of communication. In the ant species, during the foraging it leaves a substance on the ground known as *pheromone*. The other ants observe the existence of pheromone and incline toward the paths for higher concentration. The main variant of the ACO areas ant system (AS) [4], MAX-MIN ant system [5] and ant colony system (ACS) [6, 7]. A brief literature survey on ACO is presented in [8].

In 2002, Passino [9, 10, 11, 12] come with bacterial foraging optimization (BFO) algorithm by studying the foraging behavior of *E. coli* bacteria existing in our intestine. The steps involved are chemotaxis, swarming, reproduction, elimination and dispersal. During the years, many researchers investigated the stability reproduction operator [13, 14, 15], and chemotaxis dynamics [16, 17] BFO is a good optimization technique and popular for some years, but it was costlier to get an output in terms of convergence timing. Many researchers investigated the application of the algorithm in field like harmonic detection in electrical lines [18], object detection [19], face detection [20] and more detail survey was presented in [21, 22].

Karaboga D. comes with a swarm optimization based on the social behavior of the honey bee in the year 2005 [23] known as an artificial bee colony algorithm (ABC) for numerical optimization problems. The ABC algorithm consists of artificial bees that are categorized by employed bees, onlookers and scouts. The employed bee go-to food source and evaluate the nectar amount. Onlookers try to compare the nectar amount shown by different employed bee in the neighbor and employ the new bee to that source. The scouts have the responsibility to determine the new abundant food source. Then in 2007, Karaboga D. et al. proposed a variant of ABC algorithm for solving Constrained Optimization Problems [24].

Yang Xin-She et al. [25] in the year 2009 comes with a new meta-heuristic algorithm known as Cuckoo search (CS) based on the brood parasitic behavior of Cuckoo species with combination of Levy flight [26, 27] behavior of birds. A more detail survey on CS is presented in [28]. A modification to the CS algorithm without Levy flight was proposed by Naik et al. in [29, 30].

Yang Xin-She again comes with the firefly algorithm (FA) [31, 32], and bat algorithm (BA) [33] in the year 2010. The author shows that FA can solve a single objective multimodal function more efficiently. The FA was based the behavior of fireflies based mainly on the ideology of attractiveness, which is proportional to their brightness. The position of firefly in the landscape was related to the attractiveness and related to the objective function. The BA was based on the echolocation behavior of bats to sense distance between food/pray. As the bat senses the distance and adjust the velocity and position based on loudness (varying wavelength or frequency) to get into the food/pray. A brief review on FA was found in [34], and for BA in [35].

The above-said algorithm is widely used in various problem in different field. Except these also some additional bio-inspired algorithms are investigated by the researcher like artificial fish-swarm algorithm (AFSA) [36], termite algorithm [37], group search optimizer (GSO) [38], flower pollination algorithm (FPA) [39, 40], fruit fly optimization algorithm (FFOA) [41, 42], krill herd (KH) [43, 44], cuckoo optimization algorithm (COA) [45], dolphin echolocation [46], grey wolf optimizer (GWO) [47], artificial algae algorithm (AAA) [48], ant lion optimizer (ALO) [49], shark smell optimization (SSO) [50], dolphin swarm optimization algorithm (DSOA) [51], grasshopper optimization algorithm (GOA) [52], selfish herd optimizer (SHO) [53], Symbiotic Organisms Search [54], spotted hyena optimizer [55], lion's Algorithm [56], lion optimization algorithm (LOA) [57], cat swarm optimization (CSO) [58], virus colony search (VCS) [59], coral reefs optimization (CRO) algorithm [60], whale optimization algorithm (WOA) [61], mouth brooding fish algorithm[62] and competitive optimization algorithm (COOA) [63].

Also, there are some well-known nature-inspired algorithms not inspired by biological species. Some of them are simulated annealing (SA) [64], differential evolution (DE) [65, 66, 67], gravitational search algorithm (GSA) [68], mine blast algorithm (MBA) [69], lightning search algorithm (LSA) [70], exchange market algorithm (EMA) [71], atom search optimization (ASO) [72], volleyball premier league (VPL) algorithm

[73], tree growth algorithm (TGA) [74], multiverse optimizer (MVO) [75], electrosearch (ES) algorithm [76], thermal exchange optimization [77], water wave optimization (WWO) [78], weighted superposition attraction (WSA) [79, 80], Lightning Attachment Procedure Optimization (LAPO) [81] and teaching-learning-based optimization algorithm (TLBO) [82].

In this chapter, we systematically review the development of the nature-inspired algorithm in the last few years such as dragonfly algorithm (DA) [83], crow search algorithm [84], salp swarm algorithm [85], artificial butterfly algorithm (ABO) [86], sooty tern optimization algorithm (STOA) [87], seagull optimization algorithm (SOA) [88], Harris hawks optimizer (HHO) [89] and squirrel search algorithm [90].

The organization of the rest of the chapter is as follows: In Section 1.1 is the Introduction. Section 1.2 describes the brief introduction of a generic form of the optimization problem and the original principle of various nature-inspired algorithms and their application. In Section 1.3, a brief discussion is about the benchmark function related to single objective real-parameter numerical optimization and Section 1.4 is the concluding remark.

1.2 Nature-inspired optimization algorithms

The generic form of the optimization problem can be written

$$\begin{aligned}&\text{minimize}_{X\in\mathfrak{R}^n} f_i(X), \quad (i = 1, 2, \dots, M),\\&\text{subject to } h_j(X) = 0, \quad (j = 1, 2, \dots, J),\\&g_k(X) \neq 0, \quad (k = 1, 2, \dots, K),\end{aligned}$$

where $f_i(X)$, $h_j(X)$ and $g_k(X)$ are functions of the design variable x_i presented in a design vector X,

$$X = (x_1, x_{2,\dots,} x_d)^T.$$

The cost function or fitness function described by $f_i(X)$, where $i = 1, 2, \dots, M$, and M denote to the number of objective functions. When $M = 1$, the problem is a single objective. $\mathfrak{R}^d$ denote the search space for design variable x_i and solution space for fitness function f_i. The h_j and g_k, are respectively equality and inequality constraints, respectively. If $g_k < 0$, the problem is minimization problem, otherwise a maximization problem. Here, we discuss the original principle and the diversity of the biologically inspired algorithm mostly based on the single-objective optimization problem.

Now let us thoroughly discuss some more recent development optimization algorithms taking inspiration from biological species such as dragonfly, crow, salp, butterfly, sooty tern, seagull, Harris hawk and squirrel.

1.2.1 Dragonfly algorithm

Mirjalili S. proposed a dragonfly algorithm (DA) [83] in the year 2015 taking inspiration of swarming behaviors of dragonflies. The DA was based on the swarm intelligence (SI) paradigm. Dragonfly swarm for hunting and migration. The hunting of dragonflies can be related to static swarm, and migration can be related to dynamic swarm. In the static swarm, the dragonflies make small groups within a small area to hunt other small flies or insects, which can be referred to as exploitation or local search. In the dynamic swarm, the dragonflies make the larger group migrate to some other area, and can be referred as exploration (global search).

1.2.1.1 Mathematical modeling of the dragonfly algorithm

The DA used two different position update rules to find the position of the current individual X and calculate the objective value $f(X)$.

– Rule I: When a dragonfly has at least one neighbor dragonfly

The position vector can be calculated as

$$X_{t+1} = X_t + \Delta X_{t+1}, \tag{1.1}$$

where t is the current iteration, ΔX_{t+1} is the step vector for new position vector, X_{t+1} is the new position vector and X_t is the current position vector.

The step vector is calculated as

$$\Delta X_{t+1} = (sS_i + aA_i + cC_i + fF_i + eE_i) + w\Delta X_t, \tag{1.2}$$

where s is the separation weight, separation of ith individual $S_i = -\sum_{j=1}^{N} X - X_j$, a is the alignment weight, alignment of ith individual $A_i = \frac{\sum_{j=1}^{N} V_j}{N}$, V_j is the velocity of jth individual, c is the cohesion weight, cohesion of ith individual $C_i = \frac{\sum_{j=1}^{N} X_j}{N} - X$, f is the food factor, food source of ith individual $F_i = X^+ - X$, e is the enemy factor, w is the inertia weight, position of ith individual $E_i = X^- + X$. The X^+ and X^- are the position of the food source and position of the enemy, respectively.

– Rule II: When the dragonfly is alone in the surrounding

$$X_{t+1} = X_t + \text{Lévy}(d) \times X_t \tag{1.3}$$

where t is the current position and d is the search dimension. The Lévy flight is calculated using [91, 92].

The DA algorithm is extended to the binary problem (BDA) and multiobjective problem (MODA) [83]. The DA algorithm was successfully applied in the optimization of orthotropic infinite plates with a quasi-triangular cut-out [93] and feature selection [94].

1.2.2 Crow search algorithm

Askarzadeh A. proposes a novel metaheuristic algorithm known as the crow search algorithm [84, 95] based on crow intelligence. The basic principles of crow search algorithms are based on facts that the crow memorizes the position of food hiding places of other birds, mastering in thievery by following each other, protecting their catch from being pilfered by a probability and lives in a flock. Based on these rules, the crow search algorithm formulation is done.

1.2.2.1 Mathematical modeling of crow search algorithm

Let there is a d-dimensional environment with N number crows (flock size) in a search space in iteration (time) i, then position vector $x^{i,\text{iter}}$ ($i = 1, 2, \ldots, N$; $iter = 1, 2, \ldots, iter_{\max}$) where $x^{i,\text{iter}} = [x_1^{i,\text{iter}}, x_2^{i,\text{iter}}, \ldots, x_d^{i,\text{iter}}]$ and $iter_{\max}$ is the maximum number of iteration. CSA algorithm formulation is done using the following two conditions of crow movement.

- Crow movement 1: Assuming crow j does not know crow i is in the following. Then the position vector will be updated as

$$x^{i,\text{iter}+1} = x^{i,\text{iter}} + r_i \times fl^{i,\text{iter}} \times (m^{j,\text{iter}} - x^{i,\text{iter}}) \tag{1.4}$$

  where r_i is a random number between 0 and 1, and $fl^{i,\text{iter}}$ is the flight length (step size) of crow i in iteration $iter$, $m^{j,\text{iter}}$ is the memory/hiding place (best position so far) of crow j. The small value $fl^{i,\text{iter}}$ leads to local search (exploitation) and a large value leads to global search (exploration).
  The memory of crow can be updated as

$$m^{i,\text{iter}+1} = \begin{cases} x^{i,\text{iter}+1} & f(x^{i,\text{iter}+1}) \text{ is better than } f(m^{i,\text{iter}}) \\ m^{i,\text{iter}} & \text{otherwise} \end{cases} \tag{1.5}$$

- Crow movement 2: Assuming crow j knows that crow i is following: The crow j fools the crow i by going to some other position randomly.

So, by combining the crow movements 1 and 2, in the crow search algorithm, the position vector is updated as

$$x^{i,\text{iter}+1} = \begin{cases} x^{i,\text{iter}} + r_i \times fl^{i,\text{iter}} \times (m^{i,\text{iter}} - x^{i,\text{iter}}) & r_j \geq AP^{i,\text{iter}} \\ a \text{ random position} & \text{otherwise} \end{cases} \tag{1.6}$$

where r_j is a random number between 0 to 1, and $AP^{i,\text{iter}}$ is the awareness probability of crow j in the iteration $iter$ which will be initialized at the beginning.

The crow search algorithm evaluates the fitness value using the position vector represented in equation (1.6) until it reaches an optimum (best) fitness value or max iteration achieved.

The crow search algorithm has been successfully applied in constrained engineering optimization in [84], abrasive water jet (AWJ) cutting of ceramic tiles [96], economic environmental dispatch [97], fuzzy C-means clustering algorithm [98] and selection of conductor size in radial distribution networks [99].

1.2.3 Salp swarm algorithm

Mirjalili S. et al. proposed an algorithm based on the swarming behavior of salps in the ocean during navigating and foraging known as the salp swarm algorithm [85] in the year 2017. The salps have a transparent, barrel-shaped body and belongs to the family of Salpidae. They control the movement by pumping water through their body as propulsion, just like the jellyfish.

1.2.3.1 Mathematical modeling of the salp swarm algorithm

The mathematical formulation of the salp swarm algorithm was based on the behavior of salps forming swarm in the form of the salp chain that helps to achieve locomotion and foraging. The salp chain consists of a population of salps divided into two groups: leader and followers. The salp in the front of the salp chain is referred as the leader, and the rest of salps as followers. Let assume that a salp chain X consists of N number of salps in a d-dimensional search space. Then X can be written as

$$X = \begin{bmatrix} x_1^1 & \cdots & x_j^1 & \cdots & x_d^1 \\ \cdots & \cdots & \cdots & \cdots & \cdots \\ x_1^i & \cdots & x_j^i & \cdots & x_d^i \\ \cdots & \cdots & \cdots & \cdots & \cdots \\ x_1^N & \cdots & x_j^N & \cdots & x_d^N \end{bmatrix}, \quad \text{where } i = 1, 2, \ldots, N \text{ and } j = 1, 2, \ldots, d \tag{1.7}$$

The salp swarm algorithm takes the assumption that there is a target food source F in the search space. This can be referred to as the best search agent. The position of salp can be updated in two different ways: one for leader (when $i = 1$), and other for follower (when $i = 1, 2, \ldots, N$).

- Case 1: Position update for leader

$$x_j^1 = \begin{cases} F_j + c_1((UB_j - LB_j)c_2 + LB_j) & c_3 \geq 0 \\ F_j - c_1((UB_j - LB_j)c_2 + LB_j) & c_3 < 0 \end{cases} \tag{1.8}$$

where F_j is the best target in the jth dimension, UB_j and LB_j are the upper bound and lower bound of the jth dimension. The c_1, c_2 and c_3 are random numbers. The c_1 is used to balance exploitation and exploration, which depends on the current iteration l, and the maximum number of iterations L. So c_1 is defined as

$$c_1 = 2e^{-(\frac{4l}{L})^2} \tag{1.9}$$

The other random parameter c_2 is uniformly generated in the interval $[0, 1]$, and c_3 is uniformly generated in the interval $[-1, 1]$.
- Case 2: Position update for follower:

$$x_j^i = \frac{1}{2}(x_j^i + x_j^{i-1}) \tag{1.10}$$

The SSA return the best search agent X_{opt} when the optimal fitness $f(X_{\text{opt}})$ is achieved with the help of position vector X (equation (1.7)) that are updated using equations (1.8) and (1.10).

The salp swarm algorithm is extended to a multiobjective salp swarm algorithm (MSSA) [85] to solve the multi-objective problem. It is also successfully applied in engineering problems like air foil design [85], marine propeller design [85] and node localization in a wireless sensor network [100]. A more detail literature survey of the salp swarm algorithm is presented in [101].

1.2.4 Artificial butterfly optimization algorithm

Qi Xiangbo et al. came up with a new artificial butterfly optimization [86] algorithm based on the mate-finding behavior of the butterfly species. The artificial butterfly optimization algorithm uses the concept of a sunspot under the tree as a preferable mating place of the butterfly. In the artificial butterfly optimization algorithm, the artificial butterfly is divided into two groups of butterflies known as sunspot butterflies and canopy butterflies. The sunspot butterflies have greater fitness and are in the position of a sunspot where mating probability is greater, whereas the canopy butterflies are a flight above the surroundings of a sunspot and have the least amount of fitness.

1.2.4.1 Mathematical modeling of artificial butterfly optimization

Based on the behavior of the butterfly's experience, there are three types of flights to update the position X in a n-dimensional problem within a colony of population N butterflies. The types of flights are:
- Sunspot flight mode: The flight to a neighbor location experienced by the sunspot butterflies
- Canopy flight mode: The flight to a neighbor location experienced by the canopy butterflies
- Free flight mode: The flight to a neighbor location experienced by each butterfly

The artificial butterfly optimization algorithm proposes two variants coined as ABO1 and ABO2 based on the position vector updating strategy. For the simplicity of the algorithm, sunspot flight is equivalent to the canopy flight.

– Sunspot/Canopy flight mode:

In the sunspot or canopy flight mode, the sunspot or canopy butterflies search nearby neighbor locations to exploit local search, and can be referred to as exploitation. The position vector X_i^{t+1} of ith (i. e., $1 \le i \le N$) butterfly in the next iteration $t+1$, if current iteration is t (i. e., $1 \le t \le T$) with a max number of iteration T,

$$X_i^{t+1} = \begin{cases} X_i^t + (X_i^t - X_j^t) \times rand() & \text{for ABO1} \\ X_i^t + \frac{X_j^t - X_i^t}{\|X_j^t - X_i^t\|} \times (UB - LB) \times step \times rand() & \text{for ABO2} \end{cases} \tag{1.11}$$

– Free flight mode:

The butterflies experience an exploration phase during this flight mode. In this mode, the position vector X_i^{t+1} is updated as

$$X_i^{t+1} = X_j^t - 2 \times a \times rand() - a \times |2 \times rand() \times X_j^t - X_i^t| \tag{1.12}$$

In the above position, update the equations of sunspot/canopy and free flight mode, and the different variable is described as:

$$\begin{aligned} &j \text{ is a randomly chosen butterfly where } j \neq i. \\ &rand() \text{ is a random number in the range } (0,1). \\ &UB \text{ and } LB \text{ are the upper and lower boundary of the search variable.} \\ &a \text{ is linearly decreasing from 2 to 0 as the iteration } t \text{ increases.} \\ &step \text{ is the step size with a decreasing strategy from 1 to } step_e. \\ &step = 1 - (1 - step_e) \times \frac{t}{T} \end{aligned} \tag{1.13}$$

The artificial butterfly optimization algorithm updates the position vector X^t iteratively until the termination criterion met is based on the fitness function $f(X)$, and returns the optimal position vector X_{opt} as a solution vector.

1.2.5 Sooty tern and seagull optimization algorithms

Dhiman et al. proposed a bio-inspired algorithm sooty tern optimization algorithm (STOA) [87], and seagull optimization algorithm (SOA) [88] based on the attacking and migration behavior of a sea bird sooty tern and seagull. In the attacking behavior, sooty tern and seagull uses a flapping mode of flight to locate the prey, so this behavior can be referred to as exploitation. The sooty tern and seagull migrate in a group to find the richest and most abundant food source. So, the migration behavior can be referred as exploration.

1.2.5.1 Mathematical modeling of STOA and SOA

Let us define the position vector of sooty tern and seagull as $\vec{P}_{st/sg}(t)$, $t = 0, 1, \ldots, L$, where t is the current iteration, the maximum number of iterations can be L, and $\vec{P}_{\text{bst}}$ is the best search agents which are achieved from the best fitness value. The position vector $\vec{P}_{st/sg}$ of STOA and SOA can be updated using the mathematical modeling of attacking and migration behavior of the sooty tern and seagull.

– Attacking behavior

The sooty tern and seagull adjust the velocity and angle of attack. They generate spiral behavior during the attacking a pray described as

$$x' = r \times \sin(i) \tag{1.14}$$

$$y' = r \times \cos(i) \tag{1.15}$$

$$z' = r \times i \tag{1.16}$$

$$r = u \times e^{kv} \tag{1.17}$$

where r is the radius of each turn of the spiral, i represent in the range $[0 \leq i \leq 2\pi]$, u and v define the spiral shape.

– Migration behavior

The migration behavior tries to find the gap $\vec{D}_{st}$ of the sooty tern and $\vec{D}_{sg}$ of the seagull between the search agents and best search agents that depends on collision avoidance $\vec{C}_{st/sg}$, and convergence toward the best neighbors $\vec{M}_{st}$ for sooty tern and $\vec{M}_{sg}$ for seagull:

$$\vec{D}_{st}(t) = \vec{C}_{st/sg}(t) + \vec{M}_{st}(t) \tag{1.18}$$

$$\vec{D}_{sg}(t) = |\vec{C}_{st/sg}(t) + \vec{M}_{sg}(t)| \tag{1.19}$$

The collision avoidance $\vec{C}_{st/sg}$ between its neighbor search agents is represented as

$$\vec{C}_{st/sg}(t) = S_A(t) \times \vec{P}_{st/sg}(t) \tag{1.20}$$

The S_A indicates the movement of search agents and is represented as

$$S_A(t) = C_f - (t \times (C_f/L)) \tag{1.21}$$

where C_f is a controlling variable, which is linearly decreased from C_f to 0.

The convergence toward the best neighbors $\vec{M}_{st}$ and $\vec{M}_{sg}$ are defined as

$$\vec{M}_{st}(t) = B_{st} \times (\vec{P}_{\text{bst}}(t) - \vec{P}_{st/sg}(t)) \tag{1.22}$$

$$\vec{M}_{sg}(t) = B_{sg} \times (\vec{P}_{\text{bst}}(t) - \vec{P}_{st/sg}(t)) \tag{1.23}$$

where B_{st} and B_{sg} are a random variable responsible for exploration, and it can be calculated as follows:

$$B_{st} = 0.5 \times r_1 \tag{1.24}$$

$$B_{sg} = 2 \times (S_A(t))^2 \times r_2 \tag{1.25}$$

where r_1 and r_2 are a random number in the range $[0, 1]$.

Then by considering the attacking and migration behavior, the position vector $\vec{P}_{st/sg}$ of STOA and SOA will be updated as

$$\vec{P}_{st/sg}(t) = \begin{cases} (\vec{D}_{st}(t) \times (x' + y' + z')) \times \vec{P}_{\text{bst}}(t) & \text{for STOA} \\ (\vec{D}_{sg}(t) \times (x' + y' + z')) \times \vec{P}_{\text{bst}}(t) & \text{for SOA} \end{cases} \tag{1.26}$$

The STOA and SOA update the $\vec{P}_{st/sg}(t)$ iteratively and save the best optimal solution $\vec{P}_{\text{bst}}$, when $t = L$ or mate and end criterion. The STOA and SOA are successfully implemented in a constrained industrial engineering problem in [87, 88], and have a lot of potential to explore in the future.

1.2.6 Harris hawks optimizer

Heidari et al. proposed Harris hawks optimization (HHO) [89], a population-based nature-inspired optimization algorithm based on the cooperative behavior and surprise pounce of Harris' hawks. The Harris' hawk birds are one of the most intelligent creatures and use a variety of chasing patterns based on the dynamic escaping patterns of the prey in a scenario. The Harris' hawk uses the "surprise pounce," also known as the "seven kills" strategy to capture the prey, in which several hawks try to attack cooperatively from several directions and simultaneously converge to the detected prey. The mathematical model of HHO is based on the exploratory and exploitative phase taking inspiration exploring a prey, surprise pounce and the different attacking strategies of Harris' hawks.

1.2.6.1 Mathematical modeling of HHO

Let us define $X(t)$ as the position vector of hawks, $X_{\text{rabbit}}(t)$ is the position of rabbit, $X_{\text{rand}}(t)$ is a randomly selected hawk within the current population at current iteration t (i. e., $1 \leq t \leq T$), and T is the maximum number of iterations. The HHO use escaping energy E of the prey to determine exploration or exploitation phases to be established in the algorithm.

– Transition between exploration and exploitation

The transition from exploration to exploitation or vice versa depends on the escaping energy of the prey. The escaping energy of the prey can be modeled as

$$E = 2E_0\left(1 - \frac{t}{T}\right) \tag{1.27}$$

where E_0 is the initial energy which changes its states in between $[-1, 1]$ at each iteration. When E_0 is decreasing 0 to −1, the prey is physically flagging, whereas E_0 is increasing the prey is strengthening. So, when $|E| \geq 1$, the hawks search different regions to explore the prey location, that is, HHO is in the exploration phase, and when $|E| < 1$, the HHO try to exploit the neighborhood of the solution for the period of the exploitation phase.

– Exploration phase ($|E| \geq 1$)

The Harris' hawk perches arbitrarily on some location and waits to spot the prey based on two strategies. Let q be an equal chance for each perching strategy. If $q < 0.5$, then the hawk takes a position based on the other family member and the rabbit (probably close to prey and another family member), else $q \geq 0.5$ means that hawk sits on some random location in a tall tree within the group's home range. Then the next iteration position vector $X(t + 1)$ of Harris' hawk can be updated as

$$X(t+1) = \begin{cases} X_{\text{rand}}(t) - r_1|X_{\text{rand}}(t) - 2r_2X(t)| & q \geq 0.5 \\ (X_{\text{rabbit}}(t) - X_m(t)) - r_3(LB + r_4(UB - LB)) & q < 0.5 \end{cases} \tag{1.28}$$

where LB and UB shows the lower and upper bounds of variables, X_m is the mean of the position of the current generation of hawk, r_1, r_2, r_3, r_4 and q are the random numbers in the range $[0, 1]$. The X_m can be achieved in a hawk population of N by

$$X_m(t) = \frac{1}{N}\sum_{i=1}^{N} X_i(t) \tag{1.29}$$

where $X_i(t)$ indicates the position of ith hawk in iteration t.

– Exploitation phase ($|E| < 1$)

In this phase, the hawk uses the surprise pounce attacking strategy to detect the prey. Let r be the chance of escaping prey from threatening situations. Let define $r < 0.5$ for successfully escaping, and $r \geq 0.5$ for unsuccessfully escaping during the surprise pounce. Based on the chance of escaping prey, and the escaping energy, the hawks experience soft besiege or hard besiege.

– Soft besiege ($r \geq 0.5$ and $|E| \geq 0.5$)

During this phase, hawks encircle softly to make the prey exhausted and then perform a surprise pounce that can be modeled as

$$X(t+1) = \Delta X(t) - E\left|JX_{\text{rabbit}}(t) - X(t)\right| \tag{1.30}$$

$$\Delta X(t) = X_{\text{rabbit}}(t) - X(t) \tag{1.31}$$

$$J = 2(1 - r_5) \tag{1.32}$$

where r_5 is a random number (0, 1), and J is the jump strength of rabbit during the escaping procedure.

– Hard besiege ($r \geq 0.5$ and $|E| < 0.5$)

During this phase, hawks barely encircle as the prey is exhausted and has a low escaping energy, and then perform a surprise pounce that can be modeled as

$$X(t+1) = X_{\text{rabbit}}(t) - E\left|\Delta X(t)\right| \tag{1.33}$$

– Soft besiege with progressive rapid dives ($r < 0.5$ and $|E| \geq 0.5$)

During this phase, hawks encircle softly when the prey has enough energy to escape during the surprise pounce and can be modeled as

$$X(t+1) = \begin{cases} Y & \text{if } F(Y) < F(X(t)) \\ Z & \text{if } F(Z) < F(X(t)) \end{cases} \tag{1.34}$$

$$Y = X_{\text{rabbit}}(t) - E\left|JX_{\text{rabbit}}(t) - X(t)\right| \tag{1.35}$$

$$Z = Y + S \times LF(D) \tag{1.36}$$

where D is the dimension of the problem, S is a random vector of size $1 \times D$ and LF is the levy flight for the leapfrog movements experienced by the prey:

$$LF(x) = 0.01 \times \frac{u \times \sigma}{|v|^{\frac{1}{\beta}}}, \quad \sigma = \left(\frac{\Gamma(1+\beta) \times \sin(\frac{\pi\beta}{2})}{\Gamma(\frac{1+\beta}{2}) \times \beta \times 2^{(\frac{\beta-1}{2})}}\right)^{\frac{1}{\beta}}, \quad \beta = 1.5 \tag{1.37}$$

where u, v are random numbers in the range (0, 1).

– Hard besiege with progressive rapid dives ($r < 0.5$ and $|E| < 0.5$)

During this phase, hawks barely encircle when the prey has not enough energy to escape during the surprise pounce and can be modeled as

$$X(t+1) = \begin{cases} Y & \text{if } F(Y) < F(X(t)) \\ Z & \text{if } F(Z) < F(X(t)) \end{cases} \tag{1.38}$$

$$Y = X_{\text{rabbit}}(t) - E\left|JX_{\text{rabbit}}(t) - X_m(t)\right| \tag{1.39}$$

$$Z = Y + S \times LF(D) \tag{1.40}$$

The HHO algorithm updates the position vector $X(t)$ iteratively until the end criterion is met and based on the fitness function $f(X)$, and returns the optimal position vector X_{opt} as a solution vector.

The HHO is successfully tested using a benchmark function and also implemented in some of the real-world engineering problems like three-bar truss design, tension/compression spring design, pressure vessel design, welded beam design, multiplate disc clutch brake and rolling element bearing design problem in [89].

1.2.7 Squirrel search algorithm

Jain et al. proposed a squirrel search algorithm [90] inspired by the foraging behavior and gliding movement of southern flying squirrels. The flying squirrels can find acorns easily for their daily needs during autumn; after that, they search for hickory nuts as an optimal source of food to store it during the winter. During the winter, the nutrient demands are more as they are less active, so they eat the hickory nuts. When the winter passes, the flying squirrels are active again. This cyclic process is repeated throughout the life span of flying squirrels. The mathematical model of this behavior is the main source of design of the squirrel search algorithm for optimization.

1.2.7.1 Mathematical modeling of the squirrel search algorithm

For the mathematical modeling of the squirrel search algorithm, use some of the following basic assumptions:
- One squirrel in a tree at a time.
- Every single squirrel searches food on its own based on the dynamic foraging behavior.
- The search space has only three types of trees out of which are three oak trees (acorn nuts), one hickory tree (hickory nuts) and rest are normal trees.

Let X be the position vector of the N flying squirrel in a n dimensional problem. The X can be represented as

$$X = \begin{bmatrix} X_{1,1} & \cdots & X_{1,j} & \cdots & X_{1,d} \\ \cdots & \cdots & \cdots & \cdots & \cdots \\ X_{i,1} & \cdots & X_{i,j} & \cdots & X_{i,d} \\ \cdots & \cdots & \cdots & \cdots & \cdots \\ X_{N,1} & \cdots & X_{N,j} & \cdots & X_{N,d} \end{bmatrix}, \quad \text{where } i = 1, 2, \ldots, N \quad \text{and } j = 1, 2, \ldots, d. \tag{1.41}$$

The X_i can be initialized as

$$X_i = LB + rand \times (UB - LB) \tag{1.42}$$

where LB and UB are the lower and upper bound of search space, and $rand$ is random number in the range $(0,1)$.

Let $f(X) = [f(X_1), f(X_2), \ldots, f(X_N)]$ and is the fitness of the squirrel based on position X; this decides the quality of food source search by squirrels. Based on the fitness value, the quality of food resources that the squirrel search algorithm is categorized by, is the optimal food resource (hickory nut tree), normal food resource (acorn nut tree) and certainly not the food resource (normal tree). Based on this, the dynamic foraging behavior can be modeled as follows.

After the ascending order sorting of $f(X)$ during the iteration t, we classify the three sets of food sources based on their location, i. e., the optimal food resources related to hickory nut tree as X_{ht}^t, the normal food source related to acorn nut tree as X_{at}^t, and not food source related to normal tree as X_{nt}^t.

$$[sorted_f, sorted\_index] = sort(f(X)) \tag{1.43}$$

$$X_{ht}^t = X(sorted\_index(1)) \tag{1.44}$$

$$X_{at}^t = X(sorted\_index(2:4)) \tag{1.45}$$

$$X_{nt}^t = X(sorted\_index(5:N)) \tag{1.46}$$

The position of squirrels is updated using three different cases during the winter season as follows:

- Case 1: The position of the squirrel present in normal food resources (acorn nut tree) to optimal food resources (hickory nut tree) can be updated as

$$X_{at}^{t+1} = \begin{cases} X_{at}^t + d_g \times G_c \times (X_{ht}^t - X_{at}^t) & R_1 \geq P_{dp} \\ \text{Random location} & \text{otherwise} \end{cases} \tag{1.47}$$

  where $d_g (= 9 \text{ to } 20\text{ m})$ is the random gliding distance, $G_c (= 1.9)$ is the gliding constant, predator presence probability $P_{dp} (= 0.1)$ and R_1 is a random number in the range $[0,1]$.

- Case 2: The position of the squirrel presents not in a food resource (normal tree) to normal food source (acorn nut tree) can be updated as

$$X_{nt}^{t+1} = \begin{cases} X_{nt}^t + d_g \times G_c \times (X_{at}^t - X_{nt}^t) & R_2 \geq P_{dp} \\ \text{Random location} & \text{otherwise} \end{cases} \tag{1.48}$$

  where R_2 is a random number in the range $[0,1]$.

- Case 3: The position of the squirrel present in a food resource (normal tree) and already consumes the acorn nut wants to move to the optimal food resources (hickory nut tree) can be updated as

$$X_{nt}^{t+1} = \begin{cases} X_{nt}^t + d_g \times G_c \times (X_{ht}^t - X_{nt}^t) & R_3 \geq P_{dp} \\ \text{Random location} & \text{otherwise} \end{cases} \tag{1.49}$$

  where R_3 is a random number in the range $[0,1]$.

The position of squirrels is updated during the end of the winter season depends on the seasonal constant S_c for the minimum value of seasonal constant $S_{\min}$ is calculated as

$$S_c^y = \sqrt{\sum_{k=1}^{n}(X_{at,k}^y - X_{ht,k})^2}, \quad \text{where } y = 1, 2, 3 \tag{1.50}$$

$$S_{\min} = \frac{10E^{-6}}{(365)^{t/(L/2.5)}} \tag{1.51}$$

where t and L are the current and maximum iteration values.

If the seasonal monitoring condition, i. e., $S_c^y < S_{\min}$ is found true, random relocation of the position vector for squirrels present in a normal tree X_{nt}^{new} are done as follows:

$$X_{nt}^{\text{new}} = LB + Levy \times (UB - LB) \tag{1.52}$$

where *Levy* represents the Levy distribution.

The HHO algorithm updates the position vector X iteratively until the end criterion met is based on the fitness function $f(X)$, and returns the optimal position vector X_{opt} as a solution vector.

1.3 Benchmark functions for single-objective numerical optimization

The benchmark function problems are defined as

$$\min_X f(X), X = [x_1, x_2, \ldots, x_d]^T \tag{1.53}$$

where d is the search dimension of the problem, which is a set of feasible points in search spaces. The optimal solution of an optimization problem is a set of $X_{\text{opt}} \in d$ of all optimal points in d. The benchmark functions are used to validate the optimization algorithms based on the important features like dimensionality, modality and separability. The dimensionality of the problem increases, then the cost of the optimization algorithm increases as the search space increases exponentially. The modality of a benchmark function is considered as the number of peaks in search space. Based on the modality, the benchmark functions are mainly classified as unimodal and multimodal. The unimodal benchmark functions have only one peak or minima, with which the convergence rate is an important parameter to validate the optimization algorithm; while the other multimodal benchmark functions have two or more peaks or minima, with which the search of optimal minima is important. The separability refers the difficulty level offered by the benchmark functions. The separable benchmark functions

Table 1.1: Classical unimodal and separable benchmark functions.

Functions	Expression	d	Range	$f_{\min}$
Step	$f(X) = \sum_{j=1}^{d}(x_j + 0.5)^2$	30	[−5.12, 5.12]	0
Sphere	$f(X) = \sum_{j=1}^{d} x_j^2$	30	[−100, 100]	0
Sum Squares	$f(X) = \sum_{j=1}^{d} jx_j^2$	30	[−10, 10]	0
Quartic, i. e., Noise	$f(X) = \sum_{j=1}^{d} jx_j^4 + rand$	30	[−1.28, 1.28]	0

Table 1.2: Classical unimodal and nonseparable benchmark functions.

Functions	Expression	d	Range	$f_{\min}$
Beale	$f(X) = (1.5 - x_1 + x_1x_2)^2 + (2.25 - x_1 + x_1x_2^2)^2 + (2.625 - x_1 + x_1x_2^3)^2$	2	[−4.5, 4.5]	0
Easom	$f(X) = -\cos(x_1)\cos(x_2)\exp(-(x_1 - \pi)^2 - (x_2 - \pi)^2)$	2	[−100, 100]	−1
Matyas	$f(X) = 0.26(x_1^2 + x_2^2) - 0.48x_1x_2$	2	[−10, 10]	0
Colville	$f(X) = 100(x_1^2 - x_2)^2 + (x_1 - 1)^2 + (x_3 - 1)^2 + 90(x_3^2 - x_4)^2 + 10.1(x_2 - 1)^2 + (x_4 - 1)^2 + 19.8(x_2 - 1)(x_4 - 1)$	4	[−10, 10]	0
Zakharov	$f(X) = \sum_{j=1}^{d} x_j^2 + (\sum_{j=1}^{d} 0.5jx_j)^2 + (\sum_{j=1}^{d} 0.5jx_j)^4$	10	[−5, 10]	0
Schwefel 2.22	$f(X) = \sum_{j=1}^{d} \lvert x_j\rvert + \prod_{j=1}^{d} \lvert x_j\rvert$	30	[−10, 10]	0
Schwefel 1.2	$f(X) = \sum_{j=1}^{d}(\sum_{k=1}^{j} x_k)^2$	30	[−100, 100]	0
Dixon–Price	$f(X) = (x_1 - 1)^2 + \sum_{j=2}^{d} j(2x_j^2 - x_j - 1)^2$	30	[−10, 10]	0

are easier to solve as compared to the nonseparable benchmark functions. The classical benchmark functions [30, 54, 102, 103, 104] are presented in Tables 1.1, 1.2, 1.3, and 1.4 with details of functions name, expression, dimension (d), range of the search space, optimal minima ($f_{\min}$). During the years as the popularity of evolutionary computation and machine intelligence are increasing, the complexity of the optimization problem is also increasing. Based on these, researchers have come out with some new modern numerical optimization problems from the IEEE Conference on Evolutionary Computation (IEEE CEC) [105, 106, 107, 108, 109, 110]. These benchmark functions have several new features like the novel basic optimization problem, and the functions are shifted, expanded, rotated and combined more than in one basic optimization problem as hybrid and composite functions.

1.4 Conclusions

In this chapter, we systematically reviewed a recent nature bio-inspired algorithm dragonfly algorithm, crow search algorithm, salp swarm algorithm, artificial butterfly algorithm, sooty tern optimization algorithm, seagull optimization algorithm, Har-

Table 1.3: Classical multimodal and separable benchmark functions.

Functions	Expression	d	Range	f_{min}
Bohachevsky 1	$f(X) = x_1^2 + 2x_2^2 - 0.3\cos(3\pi x_1) - 04\cos(4\pi x_2) + 0.7$	2	$[-100, 100]$	0
Booth	$f(X) = (x_1 + 2x_2 - 7)^2 + (2x_1 + x_2 - 5)^2$	2	$[-10, 10]$	0
Michalewicz2	$f(X) = -\sum_{j=1}^{d} \sin(x_j)(\sin(\frac{jx_j^2}{\pi}))^{20}$	2	$[0, \pi]$	-1.8013
Michalewicz5	$f(X) = -\sum_{j=1}^{d} \sin(x_j)(\sin(\frac{jx_j^2}{\pi}))^{20}$	5	$[0, \pi]$	-4.6877
Michalewicz10	$f(X) = -\sum_{j=1}^{d} \sin(x_j)(\sin(\frac{jx_j^2}{\pi}))^{20}$	10	$[0, \pi]$	-9.6602
Rastrigin	$f(X) = \sum_{j=1}^{d} (x_j^2 - 10\cos(2\pi x_j) + 10)$	30	$[-5.12, 5.12]$	0

Table 1.4: Classical multimodal and nonseparable benchmark functions.

Functions	Expression	d	Range	f_{min}
Schaffer	$f(X) = 0.5 + \frac{\sin^2(\sqrt{x_1^2 + x_2^2}) - 0.5}{(1 + 0.001(x_1^2 + x_2^2))^2}$	2	$[-100, 100]$	0
Six Hump Camel Back	$f(X) = 4x_1^2 - 2.1x_1^4 + \frac{1}{3}x_1^6 + x_1x_2 - 4x_2^2 + 4x_2^4$	2	$[-5, 5]$	-1.03163
Bohachevsky 2	$f(X) = x_1^2 + 2x_2^2 - 0.3\cos(3\pi x_1)\cos(4\pi x_2) + 0.3$	2	$[-100, 100]$	0
Shubert	$f(X) = (\sum_{j=1}^{5} j\cos(j + 1)x_1 + j)(\sum_{j=1}^{5} j\cos(j + 1)x_2 + j)$	2	$[-10, 10]$	-186.73
Goldstein–Price	$f(X) = [1 + (x_1 + x_2 + 1)^2(19 - 14x_1 + 3x_1^2 - 14x_2 + 6x_1x_2)] \times [(2x_1 - 3x_2)^2 \times (18 - 32x_1 + 12x_1^2 + 48x_2 - 36x_1x_2 + 27x_2^2) + 30]$	2	$[-2, 2]$	3
Rosenbrock	$f(X) = \sum_{j=1}^{d-1} 100(x_{j+1} - x_j^2)^2 + (x_j - 1)^2$	30	$[-30, 30]$	0
Griewank	$f(X) = \frac{1}{4000}(\sum_{j=1}^{d}(x_j - 100)^2) - (\prod_{j=1}^{d}\cos(\frac{x_j - 100}{\sqrt{j}})) + 1$	30	$[-600, 600]$	0
Ackley	$f(X) = -20\exp(-0.2\sqrt{\frac{1}{d}\sum_{j=1}^{d} x_j^2}) - \exp(\frac{1}{d}\sum_{j=1}^{d}\cos(2\pi x_j)) + 20 + e$	30	$[-32, 32]$	0
Schwefel 2.26	$f(X) = \sum_{j=1}^{d} -x_j \sin(\sqrt{\lvert x_j \rvert})$	30	$[-500, 500]$	-12569.5

ris hawks optimizer and squirrel search algorithm, as nature has so many creatures and every one of them optimizes to survive in the environment. So, in the future, researchers may come with a new bio-inspired algorithm that performs well above as compared to the known date.

In this chapter, we also reviewed the benchmark functions used by the different researchers to validate and compare the new or/between old optimization algorithm. As the complexity increases day by day, we require more complex benchmark functions to validate the new algorithm so that it can be applied to real-life problems.

Bibliography

[1] Holland JH. Adaptation in natural and artificial systems. University of Michigan Press; 1975.
[2] Kennedy J, Eberhart R, editors. Particle swarm optimization. In: IEEE international conference on neural networks, 1995 proceedings, Nov/Dec 1995; 1995.
[3] Jain NK, Nangia U, Jain J. A review of particle swarm optimization. J Inst Eng Ser B. 2018; 99(4):407–411.
[4] Dorigo M, Maniezzo V, Colorni A Ant system: optimization by a colony of cooperating agents. IEEE Trans Syst Man Cybern, Part B, Cybern. 1996; 26(1):29–41.
[5] Stützle T, Hoos HH. MAX–MIN Ant System. Future Gener Comput Syst. 2000; 16(8):889–914.
[6] Dorigo M, Gambardella LM. Ant colony system: a cooperative learning approach to the traveling salesman problem. IEEE Trans Evol Comput. 1997; 1(1):53–66.
[7] Dorigo M, Gambardella LM. Ant colonies for the travelling salesman problem. Biosystems. 1997; 43(2):73–81.
[8] Dorigo M, Birattari M, Stutzle T. Ant colony optimization. IEEE Comput Intell Mag 2006; 1(4):28–39.
[9] Passino KM. Biomimicry of bacterial foraging for distributed optimization and control. IEEE Control Syst. 2002; 22(3):52–67.
[10] Gazi V, Passino KM, editors. Stability analysis of swarms in an environment with an attractant/repellent profile. In: Proceedings of the 2002 American control conference (IEEE Cat NoCH37301); 2002.
[11] Gazi V, Passino KM, editors. Stability analysis of swarms. In: 2002 Proceedings of the American control conference; 2002.
[12] Liu Y, Passino KM. Biomimicry of social foraging bacteria for distributed optimization: models, principles, and emergent behaviors. J Optim Theory Appl. 2002; 115(3):603–628.
[13] Abraham A, Biswas A, Dasgupta S, Das S, editors. Analysis of reproduction operator in Bacterial Foraging Optimization Algorithm. In: IEEE congress on evolutionary computation, CEC 2008 (IEEE world congress on computational intelligence), 1–6 June 2008; 2008.
[14] Biswas A, Das, S, Abraham, A, Dasgupta S. Stability analysis of the reproduction operator in bacterial foraging optimization. Theor Comput Sci. 2010; 411(21):2127–2139.
[15] Biswas A, Das, S, Abraham, A, Dasgupta S. Analysis of the reproduction operator in an artificial bacterial foraging system. Appl Math Comput. 2010; 215(9):3343–3355.
[16] Das S, Dasgupta S, Biswas A, Abraham A, Konar A. On stability of the chemotactic dynamics in bacterial-foraging optimization algorithm. IEEE Trans Syst Man Cybern, Part A, Syst Hum. 2009; 39(3):670–679.
[17] Dasgupta S, Das, S, Abraham, A, Biswas A. Adaptive computational chemotaxis in bacterial foraging optimization: an analysis. IEEE Trans Evol Comput. 2009; 13(4):919–941.
[18] Mishra S. A hybrid least square-fuzzy bacterial foraging strategy for harmonic estimation. IEEE Trans Evol Comput. 2005; 9(1):61–73.
[19] Dasgupta S, Das, S, Biswas A, Abraham A. Automatic circle detection on digital images with an adaptive bacterial foraging algorithm. Soft Comput. 2010; 14(11):1151–1164.
[20] Panda R, Naik MK, Panigrahi BK. Face recognition using bacterial foraging strategy. Swarm Evol Comput. 2011; 1(3):138–146.
[21] Hernández-Ocaña B, Mezura-Montes E, Pozos-Parra P, editors. A review of the bacterial foraging algorithm in constrained numerical optimization. In: 2013 IEEE congress on evolutionary computation, 20–23 June 2013; 2013.
[22] Das S, Biswas A, Dasgupta S, Abraham A. Bacterial foraging optimization algorithm: theoretical foundations, analysis, and applications. In: Abraham A, Hassanien A-E, Siarry P, Engelbrecht A, editors. Foundations of computational intelligence, volume 3. Studies in

computational intelligence. vol. 203. Berlin, Heidelberg: Springer; 2009. 23–55.
[23] Karaboga D. An idea based on honey bee swarm for numerical optimization. Turkey: Erciyes University; 2005.
[24] Karaboga D, Basturk B. Artificial Bee Colony (ABC) optimization algorithm for solving constrained optimization problems. In: Proceedings of the 12th international fuzzy systems association world congress on foundations of fuzzy logic and soft computing, Cancun, Mexico. Springer-Verlag; 2007. 789–798.
[25] Yang X-S, Deb S, editors. Cuckoo search via Lavy flights. In: NaBIC 2009 World Congress on Nature & Biologically Inspired Computing; 9–11 Dec. 2009.
[26] Barthelemy P, Bertolotti J, Wiersma DS. A Levy flight for light. Nature. 2008; 453(7194):495–498.
[27] Gutowski M. Levy flights as an underlying mechanism for global optimization algorithms. ArXiv Mathematical Physics e-Prints. 2001.
[28] Yang X-S. Cuckoo search and firefly algorithm: overview and analysis. In: Yang X-S, editor. Cuckoo search and firefly algorithm. Studies in computational intelligence. 516. Springer International Publishing; 2014. 1–26.
[29] Naik M, Nath MR, Wunnava A, Sahany S, Panda R, editors. A new adaptive cuckoo search algorithm. In: 2nd IEEE international conference on recent trends in information systems (ReTIS-15), 9–11 July 2015; Jadavpur University, Kolkata, India; 2015.
[30] Naik M, Panda, R. A novel adaptive cuckoo search algorithm for intrinsic discriminant analysis based face recognition. Appl Soft Comput. 2016; 38:661–675.
[31] Yang X-S. Firefly algorithms for multimodal optimization. In: Watanabe O, Zeugmann T, editors. Stochastic algorithms: foundations and applications. Lecture notes in computer science. vol. 5792. Berlin, Heidelberg: Springer; 2009. 169–178.
[32] Yang X-S. Cuckoo search and firefly algorithm. Springer International Publishing; 2014.
[33] Yang X-S. A new metaheuristic bat-inspired algorithm. In: González J, Pelta D, Cruz C, Terrazas G, Krasnogor N, editors. Nature inspired cooperative strategies for optimization (NICSO 2010). Studies in computational intelligence. 284. Berlin, Heidelberg: Springer; 2010. 65–74.
[34] Fister I, Yang X-S, Fister D, Fister I. Firefly algorithm: a brief review of the expanding literature. In: Yang X-S, editor. Cuckoo search and firefly algorithm: theory and applications. Cham: Springer International Publishing; 2014. 347–360.
[35] Yang X-S, He X. Bat algorithm: literature review and applications. Int J Bio-Inspir Comput 2013; 5(3):141–149.
[36] Neshat M, Sepidnam G, Sargolzaei M, Toosi AN. Artificial fish swarm algorithm: a survey of the state-of-the-art, hybridization, combinatorial and indicative applications. Artif Intell Rev. 2014; 42(4):965–997.
[37] Roth M, Wicker S. Termite: a swarm intelligent routing algorithm for mobile wireless ad-hoc networks. In: Ajith A, Crina G, Vitorino R, editors. Stigmergic optimization: Berlin, Heidelberg: Springer; 2006. 155–184.
[38] He S, Wu QH, Saunders JR. Group search optimizer: an optimization algorithm inspired by animal searching behavior. IEEE Trans Evol Comput. 2009; 13(5):973–990.
[39] Yang X-S, Karamanoglu M, He X. Flower pollination algorithm: a novel approach for multiobjective optimization. Eng Optim. 2014; 46(9):1222–1237.
[40] Chiroma H, Shuib NLM, Muaz SA, Abubakar AI, Ila LB, Maitama JZ. A review of the applications of bio-inspired flower pollination algorithm. Proc Comput Sci. 2015; 62:435–441.
[41] Dai H, Zhao G, Lu J, Dai S. Comment and improvement on "A new Fruit Fly Optimization Algorithm: Taking the financial distress model as an example". Knowl-Based Syst. 2014; 59:159–160.
[42] Pan W-T. A new Fruit Fly Optimization Algorithm: taking the financial distress model as an

example. Knowl-Based Syst. 2012; 26:69–74.
[43] Bolaji ALa, Al-Betar MA, Awadallah MA, Khader AT, Abualigah LM. A comprehensive review: Krill Herd algorithm (KH) and its applications. Appl Soft Comput. 2016; 49:437–446.
[44] Gandomi AH, Alavi AH. Krill herd: a new bio-inspired optimization algorithm. Commun Nonlinear Sci Numer Simul. 2012; 17(12):4831–4845.
[45] Rajabioun R. Cuckoo optimization algorithm. Appl Soft Comput. 2011; 11(8):5508–5518.
[46] Kaveh A, Farhoudi N. A new optimization method: dolphin echolocation. Adv Eng Softw. 2013; 59:53–70.
[47] Mirjalili S, Mirjalili SM, Lewis A. Grey wolf optimizer. Adv Eng Softw. 2014; 69(0):46–61.
[48] Uymaz SA, Tezel G, Yel E. Artificial algae algorithm (AAA) for nonlinear global optimization. Appl Soft Comput. 2015; 31:153–171.
[49] Mirjalili S. The ant lion optimizer. Adv Eng Softw. 2015; 83:80–98.
[50] Abedinia O, Amjady N, Ghasemi A. A new metaheuristic algorithm based on shark smell optimization. Complexity. 2016; 21(5):97–116.
[51] Yong W, Tao W, Cheng-Zhi Z, Hua-Juan H. A new stochastic optimization approach — dolphin swarm optimization algorithm. Int J Comput Intell Appl. 2016; 15(02):1650011.
[52] Saremi S, Mirjalili S, Lewis A. Grasshopper optimisation algorithm: theory and application. Adv Eng Softw. 2017; 105:30–47.
[53] Fausto F, Cuevas E, Valdivia A, González A. A global optimization algorithm inspired in the behavior of selfish herds. Biosystems. 2017; 160:39–55.
[54] Cheng M-Y, Prayogo D. Symbiotic Organisms Search: a new metaheuristic optimization algorithm. Comput Struct. 2014; 139:98–112.
[55] Dhiman G, Kumar V. Spotted Hyena Optimizer: a novel bio-inspired based metaheuristic technique for engineering applications. Adv Eng Softw. 2017; 114:48–70.
[56] Rajakumar BR. The Lion's Algorithm: a new nature-inspired search algorithm. Proc Technol. 2012; 6:126–135.
[57] Yazdani M, Jolai F. Lion Optimization Algorithm (LOA): a nature-inspired metaheuristic algorithm. J Comput Des Eng. 2016; 3(1):24–36.
[58] Chu S-C, Tsai P-W, Pan J-S. Cat swarm optimization. In: Yang Q, Webb G, editors. PRICAI 2006: trends in artificial intelligence: 9th Pacific Rim international conference on artificial intelligence, 2006 proceedings, Guilin, China, August 7–11. Berlin, Heidelberg: Springer; 2006. 854–858.
[59] Li MD, Zhao H, Weng XW, Han T. A novel nature-inspired algorithm for optimization: virus colony search. Adv Eng Softw. 2016; 92:65–88.
[60] Salcedo-Sanz S, Del Ser J, Landa-Torres I, Gil-Lopez S, Portilla-Figueras JA. The Coral Reefs Optimization Algorithm: a novel metaheuristic for efficiently solving optimization problems. Sci World J. 2014;2014:15.
[61] Mirjalili S, Lewis A. The whale optimization algorithm. Adv Eng Softw. 2016; 95:51–67.
[62] Jahani E, Chizari M. Tackling global optimization problems with a novel algorithm – Mouth Brooding Fish algorithm. Appl Soft Comput. 2018; 62:987–1002.
[63] Sharafi Y, Khanesar MA, Teshnehlab M. COOA: competitive optimization algorithm. Swarm Evol Comput. 2016; 30:39–63.
[64] Kirkpatrick S, Gelatt CD, Vecchi MP. Optimization by simulated annealing. Science. 1983; 220(4598):671.
[65] Feoktistov V. Differential evolution. Springer; 2006.
[66] Storn R, Price K. Differential evolution – a simple and efficient heuristic for global optimization over continuous spaces. J Glob Optim. 1997; 11(4):341–359.
[67] Storn R, editor On the usage of differential evolution for function optimization. In: 1996 biennial conference of the North American Fuzzy Information Processing Society (1996

NAFIPS), 19–22 Jun 1996; 1996.
[68] Rashedi E, Nezamabadi-pour H, Saryazdi S. GSA: a gravitational search algorithm. Inf Sci. 2009; 179(13):2232–2248.
[69] Sadollah A, Bahreininejad A, Eskandar H, Hamdi M. Mine blast algorithm: a new population based algorithm for solving constrained engineering optimization problems. Appl Soft Comput. 2013; 13(5):2592–2612.
[70] Shareef H, Ibrahim AA, Mutlag AH. Lightning search algorithm. Appl Soft Comput. 2015; 36:315–333.
[71] Ghorbani N, Babaei E. Exchange market algorithm. Appl Soft Comput. 2014; 19:177–187.
[72] Zhao W, Wang L, Zhang Z. Atom search optimization and its application to solve a hydrogeologic parameter estimation problem. Knowl-Based Syst. 2019; 163:283–304.
[73] Moghdani R, Salimifard K. Volleyball Premier League Algorithm. Appl Soft Comput. 2018; 64:161–185.
[74] Cheraghalipour A, Hajiaghaei-Keshteli M, Paydar MM. Tree Growth Algorithm (TGA): a novel approach for solving optimization problems. Eng Appl Artif Intell. 2018; 72:393–414.
[75] Mirjalili S, Mirjalili SM, Hatamlou A. Multi-Verse Optimizer: a nature-inspired algorithm for global optimization. Neural Comput Appl 2016; 27(2):495–513.
[76] Tabari A, Ahmad A. A new optimization method: electro-search algorithm. Comput Chem Eng. 2017; 103:1–11.
[77] Kaveh A, Dadras A. A novel meta-heuristic optimization algorithm: thermal exchange optimization. Adv Eng Softw. 2017; 110:69–84.
[78] Zheng Y-J. Water wave optimization: a new nature-inspired metaheuristic. Comput Oper Res. 2015; 55:1–11.
[79] Baykasoğlu A, Akpinar Ş. Weighted Superposition Attraction (WSA): a swarm intelligence algorithm for optimization problems – Part 1: Unconstrained optimization. Appl Soft Comput. 2017; 56:520–540.
[80] Baykasoğlu A, Akpinar Ş. Weighted Superposition Attraction (WSA): a swarm intelligence algorithm for optimization problems – Part 2: Constrained optimization. Appl Soft Comput. 2015; 37:396–415.
[81] Nematollahi AF, Rahiminejad A, Vahidi B. A novel physical based meta-heuristic optimization method known as Lightning Attachment Procedure Optimization. Appl Soft Comput. 2017; 59:596–621.
[82] Črepinšek M, Liu S-H, Mernik L. A note on teaching–learning-based optimization algorithm. Inf Sci. 2012; 212(0):79–93.
[83] Mirjalili S. Dragonfly algorithm: a new meta-heuristic optimization technique for solving single-objective, discrete, and multi-objective problems. Neural Comput Appl 2016; 27(4):1053–1073.
[84] Askarzadeh A. A novel metaheuristic method for solving constrained engineering optimization problems: crow search algorithm. Comput Struct. 2016; 169:1–12.
[85] Mirjalili S, Gandomi AH, Mirjalili SZ, Saremi S, Faris H, Mirjalili SM. Salp Swarm Algorithm: a bio-inspired optimizer for engineering design problems. Adv Eng Softw. 2017; 114:163–191.
[86] Qi X, Zhu Y, Zhang H. A new meta-heuristic butterfly-inspired algorithm. J Comput Sci. 2017; 23:226–239.
[87] Dhiman G, Kaur A. STOA: a bio-inspired based optimization algorithm for industrial engineering problems. Eng Appl Artif Intell. 2019; 82:148–174.
[88] Dhiman G, Kumar V. Seagull optimization algorithm: theory and its applications for large-scale industrial engineering problems. Knowl-Based Syst. 2019; 165:169–196.
[89] Heidari AA, Mirjalili S, Faris H, Aljarah I, Mafarja M, Chen H. Harris hawks optimization: algorithm and applications. Future Gener Comput Syst. 2019; 97:849–872.

[90] Jain, M, Singh V, Rani A. A novel nature-inspired algorithm for optimization: squirrel search algorithm. Swarm Evol Comput. 2019; 44:148–175.
[91] Soneji H, Sanghvi RC, editors. Towards the improvement of Cuckoo search algorithm. In: 2012 world congress on information and communication technologies (WICT), Oct. 30 2012–Nov. 2 2012; 2012.
[92] Yang X-S, Ting TO, Karamanoglu M. Random walks, Lévy flights, Markov chains and metaheuristic optimization. In: Jung H-K, Kim JT, Sahama T, Yang C-H, editors. Future information communication technology and applications. Lecture notes in electrical engineering. 235. Springer Netherlands; 2013. 1055–1064.
[93] Jafari M, Bayati Chaleshtari MH. Using dragonfly algorithm for optimization of orthotropic infinite plates with a quasi-triangular cut-out. Eur J Mech A, Solids. 2017; 66:1–14.
[94] Mafarja M, Heidari AA, Faris H, Mirjalili S, Aljarah I. Dragonfly Algorithm: theory, literature review, and application in feature selection. In: Mirjalili S, Song Dong J, Lewis A, editors. Nature-inspired optimizers: theories, literature reviews and applications. Cham: Springer International Publishing; 2020. 47–67.
[95] Zolghadr-Asli B, Bozorg-Haddad O, Chu X. Crow Search Algorithm (CSA). In: Bozorg-Haddad O, editor. Advanced optimization by nature-inspired algorithms. Singapore: Springer Singapore; 2018. 143–149.
[96] Tamilarasan A, Renugambal A, Manikanta D, Sekhar Reddy GBC, Sravankumar K, Sreekar B, et al. Application of crow search algorithm for the optimization of abrasive water jet cutting process parameters. In: IOP conference series: materials science and engineering. 2018; 390:012034.
[97] Ela AAAE, El-Sehiemy RA, Shaheen AM, Shalaby AS, editors. Application of the crow search algorithm for economic environmental dispatch. In: 2017 nineteenth international Middle East power systems conference (MEPCON), 19–21 Dec. 2017; 2017.
[98] Parvathavarthini S, Karthikeyani Visalakshi N, Shanthi S, Lakshmi K. Crow-search-based intuitionistic fuzzy C-means clustering algorithm. In: Vijayan S, editor. Developments and trends in intelligent technologies and smart systems. Hershey, PA, USA: IGI Global; 2018. 129–150.
[99] Abdelaziz, AY, Fathy, A. A novel approach based on crow search algorithm for optimal selection of conductor size in radial distribution networks. Int J Eng Sci Technol. 2017; 20(2):391–402.
[100] Kanoosh HM, Houssein EH, Selim MM. Salp swarm algorithm for node localization in wireless sensor networks. J Comput Netw Commun. 2019; 2019:12.
[101] Faris H, Mirjalili S, Aljarah I, Mafarja M, Heidari AA. Salp Swarm Algorithm: theory, literature review, and application in extreme learning machines. In: Mirjalili S, Song Dong J, Lewis A, editors. Nature-inspired optimizers: theories, literature reviews and applications. Cham: Springer International Publishing; 2020. 185–199.
[102] Yao X, Yong L, Guangming L. Evolutionary programming made faster. IEEE Trans Evol Comput. 1999; 3(2):82–102.
[103] Cheng M-Y, Lien L-C. Hybrid artificial intelligence-based PBA for benchmark functions and facility layout design optimization. J Comput Civ Eng. 2012; 26(5):612–624.
[104] Jamil M, Yang X-S. A literature survey of benchmark functions for global optimization problems. Int J Math Model Numer Optim. 2013; 4(2):150–194.
[105] Suganthan P, Hansen N, Liang J, Deb K, Chen Y-P, Auger A, et al. Problem definitions and evaluation criteria for the CEC 2005 special session on real-parameter optimization. Nat Comput. 2005; 341–357.
[106] Tang K, Li X, Suganthan P, Yang Z, Weise T. Benchmark functions for the CEC'2008 special session and competition on large scale global optimization; 2009.

[107] Tang K, Li X, Suganthan PN, Yang Z, Weise T, editors. Benchmark functions for the CEC'2010 special session and competition on large-scale global optimization; 2010.
[108] Li X, Tang K, Omidvar MN, Yang Z, Qin K, editors. Benchmark functions for the CEC'2013 special session and competition on large-scale global optimization; 2013.
[109] Liang JJ, Qu BY, Suganthan PN, editors. Problem definitions and evaluation criteria for the CEC 2014 special session and competition on single objective real-parameter numerical optimization; 2013.
[110] Chen Q, Liu B, Zhang Q, Liang JJ, Suganthan PN, Qu BY, editors. Problem definitions and evaluation criteria for CEC 2015 special session on bound constrained single-objective computationally expensive numerical optimization; 2015.

Gunjan Goyal, Pankaj Kumar Srivastava, and Dinesh C. S. Bisht

2 Genetic algorithm: a metaheuristic approach of optimization

Abstract: Genetic algorithms are the search algorithms inspired from natural genetics. The genetic algorithm was proposed by John Holland, for the search and optimization problems to find the best solution among a population of solutions that works on the basic genetic operators namely encoding, selection, crossover and mutation and variants of a genetic algorithm are based on the different types of these operators, which are explored to minimize the time needed to find a solution. In this chapter, genetic algorithms and variants of genetic algorithms will be discussed with their advantages and disadvantages.

Keywords: Evolutionary algorithm, genetic algorithm, encoding, selection, crossover, mutation

2.1 Introduction

Nature is extremely smart. Through millions and billions of years, nature find ways to optimize everything it does. To learn and use the methods that had worked before naturally, Charles Darwin introduced the theory of evolution and the process of natural selection. An evolutionary algorithm is an optimization tool which works on the principle of Darwin's theory. This chapter focuses on a particular type of evolutionary algorithm, i. e., genetic algorithm and some of its variants.

> It is not the strongest of the species that survives, nor the most intelligent that survives. It is the one that is the most adaptable to change, that lives within the means available and works cooperatively against common threats.
>
> Charles Darwin

Evolution is the optimization process and the genetic algorithm simulates the process of evolution. Genetic algorithms are the metaheuristic search algorithms inspired from natural genetics often applied to the optimization or learning. These algorithms are used because they are more robust over conventional algorithms. The genetic algorithm has many advantages and it can overcome many limitations of traditional tools but it has a few disadvantages as well.

Gunjan Goyal, Pankaj Kumar Srivastava, Dinesh C. S. Bisht, Department of Mathematics, Jaypee Institute of Information Technology, 201304 Noida, India, e-mails: gunjangoyal.04@gmail.com, panx.16j@gmail.com, drbisht.math@gmail.com

https://doi.org/10.1515/9783110671353-002

Advantages
- It can deal with a large number of variables (discrete or continuous).
- It can provide us with a list of fit solutions.
- It is appropriate for parallelism.
- It is a versatile tool for optimization.

Disadvantages
- This optimization tool works like a black-box tool.
- It does not guarantee the optimal solution.
- It is computationally high-priced.
- Finding appropriate parameters like fitness function, population size, etc. is a difficult task.

The theoretical basis for a genetic algorithm was given by J. Holland [1], for the search and optimization problems to find the suitable solution among a list of solutions and it was popularized by David Goldberg [2], one of the students of Holland, who solved the problem, which involved the control of a gas pipeline transmission. The genetic algorithm works on the four basic genetic operators:
- Encoding
- Selection
- Crossover
- Mutation

Before going into the details of these operators, first of all let us have a glimpse of the background of genetics. The basic unit of genetics is a gene. The chromosomes carry the genes by the double-helix structure called deoxyribonucleic acid (DNA). DNA carries the genetic information in a specific form, known as a genetic code. Every individual has a specific genetic code. Genes basically represent the characteristics of an individual and possible combinations of genes for one property and are known as alleles. Also, a gene can take various alleles. The set of all of the possible alleles is known as a gene pool and this gene pool gives the idea of variants in future generations. For example, if a person has brown hair, he may have an allele for black hair as well. However in this case, the dominant allele is for brown hair. So, there is a possibility that the next generation may have black hair or a combination of black-brown hair. This study of genetics started with a pea plant experiment done by Gregor Mendel in 1822 [3, 4].

2.2 Working of genetic algorithm

The basic terminologies used in the process are population, genes, chromosome and fitness value. The population is the set of individuals representing the feasible so-

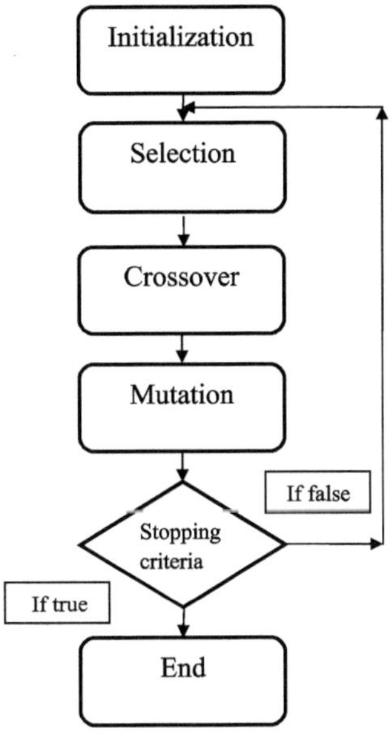

Figure 2.1: Flowchart for genetic algorithm.

lution for a specified problem and a solution is the set of parameters recognized as genes. A string of values is formed by joining these genes known as chromosomes. Every chromosome has a fitness score which is evaluated using the fitness function.

Let us understand the working principle of genetic algorithm (Figure 2.1) in brief:

Step 1. First of all, a random population of solutions is generated in the genes representation.
Step 2. The fitness value of each of the solution is evaluated and then selection scheme is used to select good solutions based on the fitness value. With the help of this selection scheme, mating pool is obtained.
Step 3. Crossover is applied to the above mating pool. In the process of crossover, the properties of parent solutions are exchanged and new child solutions are obtained.
Step 4. Mutation: Sudden change in some of the genes depending upon the mutation schemes to obtain a local change in the solution.
Step 5. Once the selection, crossover and mutation have occurred, this is known as one generation of GA. The whole process is repeated until the desired output is obtained or the process can be terminated using different termination criteria.

Now, to understand the genetic algorithm, let us consider an example in which one coin is tossed 60 times. Our objective is to maximize the number of tails. In the first 10 tosses, the result obtained is 1 1 1 1 0 1 0 1 0 1. Here, 1 represents the occurrence of tail and 0 represents occurrence of tails. Similarly, the results are obtained in 6 slots ($T_1, T_2, \ldots, T_6$) of 10 tosses each. Here, fitness function is occurrence of 1's in a chromo-

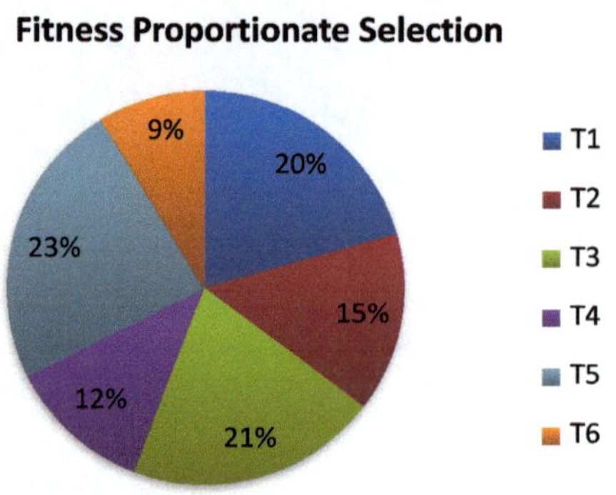

Figure 2.2: Roulette wheel sampling.

some and population size is 6. So the process of initialization is done (Table 2.1 shows initial population).

Table 2.1: Fitness value for Initial population.

Chromosome		Fitness value
T_1	1 1 1 1 0 1 0 1 0 1	7
T_2	0 1 1 1 0 0 0 1 0 1	5
T_3	1 1 1 0 1 1 0 1 0 1	7
T_4	0 1 0 0 0 1 0 0 1 1	4
T_5	1 1 1 0 1 1 1 1 0 1	8
T_6	0 1 0 0 1 1 0 0 0 0	3
Total		34

Now for the selection technique, roulette wheel sampling method (Figure 2.2) is applied and (T_1, T_3) and (T_4, T_6) are selected for single point crossover:

T_1	1111010101
T_3	1110110101
	↓
T_1'	1111110101
T_3'	1110010101

T_4	0100010011
T_6	01001 10000
	↓
T_4'	0100010000
T_6'	01001 10011

From the process of crossover, four new chromosomes are obtained as shown above. After applying mutation on T_2 and T_4', new population is obtained shown in Table 2.2 and the new fitness value is calculated, i. e., 36, which is higher than initial fitness value.

Table 2.2: Fitness value for new population.

New Chromosome		Fitness value
T_1'	1 1 1 1 1 1 0 1 0 1	8
T_2	0 1 1 1 0 1 0 1 0 1	6
T_3'	1 1 1 0 0 1 0 1 0 1	6
T_4'	0 1 0 0 0 1 0 0 1 0	3
T_5	1 1 1 0 1 1 1 1 0 1	8
T_6'	0 1 0 0 1 1 0 0 1 1	5
Total		36

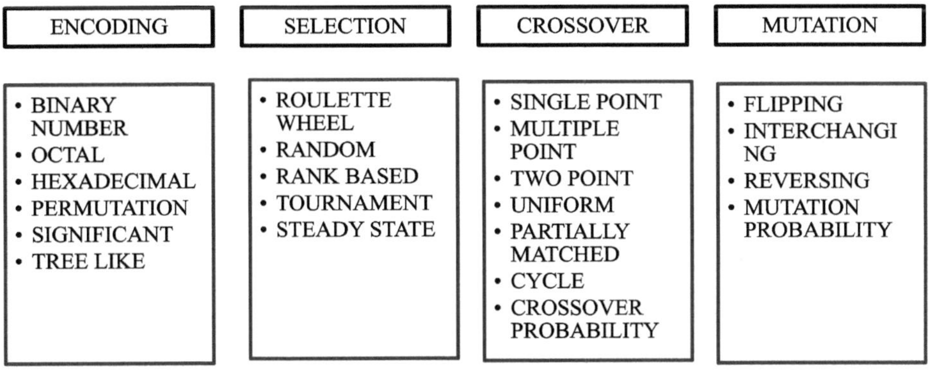

Figure 2.3: Different schemes of basic operators of genetic algorithm.

The above mentioned steps (encoding, selection, crossover and mutation) can be done using different schemes (shown in Figure 2.3). A variety of schemes of these operators had been introduced in literature. Some of them are explained below with the help of examples for each scheme.

2.3 Different schemes of basic operators of genetic algorithm

2.3.1 Encoding

The process of converting genes into a coded form is known as encoding. Encoding refers to the kind of solution set which is to be optimized. These coded forms can be of different types like numbers, bits, trees or values, depending upon the problem. They are classified into two categories: 1-dimensional encoding (binary, octal, hexadecimal, permutation, value) and 2-dimensional encoding [5].

2.3.1.1 Binary number encoding

In the binary number encoding, every chromosome is presented in binary numbers (0 & 1). It is commonly used 1-dimensional encoding method due to its simplicity. Each string consists of some bits, where each bit can represent the characteristics of solution and the string length depends upon accuracy.

For example: Chromosome A and chromosome B are 10-bit string:

Chromosome A	1 1 0 0 1 1 0 1 0 1 0 0
Chromosome B	0 1 0 1 1 0 0 1 1 0 0 1

2.3.1.2 Octal encoding

In this scheme of encoding, strings are made from the octal numbers, i. e., 0 to 7, known as octal encoding. Due to small size, octal encoding is advantageous over binary encoding.

For example:

Chromosome A	20346151
Chromosome B	12670231

2.3.1.3 Hexadecimal encoding

Hexadecimal encoding is 1-dimensional type of encoding in which the strings are made of hex (0–9, A–F).

For example:

Chromosome A	97AE
Chromosome B	A2C6

2.3.1.4 Permutation encoding

Each chromosome in permutation encoding is composed of a sequence of integers/real numbers. Permutation encoding is commonly applied in ordering problems, where the string represents the sequence of cities visited by a salesman.

For example:

Chromosome A	8 4 6 7 2 3 1 5 9
Chromosome B	1 5 3 6 2 4 7 9 8

2.3.1.5 Significant encoding

This type of encoding represents the chromosome as the string of complicated values such as integer or object or character depending upon the specific problem. It is beneficial for some specific problems where binary encoding would be very difficult such as finding weights for neural network.

For example:

Chromosome A	ABDJEIFJHDDLDFLFEGT
Chromosome B	(right), (left), (back), (forward), (left)

2.3.1.6 Tree-like encoding

Tree-like encoding is a 2-dimensional type of encoding that is used for evolving expressions or programs, mainly used in genetic programming. Each chromosome is represented as a tree of functions or commands of programming language.

For example:

Chromosome A

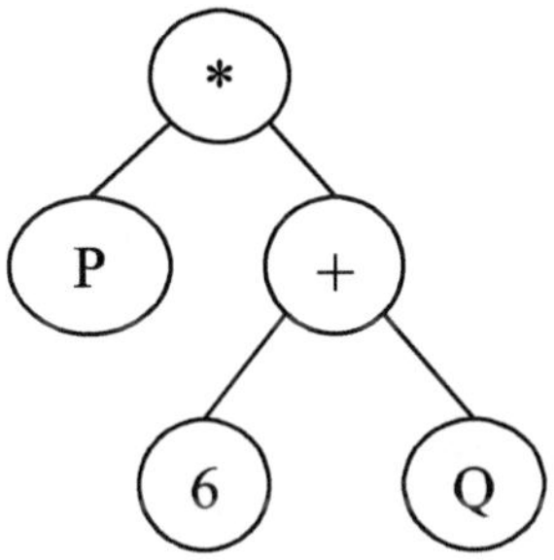

(* P (+ 6 Q))

Chromosome B

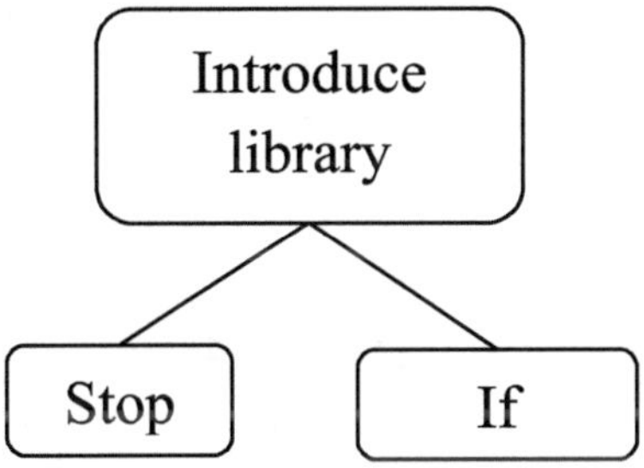

(Introduce library Stop If)

2.3.2 Selection

In the process of genetic algorithm, selection of parent chromosomes is done from population for crossing. The process of selection is done to obtain the offspring of individuals having higher fitness. From the initial population, individuals (parents) are selected for reproduction according to their evaluation function. Selection scheme is of the two kinds: proportionate based selection (based upon fitness value) and ordinal based selection (depending upon the rank within individuals). A variety of selection methods are discussed below [6, 7].

2.3.2.1 Proportionate selection

In proportionate selection, probability of chromosome getting selected in mating pool and the fitness are proportional to each other. The aim of this type of selection is the search of chromosomes using roulette wheel due to which it is also known as roulette wheel selection. Let us consider a roulette wheel where every individual is placed in a wheel segment where area of each segment is relative to fitness. The wheel is rotated either in clockwise or anticlockwise direction and number of rotations will be equal to population size (n). For selecting a chromosome, a marble is thrown there. On every spin, a chromosome is selected by roulette wheel pointer. For the selection, the probability p_i of each chromosome is calculated and then cumulative probability is evaluated. The range of cumulative frequency will be 0 to 1. A random number is chosen from 0 to 1. If this random number lies in cumulative probability range p_{i-1} to p_i, then that chromosome is selected. This process is repeated n times. Therefore, the chromosome having a higher fitness value will have more probability of getting selected. In this process, the selection is easier but the rate of evolution and variance of fitness is dependent, due to which one can face problems if the fitness varies too much.

2.3.2.2 Random selection

In random selection, parents are randomly chosen from population. This type of selection is more complicated than the roulette wheel selection.

2.3.2.3 Rank based selection

In 1985, rank based selection was given by Baker [8] in which population is ranked first and then fitness of each chromosome is received from the ranking. In the process, the chromosomes receive the fitness in decreased order, i. e., the worst chromosome has fitness one and the best chromosome has the fitness N, where N is total chromosomes in population. It has a slower convergence. The advantage of the process is that it prevents too quick convergence which may lead to poor selection.

2.3.2.4 Tournament selection

This type of strategy holds a tournament competition [7] among the population, i. e., randomly choose some chromosomes and from that set, select best individuals for further genetic procedure. Generally, the tournament is held between two individuals but it may occur in more individuals as well. The method of tournament selection is more proficient which leads to a most favorable solution.

2.3.2.5 Steady state based selection

The main inspiration of steady state based selection is that major part of chromosome should survive in next generation. It is not a specific method of selecting parents. From every generation, some of the chromosomes having high fitness value are chosen to create new offspring. Then these new offspring replace the bad chromosomes and remaining population carries on to next generation.

2.3.3 Crossover

The process of producing a child solution from a two-parent solution is known as crossover. The crossover is done by applying the operator to the mating pool, hoping for the better child solution (high fitness value) to be produced. In the process, properties are exchanged between the parents. There are various crossover techniques in genetic algorithm literature [9, 10, 11, 12, 13, 14]. Some of them are discussed below.

2.3.3.1 Single point crossover

Parent chromosomes are slashed only once and the corresponding parts are exchanged. This type of crossover is mainly used by the traditional genetic algorithm. Crossover position is selected at random and bits after that position is exchanged, which implies that if appropriate position is chosen, enhanced child chromosomes can be achieved.

For example: The crossover position, i. e., the fifth position is chosen at random. Therefore, the bits after the fifth position are exchanged:

First Parent	1 0 1 1 1 0 0 1
Second Parent	0 1 0 1 0 0 1 1

First Child	1 0 1 1 1 0 1 1
Second Child	0 1 0 1 0 0 0 1

2.3.3.2 Two-point crossover

In the two-point crossover, chromosomes are slashed at two positions and bits in between the cuts are exchanged. The introduction of more cuts reduces the performance of genetic algorithm and the reason behind the reduction of performance may be the disruption in the building blocks. Besides the above disadvantage, it has one advantage that problem space can be searched more effectively. Two-point crossovers is preferred over single point crossover because if the chromosome is having good genetic information in the head and the crossover is occurring at the tail, then chromosome produced may not be the good one.

For example:

First Parent	1 0 1 1 1 0 0 1
Second Parent	0 1 0 1 0 0 1 1

First Child	1 0 0 1 0 0 0 1
Second Child	0 1 1 1 1 0 1 1

2.3.3.3 Multiple point crossover

In multiple point crossovers, multiple cuts are introduced and the alternative sections are swapped. Multiple point crossovers are equivalent to two-point crossover.

For example:

First Parent	1 0 1 1 1 0 0 1
Second Parent	0 1 0 1 0 0 1 1

First Child	0 1 1 1 0 0 0 1
Second Child	1 0 0 1 1 0 1 1

2.3.3.4 Uniform crossover

In the uniform crossover [15], the cuts are not introduced in the chromosomes; instead, the bits are swapped independently depending upon the distribution. By flipping the coin, it is decided whether an individual gene should be included in child chromosome or not.

For example:

First Parent	0 1 2 3 4 5 6 7 8 9
Second Parent	5 8 9 4 2 3 5 7 5 8

First Child	5 1 9 4 4 5 5 7 5 9
Second Child	0 8 2 3 2 3 6 7 8 8

2.3.3.5 Crossover probability

Crossover probability (P_c) is the basic parameter used in crossover, which describes how frequently crossover will be performed.

2.3.3.6 Partially matched crossover

In partially matched crossover (PMX) [4, 5], information that is transferred from parent chromosome to the child chromosome is based on value and order, i. e., mapping is done from one portion of first parent string onto second parent string and rest information is exchanged. This type of crossover is mainly used in traveling salesman problem (TSP) [16] because in TSP, chromosomes are the values which represent the different cities and order corresponds to the time of visiting a city.

For example:

First Parent	1 2 3 4 5 6 7 8
Second Parent	3 7 5 1 6 8 2 4

First Child	4 2 3 1 6 8 7 5
Second Child	2 7 8 4 5 6 2 1

2.3.3.7 Cycle crossover

In the cycle crossover (CX) [9, 10], the child chromosome is obtained by occupying the corresponding element from the parent chromosome and remaining bits of the string are exchanged.

For example:

First Parent	8 4 7 3 6 2 5 1 9 0
Second crossover	0 1 2 3 4 5 6 7 8 9

First Child	8 1 2 3 4 5 6 7 9 0
Second Child	0 4 7 3 6 2 5 1 8 9

2.3.4 Mutation

Mutation is the process used to introduce and maintain the diversity in the population. Mutation is defined as the process of alteration of one or more bits in the string of chromosomes from its initial population to get a new solution. This process is essential to the convergence of the genetic algorithm.

2.3.4.1 Flipping mutation

Flipping mutation inverts the bits of the generated mutation chromosome. The flipping of bits generally occurs in binary representation either from 0 to 1 or 1 to 0.

For example:

Parent Chromosome	1 0 1 0 0 1 1 0
Child Chromosome	1 0 0 0 0 1 1 0

2.3.4.2 Interchanging mutation

In this mutation, randomly two positions are taken and the bits are interchanged corresponding to those positions.

For example:

Parent Chromosome	1 0 1 1 0 1 0 0
Child Chromosome	1 0 0 1 0 1 1 0

2.3.4.3 Reversing mutation

In this process, a position is selected randomly and bits from next position are swapped to obtain the child chromosome.

For example: In the given example, sixth position is selected and bits from seventh position are swapped:

Parent Chromosome	1 0 1 1 0 1 1 0
Child Chromosome	2 0 1 1 0 1 0 1

2.3.4.4 Mutation probability

It is the most commonly used mutation operator. Usually, mutation probability is applied with the low probability because if the high probability is used, then genetic algorithm reduces to random search.

2.4 Variants of a genetic algorithm

A large range of the variants of genetic algorithm is there, based on the different schemes. Some of them are discussed below.

2.4.1 Traditional genetic algorithm

In traditional genetic algorithm, all the genetic operators are ruled by fixed probabilities due to which each generation consists of fixed average number of local and global searches and reveals a fixed convergence rate. Therefore, also known as fixed rate genetic algorithm.

2.4.2 Binary coded genetic algorithm

The building block of binary coded genetic algorithms are genes and the chromosomes where the chromosomes are encoded into binary code string [2].

2.4.3 Real coded genetic algorithm

This type of genetic algorithm is advantageous over binary coded genetic algorithms where more numerical accuracy is required and to deal with the search spaces with many dimensions. In this type of algorithm, the chromosomes are represented in the form of vector of real numbers and any selection schemes can be applied. However, for crossover and mutation, schemes like a flat crossover, BLX-α crossover, random mutation, uniform mutation and many more are used [17, 18, 19].

2.4.4 Messy genetic algorithm

This algorithm was invented by Goldberg et al. in 1989 to speed up the convergence of the genetic algorithm where chromosomes are of variable length. In a messy genetic algorithm (MGA) [20], a position number is assigned to each binary digit because the genes are position independent. For example, (1 0 1 1 0 0) is a chromosome which is normally used in genetic algorithm in binary representation but in a messy genetic algorithm, this chromosome is represented as $[(1, 1)\ (2, 0)\ (3, 1)\ (4, 1)\ (5, 0)\ (6, 0)]$. In the ordered pair (2, 0), 2 is the position and the 0 is the bit. There are two phases in MGA: primordial phase and the juxtaposition phase. In primordial phase, initial population is generated which includes all the possible building blocks of a specified length and the GA is applied using only crossover for some of the generations where half the chromosomes are eliminated after few generations. In the second phase, genetic operators (cutting and splicing) are applied until and unless an acceptable solution is found. For cutting and splicing, two chromosomes are selected and a cut is randomly made dividing it into four sections. Then randomly two sections are selected for splicing together to obtain new chromosomes. The chromosome obtained using these two operators vary in length. Various applications using MGA are explained in detail by Knjazew in 2002 [21].

2.4.5 Microbial genetic algorithm

In the microbial genetic algorithm, two chromosomes are randomly selected from the initial population and their fitness value is determined. According to the fitness value, chromosome with a higher fitness value is called as winner and the worst of them is called as loser. In this process, only loser gets changed, i. e., crossover (copy from winner to the loser) and mutation are performed on the loser chromosome to ensure that the best chromosomes remain in the population. Lastly, the loser chromosome is replaced by the new chromosome generated in the population and continues until the stopping criterion is satisfied.

2.4.6 Parallel genetic algorithm

In parallel genetic algorithm [22], after creating initial population, decisions for selection are made by individuals. Every individual does local hill climbing and selects their partner themselves for crossover in the neighborhood and during the process, they may improve their fitness. Then the offspring produced does local hill climbing and the parent is replaced by offspring if it is better than parent. The process is continued until the convergence is obtained. The parallel genetic algorithm can be applied to problems like the traveling salesman and m-graph partitioning [23] and it can be further classified depending upon the single population and multiple populations [24].

2.4.7 Steady state based genetic algorithm

In the steady state genetic algorithm [25], the size of the population maintained by adding new members in the population one at a time and eliminates a member according to replacement strategy. This algorithm is also known as an incremental genetic algorithm.

2.4.8 Adaptive Genetic Algorithm (AGA)

A significant variant of genetic algorithm is the genetic algorithm with adaptive parameters, known as AGA. In AGA, the like population size, crossover probability and mutation probability are varied in the process. Initially convergence of GA is identified so that crossover probability and mutation probability can vary adaptively to maintain the population diversity [26].

2.4.9 Miscellaneous genetic algorithm variants

There are many more variants of genetic algorithms existing in literature. For instance, the micro genetic algorithm [19] is a variant of genetic algorithm proposed by Goldberg for small population size but it was first implemented by Krishnakumar [27] on population size 5. As an extension of micro genetic algorithm, Kuomousis and Katsaras introduced the saw-tooth genetic algorithm [28] in which the saw-tooth function is used for variable population size.

2.5 Applications of genetic algorithm and its variants

- Optimization problems like the job shop problem are effectively solved using the genetic algorithm [12].
- The genetic algorithm can be used to solve problems based on resource constrained scheduling [29].
- Genetic algorithm found its application in reduction of the additive noise in single trial evoked potentials [30].
- Genetic algorithm is successfully applied to optimize fuel consumption and range for flying altitude and for path planning for unmanned aerial vehicles (UAVs) [31, 32, 33].
- To reduce the data transmissions, genetic algorithm has been used for data replica placement strategy in cloud computing [34].
- In the power market, different variants of genetic algorithms are applied to obtain suitable solutions [35, 36].
- Identification of soil parameters can be done using real coded genetic algorithm [37].
- The improved genetic algorithm was used for tunnel network optimization in which pipe network is designed using gray codes [38].
- Trajectory planning has been done using genetic algorithm for robots [39].

- The use of genetic algorithms is also seen in the field of medical sciences to identify the ECG signals [40].
- This algorithm is found helpful in developing the model for the prediction of air blast [41].
- A new approach can be formed to model the bankruptcy prediction using genetic algorithm for the extraction of rules [42].

Bibliography

[1] Holland JH. Adaptation in natural and artificial systems. Ann Arbor: University of Michigan Press; 1975.

[2] Goldberg DE. Genetic algorithms in search, optimization, and machine learning. Addison-Wesley, Reading, MA; 1989.

[3] Andrei A. "Experiments in plant hybridization" (1866), by Johann Gregor Mendel. Embryo Proj Encycl.; 2013.

[4] Iltis H. Gregor Mendel and his work. Sci Mon. 1943;56(5):414–423.

[5] Kumar A. Encoding schemes in genetic algorithm. Int J Adv Res IT Eng. 2013;2(3):1–7.

[6] Blickle T, Thiele L. A comparison of selection schemes used in evolutionary algorithms. Evol Comput. 1996;4(4):361–394.

[7] Goldberg, DE, Deb, K. A comparative analysis of selection schemes used in genetic algorithms. In: Foundations of genetic algorithms. Elsevier; 1991. p. 69–93.

[8] Baker JE. Adaptive selection methods for genetic algorithms. In: Proceedings of an international conference on genetic algorithms and their applications. Hillsdale, New Jersey; 1985. p. 101–111.

[9] Larranaga P, Kuijpers CM, Murga RH, Yurramendi Y. Learning Bayesian network structures by searching for the best ordering with genetic algorithms. IEEE Trans Syst Man Cybern, Part A, Syst Hum. 1996;26(4):487–493.

[10] Oliver IM, Smith DJ, Holland JR. Study of permutation crossover operators on the traveling salesman problem. In: Genetic algorithms and their applications: proceedings of the second international conference on genetic algorithms, July 28–31, 1987, The Massachusetts Institute of Technology, Cambridge, MA. Hillsdale, NJ: L. Erlhaum Associates; 1987.

[11] Davis L. Applying adaptive algorithms to epistatic domains. In: IJCAI; 1985. p. 162–164.

[12] Qi JG, Burns GR, Harrison DK. The application of parallel multipopulation genetic algorithms to dynamic job-shop scheduling. Int J Adv Manuf Technol. 2000;16(8):609–615.

[13] Larrañaga P, Kuijpers CMH, Poza M, Murga RH. Optimal decomposition of Bayesian networks by genetic algorithms. Dept Com Sci Art Intel Univ Basque Ctry Int Rep EHU-KZAA-IKT-3-94; 1994.

[14] Mühlenbein H. Parallel genetic algorithms, population genetics and combinatorial optimization. In: Workshop on parallel processing: logic, organization, and technology. Springer; 1989. p. 398–406.

[15] Spears WM, De Jong KD. On the virtues of parameterized uniform crossover. Naval Research Lab, Washington DC; 1995.

[16] Goldberg DE, Lingle R. Alleles, loci, and the traveling salesman problem. In: Proceedings of an international conference on genetic algorithms and their applications. Hillsdale, NJ: Lawrence Erlbaum; 1985. p. 154–159.

[17] Goldberg DE. Real-coded genetic algorithms, virtual alphabets, and blocking. Complex Syst. 1991;5(2):139–167.
[18] Eshelman LJ, Schaffer JD. Real-coded genetic algorithms and interval-schemata. In: Foundations of genetic algorithms. Elsevier; 1993. p. 187–202.
[19] Pratihar DK. Soft computing. Alpha Science International, Ltd; 2007.
[20] Goldberg DE. Messy genetic algorithms: Motivation analysis, and first results. Complex Syst. 1989;4:415–444.
[21] Knjazew D. OmeGA: a competent genetic algorithm for solving permutation and scheduling problems, vol. 6. Springer Science & Business Media; 2012.
[22] Haupt RL, Haupt SE. Practical genetic algorithms, vol. 2. Wiley, New York; 1998.
[23] Mühlenbein H. Parallel genetic algorithms in combinatorial optimization. In: Computer science and operations research. Elsevier; 1992. p. 441–453.
[24] Sivanandam SN, Deepa SN. Principles of soft computing (with CD). John Wiley & Sons; 2007.
[25] Whitley D. GENITOR: a different genetic algorithm. In: Proceedings of the Rocky Mountain conference on artificial intelligence; 1988. p. 118–130.
[26] Srinivas M, Patnaik LM. Adaptive probabilities of crossover and mutation in genetic algorithms. IEEE Trans Syst Man Cybern. 1994 Apr;24(4):656–667.
[27] Krishnakumar K. Micro-genetic algorithms for stationary and non-stationary function optimization. In: Intelligent control and adaptive systems. International Society for Optics and Photonics; 1990. p. 289–297.
[28] Koumousis VK, Katsaras CP. A saw-tooth genetic algorithm combining the effects of variable population size and reinitialization to enhance performance. IEEE Trans Evol Comput. 2006;10(1):19–28.
[29] Kadri RL, Boctor FF. An efficient genetic algorithm to solve the resource-constrained project scheduling problem with transfer times: the single mode case. Eur J Oper Res. 2018;265(2):454–462.
[30] Pougnet MJ, Lovely DF, Parker PA. A genetic algorithm approach to the reduction of additive noise in single trial evoked potentials. CMBES Proc. 2018;33(1).
[31] Fu Y, Ding M, Zhou C, Hu H. Route planning for unmanned aerial vehicle (UAV) on the sea using hybrid differential evolution and quantum-behaved particle swarm optimization. IEEE Trans Syst Man Cybern Syst. 2013;43(6):1451–1465.
[32] Roberge V, Tarbouchi M, Labonté G. Fast genetic algorithm path planner for fixed-wing military UAV using GPU. IEEE Trans Aerosp Electron Syst. 2018.
[33] Eun Y, Bang H. Cooperative task assignment/path planning of multiple unmanned aerial vehicles using genetic algorithm. J Aircr. 2009;46(1):338–343.
[34] Cui L, Zhang J, Yue L, Shi Y, Li H, Yuan D. A genetic algorithm based data replica placement strategy for scientific applications in clouds. IEEE Trans Serv Comput. 2018;11(4):727–739.
[35] Da Silva EL, Gil HA, Areiza JM. Transmission network expansion planning under an improved genetic algorithm. IEEE Trans Power Syst. 2000;15(3):1168–1174.
[36] Mahmoudabadi A, Rashidinejad M, Zeinaddini-Maymand M. A new model for transmission network expansion and reactive power planning in a deregulated environment. Engineering. 2012;4(02):119.
[37] Jin Y-F, Yin Z-Y, Shen S-L, Zhang D-M. A new hybrid real-coded genetic algorithm and its application to parameters identification of soils. Inverse Probl Sci Eng. 2017;25(9):1343–1366.
[38] Dandy GC, Simpson AR, Murphy LJ. An improved genetic algorithm for pipe network optimization. Water Resour Res. 1996;32(2):449–458.
[39] Tian L, Collins C. An effective robot trajectory planning method using a genetic algorithm. Mechatronics. 2004;14(5):455–470.
[40] Diker A, Cömert Z, Avci E, Velappan S. Intelligent system based on Genetic Algorithm and

support vector machine for detection of myocardial infarction from ECG signals. In: 2018 26th signal processing and communications applications conference (SIU). IEEE; 2018.
[41] Armaghani DJ, Hasanipanah M, Mahdiyar A, Majid MZA, Amnieh HB, Tahir MM. Airblast prediction through a hybrid genetic algorithm – ANN model. Neural Comput Appl. 2018;29(9):619–629.
[42] Shin K-S, Lee Y-J. A genetic algorithm application in bankruptcy prediction modeling. Expert Syst Appl. 2002;23(3):321–328.

Adam Price, Thomas Joyce, and J. Michael Herrmann

3 Ant colony optimization and reinforcement learning

Abstract: Ant Colony Optimization (ACO) used to be one of the most frequently used algorithms in metaheuristic optimization. It can be interpreted as a Bayesian accumulation of information about an optimization problem, and bears some similarity to reinforcement learning. With this parentage, given a better theoretical understanding, ACO could be advanced into a practically competitive optimization algorithm. We are trying to support this development by studying the complex effects of parameters in ACO. For a toy example, which is, however, highly instructive in regard to the ability of the algorithm to overcome deception, we investigate how the parameter such as the number of ants and the rate of pheromone evaporation influences the performance. In addition, we investigate the suggestive link between ACO and reinforcement learning (RL) which does not only have the potential to facilitate the analysis of ACO, but which provides also options for the generation of new algorithms by the transfer of innovations across categories of algorithms. Based on these ideas, we propose an embedding of both ACO and RL into an joint evolutionary framework.

Keywords: Ant colony optimization, reinforcement learning, Fisher–Eigen equation, criticality, random dynamical systems, parameter selection, deceptiveness

3.1 Introduction

Ant colony optimization (ACO) refers to a class of algorithms for combinatorial optimization problems that adopts a stochastic strategy to find near-optimal solutions for search problems [1, 2]. A typical problem for ACO is the traveling salesperson problem (TSP), where the task is to find a shortest route that passes through all of a number of cities. In ACO, a small population of "ants" is used to sample several tours in each trial. Over many such trials, the information about the lengths of many tours is aggregated and eventually, if nothing goes wrong, near-optimal tours will be found.

For more than two decades, ACO has been one of the most popular algorithms in metaheuristic optimization, which is in part due to the convenient way specific knowledge can be included into the algorithm and the seemingly straightforward way it functions as an optimizer. On the other hand, while several metaheuristic algorithms are currently receiving renewed interest (see, e. g., [3]), ACO is at risk of falling behind, although it bears an immense potential which can be unlocked if a better theoretical

Adam Price, Thomas Joyce, J. Michael Herrmann, University of Edinburgh, Institute for Perception, Action and Behaviour Informatics Forum, 10 Crichton St, Edinburgh, EH8 9AB, Scotland, U.K., e-mail: michael.herrmann@ed.ac.uk

https://doi.org/10.1515/9783110671353-003

understanding of this approach can be achieved. ACO has a special position among metaheuristic algorithms for several reasons:

1. Solutions are produced stepwise by a series of decisions rather than in parallel. This is particularly useful if there is a natural order in the search space such as in dynamical optimization problems for which also reinforcement learning provides good approximations even in challenging practical applications.

2. Local and global information is combined in a multiplicative form rather than additively as usually in metaheuristic algorithms of the biased-random-walk type. In this way, ACO has an obvious Bayesian interpretation that remains to be exploited to reveal optimality properties of ACO.

3. The behavior of ACO is not easily controllable, which could be an consequence of the above two points. ACO tends to converge prematurely unless random noise is included (e. g., in the MIN–MAX version [4]). It might be preferable to adapt the parameters of ACO, but although ACO has similarities to existing machine-learning approaches where such adaptation is not uncommon, there are only a few cautious attempts to achieve this in ACO.

Our aim in this chapter is not to present a new algorithm that outperforms particularly parameterized versions of known algorithms for particularly selected benchmark functions: Optimal parameterization and fair comparison of algorithms is a problem beyond the scope of this chapter. Moreover, no-free-lunch results [5] mean that such comparison must be with regard to a specific task, rather than in a general setting. Interestingly, many current benchmark functions have little practical relevance and seem to encourage random-search algorithms rather than algorithms that can learn to represent the structure of a certain class of problems, but also the simulations included here in Section 3.3 have a purely illustrative purpose.

We will instead try to provide a better understanding of ACO algorithms, including their genuine relation to reinforcement learning (RL) algorithms. For this purpose, we will first discuss a standard ACO algorithm (Section 3.2), and then, in Section 3.3, consider a simple example which has interesting implications for the resilience of ACO to deception as well as to the parameter choice. In the remainder of this chapter, we will compare two main RL algorithms (SARSA and Q-learning) with variants of ACO and reformulate all of these in the context of an evolutionary dynamical system. This relationship can be made more fruitful on both sides, to which the main interest of this chapter is devoted.

3.2 Ant colony optimization

3.2.1 ACO algorithms

ACO is part of the larger class of Monte Carlo algorithms, which means that it uses a global criterion (such as the tour length in the TSP) in order to decide which local decisions are to be taken. We will not be able to treat here a substantial fraction of the existing ant-based algorithms; for an up-to-date and comprehensive overview, see [2]. Instead, we will restrict ourselves to a simple version of the Ant Colony System [6] of which will consider two variants. Moreover, we will discuss ACO only in the context of the Traveling Salesperson Problem (TSP) although the claims presented here are meant to apply also to other problems that are suitable for ACO.

We assume that each ant starts in one of the cities and decides to which city it will go next. The decision is based on the distances to all reachable cities and on information regarding the total length of tours that pass through both the current city and the next city. Pheromone trails are initialized by $\tau_{ij} = \tau_0$, and the pheromone update can be performed as a sum over all of the n_{ij} ants that traversed the solution component $(i.j)$

$$\tau_{ij} \leftarrow \rho\tau_{ij} + (1-\rho)\sum_{k=1}^{n_{ij}} \Delta\tau_0 \tag{3.1}$$

where $\Delta\tau_0$ is the amount of pheromone added per ant and ρ is the pheromone decay rate which means here that a fraction of $1-\rho$ of the pheromone evaporates within a time step.

Alternatively, the algorithm can use the pheromone from the best ant only,

$$\tau_{ij} \leftarrow \rho\tau_{ij} + (1-\rho)\Delta\tau_{ij}^{\text{best}} \tag{3.2}$$

where

$$\Delta\tau_{ij}^{\text{best}} = \begin{cases} \frac{1}{L} & \text{if best ant used edge } (i,j) \text{ in its tour} \\ 0 & \text{otherwise} \end{cases} \tag{3.3}$$

and L is the length of the tour the best ant has taken.

The tour of an individual ant consists of individual steps which are decided independently, apart from the effects of certain explicit conditions, such as that cities cannot be visited twice, and of the pheromone trail that also may cause problem-dependent dependencies among the decisions. This local and global information enters the probability rule by the factors η_{ij} and τ_{ij}, respectively, where i denotes the current city and j any potential next city. The probability rule [1] can be stated as

$$p(j|s^k(t-1)) = \begin{cases} \frac{\tau_{ij}^{\alpha}\eta_{ij}^{\beta}}{\sum_{m\in N(i)} \tau_{im}^{\alpha}\eta_{im}^{\beta}} & \text{if } j \in N(i) \\ 0 & \text{otherwise} \end{cases} \tag{3.4}$$

where $s^k(t-1) = (s_1^k(t-1), \dots, s_{t-1}^k(t-1))$ denotes the path taken by ant k so far, with $s_{t-1}^k(t-1) = i$, and $N(i)$ is the set of reachable nodes after arriving at node i. The local heuristic η_{ij} can be as simple as a length of the next leg in the traveling salesperson problem, or can include any knowledge that is accessible to the ant at the current step of the construction of the solution. After the transition has been made, the ant's new location $j(t)$ that was chosen based on equation (3.4) is concatenated to the path, i. e., $s^k(t) = (s_1^k(t-1), \ldots, s_{t-1}^k(t-1), j(t))$. The exponents α and β are discussed below. Equation (3.4) is often used with a probability q_0 only, while with a probability $1 - q_0$ the ant takes a random (admissible) transition, but for simplicity, we will assume here $q_0 = 1$.

3.2.2 ACO parameters

The exponent of the local heuristic in the probability rule (3.4) can be set to $\beta = 1$ without loss of generality, because instead of choosing a different value of β, it is always possible to choose a different local heuristic. In other words, the choice of values β refers to different forms of information about a certain problem or, in fact, to different problems, rather than to a different algorithm. Nevertheless, it can be useful to consider different values of β, e. g., in order to study the effect of a heuristics systematically.

The parameter α is often set to unity, but it should be noted, however, that it cannot be normalized simply by redefining the increment of the pheromone trail (3.3) or the evaporation parameter ρ, because different values of α correspond to different forms of the update rule (3.1), i. e., the algorithm does not remain the same, if α is changed. In principle, the effect of α is a combination of the accumulation and the shape of the fitness function such that a transformation of the fitness function cannot fully account for a change of α.

The exploration parameter q_0, or any thresholds values for the pheromone trail will not be considered here. The complement $1 - q_0$ plays a role similar to ε-greedy exploration in reinforcement learning. For a finite value of $1 - q_0$ (or likewise, for a nonzero minimal pheromone level) it is easy to show that the algorithm finds the global optimum [7], but this result is of little practical value because it is usually implying an exponential time complexity. Interestingly, convergence proofs for discrete-space discrete-time reinforcement learning use the same argument [8].

A similar effect is achieved by minimal and maximal pheromone levels [9, 4, 10, 11]: If rules (3.1) or (3.2) are used, only if the resulting pheromone level does not fall below a certain threshold, then the algorithm does not converge, but retains instead its flexibility, such that ongoing exploration is possible, although large scale exploration is exponentially unlikely. Moreover, the minimal pheromone level will typically be low in order to guarantee some improvement of the solutions over time. The introduction of a maximal value is less intuitive, but can be used to continue to explore several promising branches (rather than just one) with a larger probability than enabled by only guaranteeing a minimum pheromone level.

We will see below (Section 3.3) that the optimal rate of pheromone evaporation is not immediately clear, even on a quantitative level. The reason is that the persistence of the pheromones should depend on whether the ants provide useful or misleading information with respect to the achievable optimum, which is difficult to decide while the algorithm is still far from a good solution.

Optimal parameter values for ACO have been studied numerically, e. g., in [12, 13], although it is unlikely that their results are scalable with problem size [14]. Adaptive scheme for parameter adaptation remain largely an open issue [15].

The theory of ant colony algorithms has been addressed in many papers, most prominently in [16], but it has found its place also in reviews; see, e. g., [17, p. 20]. Particular results include the analysis of the convergence of the algorithms in terms of efficiency [18] and the relation to optimization by gradient descent [19]. In the following section, we will in particular consider the effect of deception, an issue that was first addressed in [20]. We will study this effect in dependence of the pheromone evaporation rate ρ, the number of ants n in a very simple example, which will, however, provide some interesting insights into ACO algorithms also in a more general sense.

3.2.3 Bayesian interpretation

Equation (3.4) has an intuitive Bayesian interpretation which, however, is not used regularly in the analysis of the algorithm, although this relation has been emphasized, e. g., in papers on learning the structure of a Bayesian network using ACO [21].

In this formulation, τ_{ij}^{α} is a prior and η_{ij}^{β} represents the likelihood of current evidence. Normalization of the two factors can be trivially include into equation (3.4),

$$\frac{\frac{\tau_{ij}^{\alpha}}{\sum_{r\in N(i)}\tau_{ir}^{\alpha}}\frac{\eta_{ij}^{\beta}}{\sum_{q\in N(i)}\eta_{iq}^{\beta}}}{\sum_{m\in N(s_i^k)}\frac{\tau_{im}^{\alpha}}{\sum_{r\in N(i)}\tau_{ir}^{\alpha}}\frac{\eta_{im}^{\beta}}{\sum_{q\in N(i)}\eta_{iq}^{\beta}}}$$

We can therefore assume that τ_{ij}^{α} and η_{ij}^{β} are already normalized which can be achieved easily by a redefinition of the local heuristic η_{ij}^{β}, and requires also a change of the update rule (3.3).

Together and after normalization, η_{ij}^{β} and τ_{ij}^{α} assign a posterior probability to the outgoing edges, expressing the posterior belief that the link (i,j) is part of the solution. Neither η_{ij}^{β} nor τ_{ij}^{α} is genuinely a probability, but as the pheromone level is positive and has an upper bound, and the local heuristics can be chosen appropriately. Instead of the inverse of the path length a regularized inverse should be used. Both quantities can be normalized and then be treated as a probability.

The optimality of Bayes law in accumulating evidence to improve a given event implies that η_{ij}^{β} and τ_{ij}^{α} and should reflect underlying probabilities as closely as possible.

3.3 Deceptiveness in ACO

3.3.1 A simple example

As a toy problem, we consider the TSP on four cities (A, B, C, D) with distances $|AB| = |BC| = |CD| = |DA| = 1$ and $|AC| = a$ and $|BD| = 2\sqrt{1-(\frac{a}{2})^2}$, i. e., the cities are located at the corners of a diamond-shaped quadrilateral; see Figure 3.1.

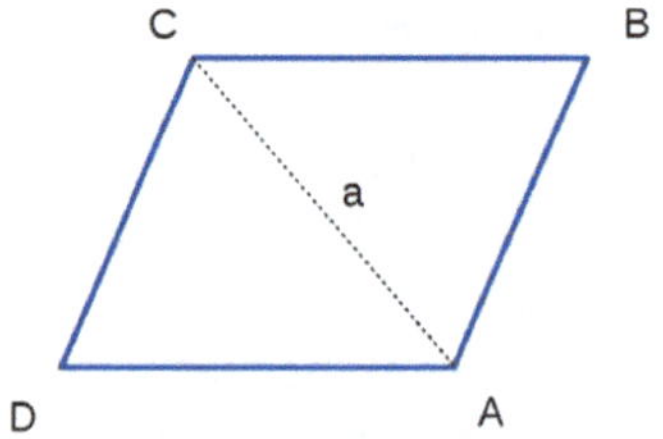

Figure 3.1: Example of a four-city TSP; see Section 3.3.1.

Paths can have two different lengths in this example: *ABCDA* is of length $L_0 = 4$ and *ACBDA* of length $L_1 = 2+a+2\sqrt{1-(\frac{a}{2})^2}$. We consider the range $a \in [0, 1]$, i. e., the angle at A is in the interval $[120°, 180°]$. Starting at A, the ant can continue either toward B leading to a shorter total length because $L_1 \geq L_0$, or diagonally toward C which is implied by the local heuristics for $a < 1$. We do not allow a first step from A toward D in order to give both types of paths the same initial prior probability.

As for the indicated range of a, the shorter initial segment leads to a longer total path length, the problem is *deceptive*. We will study how ACO deals with deceptiveness in this simple example.

Figure 3.2 shows the probability of the two options as a function of the pheromone evaporation rate ρ and the deceptiveness a for a single ant. For $\rho = 1$, i. e., instantaneous evaporation, the ant cannot build a memory of which total path length is shorter, such that even for a deceptiveness just below 1 it will be drawn toward the longer path. If the evaporation is slower, than exploration of both paths will lead eventually to a preference of the short tour in spite of some initial deceptiveness. If the deception is larger (e. g., $a < 0.9$), then the effect of the shorter tour length will be compensated by the tempting short initial segment of the tour. The actual outcome is, however, a bit more complicated as it also depends on the frequency each of the branches is taken by the ant (see, e. g., [20] for a more general discussion of this effect). Considering the boundary (see base contour in Figure 3.2) between the regions of a deceived ant and a successful solution of the problem, it seems obvious that the boundary improves (less deception) for slower evaporation (lower ρ).

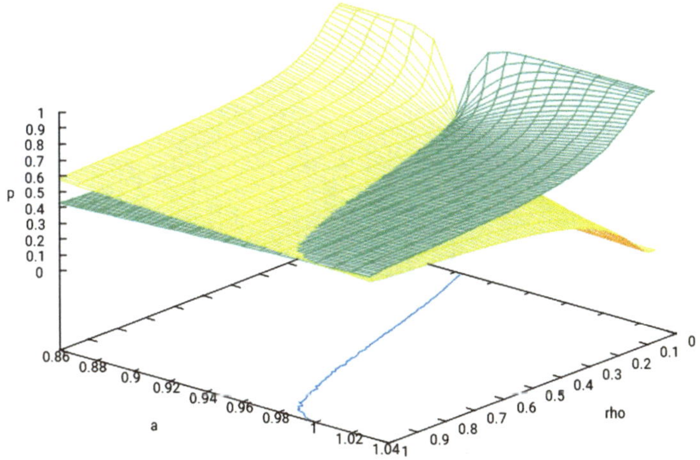

Figure 3.2: The pobability of (not) being decevied is shown in yellow (green) for one ant in the diamond problem (see Figure 3.1). The parameter ρ is the evaporation rate and a is a measure of the deceptiveness of the problem. For $a \geq 1$, the problem is not deceptive. If $a < 1$, then the first leg of the tour is shorter than the alternative legs, but will lead to a longer tour. For example, if $a = 1$ and $\rho = 1$ (immediate evaporation), then both alternatives occur with the same probability. The vertical axis shows the probability that the ant falls prey (or does not fall prey) to the deception. If there is some built-up of the pheromone, then the ant has a better chance to escape the deception, but for a single ant the advantage of using a pheromone trail is small.

Figure 3.3 includes the result from Figure 3.2 for comparison with the case of two ants. Like in the single-ant case, the effect of the evaporation rate is rather small also for more than one ant, but interestingly is qualitatively different from the single-ant case.

3.3.2 Discussion

The performance of a metaheuristic algorithm depends (i) on its ability to perform hill climbing, i. e., the exploitation of an implicit smoothness prior, (ii) on the avoidance of resampling, which can often satisfactorily achieved by a substantial random noise component and (iii) on the ability to perform global optimization, which implies the avoidance of deception. The latter requires some form of cooperation in order to efficiently combine information from various sources, which can be understood as a compositionality prior (e. g., as stated in the building block hypothesis). We need to ask thus why a single ant is hardly able to withstand deception, whereas more than one ant does much better.

The above example shows that this is the case for either way of their information being combined into the pheromone trail and independently of the evaporation rate. It clear from many experiments that multiple ants produce better results than a single

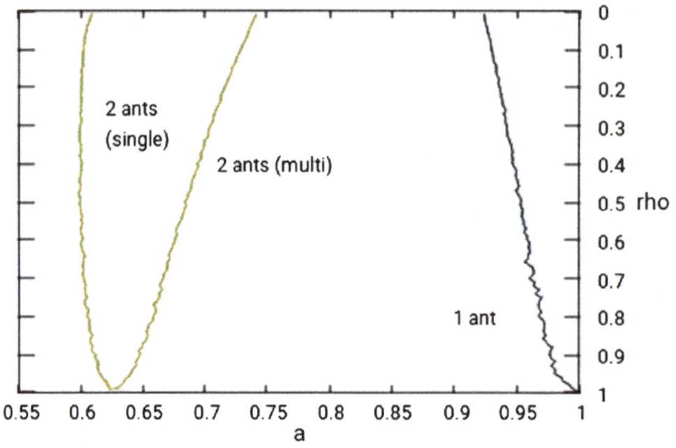

Figure 3.3: The lines show above what level of deceptiveness (x-axis) the algorithm is more likely to find the optimal solution in the diamond problem (Figure 3.1). The curve on the right (1 ant) is extracted from Figure 3.2. For the left curve only, the best ant contributes to the pheromone trace, for the middle curve, both ants add to the pheromone according to path lengths. The curves show the effect of the pheromone evaporation rate (y-axis), which exemplifies that the effect of this parameter (ρ) can depend essentially on the set-up of the algorithm. Whereas it is obvious that two ants are less easily deceived than a single one, the evaporation rate may have the opposite (or no) effect in case of two ants.

one, even when compensated by appropriately longer runtime. The reason is simply that for moderate deception it is less likely that two ants are both deceived at the same time. For example, if equation (3.4) assigns a probability p_{dec} to the deceptive leg of the tour, then the probability for both ants being deceived is $p_{\text{dec}}^2 \leq p_{\text{dec}}$. The case where one ant takes the deceptive path and the other one the optimal one, it still leads to a relative increase of the pheromone on the nondeceptive, i. e., shorter tour, although less so in the case when both ants add pheromone to the trail (3.1) than when only the best is incrementing the trail (3.2). It is a puzzling outcome that for the considered evaluation criterion of equally likely deception and nondeception, the effect of the pheromone evaporation rate ρ is different for the different cases. As we have seen for one ant, a lower rate leads to more resilience toward deception, while for two ants which both lay pheromone the opposite effect is observed, whereas for the best-only pheromone laying case of two ants, the effect of ρ is negligible. This may, however, be an effect of the particularly simple task. If the number of runs is small or if the ants change their paths quickly, the pheromone evaporation rate may have a bigger or different effect. It should be considered, however, that based on the results for the simple example, ρ is not a trivial parameter, i. e., good parameter values cannot easily be guessed.

Obviously, the result depends on the initialization. Whereas we have started here with a minimal level of pheromone, a large initial amount can improve the exploration, and thus reduce the effect of deception; see [22, Section 3.5].

The number of ants is a main factor determining the general properties of the algorithm. For more than two ants, it is obvious that the capabilities of coping with deception continue to increase, but it seems more interesting to study problems where deception occurs in more than one place.

For more complex problems, we can expect that a similar mismatch between the local heuristic and the total tour length occurs more and more often for larger problems such that more and more ants will be needed. They often made claim that a few dozens of ants are sufficient and clearly applies to problems of medium complexity, but certainly not in general. Large numbers of ants can lead to more redundancy and to a lower rate of information accumulation, but the level at which this happens depends on the complexity of the problem.

3.4 Reinforcement learning

3.4.1 RL and ACO

Reinforcement learning (RL) uses only a single agent. Multiagent RL (MARL) does not usually exploit the advantages of redundant search, but studies the cooperation of agents. The similarity between ACO and RL has been emphasized starting from the Ant-Q algorithm [1, 23] which among others introduced a local version of the fitness function; similarly, as in RL [24] the rewards are accumulated over a certain time horizon, which is determined by the parameter γ. It is of a typical length of $(1-\gamma)^{-1}$ for $\gamma < 1$, with γ usually being close to 1. The effect of γ in Ant-Q as in RL is twofold: On the one hand, large γ give a better approximation of the actual goal of the search, while smaller γ (less close to 1) leads to an effectively reduced search space (that may or may not contain the optimal solution), and thus to an improved performance for problems with a strong limitation of the number of fitness evaluations.

Other studies have proposed various hybrids of ACO and RL. References [25, 26] adapted Q-learning to match ACO, where the pheromone introduces a nonlinearity in the value function which can be useful if background knowledge on the problem is available. Sometimes the relation consists merely in a naive identification of RL and ACO [27]. A different connection between ACO and reinforcement learning has been attempted in *ant programming* [28].

We agree to the conclusion in [1] that the set of RL algorithms and the set of ACO algorithms intersect, but we want to restrict ourselves to a few algorithms that are very well known and (in Section 3.5) to an less well-known algorithm that has the potential not only to form a bridge between ACO and RL, but also to contribute to a quantitative understanding of both types of algorithms.

3.4.2 SARSA

SARSA [24] is an on-policy reinforcement learning algorithm. It means simply *state – action – reward – state – action*, indicating the sequence of steps in the interaction of the agent with the learning algorithm. It uses a stochastic policy $\pi(a|s)$ to select an action a depending on the current state s of an agent. Each action moves the agent to a new state s'. In connection with the state transition, a reward signal r becomes available which is then used to improve the estimate $Q(s,a)$ of the value function, i. e., of the expectation of the total discounted future reward conditioned on the execution of action a in state s. This is done by the following update rule, which is a stochastic approximation of the Bellman equation [29],

$$Q'(s,a) = (1-\varepsilon)Q(s,a) + \varepsilon(r + \gamma Q(s',a')), \tag{3.5}$$

where $\varepsilon \geq 0$ is a learning rate and $\gamma < 1$ is the above-mentioned discount factor.

The estimate (3.5) can be used to define a stochastic policy $\pi(a|s)$, e. g., using the Boltzmann strategy

$$\pi(a|s) = \frac{\exp \omega Q(s,a)}{\sum_b \exp \omega Q(s,b)}, \tag{3.6}$$

which can be tuned by varying the parameter ω. High ω causes the algorithm to choose essentially only the best action, while at low ω all actions are sampled. In this sense, the role of ω as an inverse temperature, which is inherited from statistical physics, is also intuitive. It makes sense to increase ω over the run time of the algorithm, but the time course of this parameters is not easy to decide in general.

The relation of SARSA (3.5) and ACO (3.4) becomes clear when considering that equation (3.5) represents an exponential average of the sum of the immediate reward r and the expected future reward $Q(s',a')$ which enters the typical choice of a policy (3.6) as an exponential, while equation (3.4) directly features a product, i. e., the Q-function is similar to the logarithm of the fitness in ACO; see also [30, p. 48]. However, in RL the reward is often exponentially averaged over a number of steps rather than over a number of trials as in ACO. We will therefore consider a nondiscounted SARSA algorithm ($\gamma = 1$ in equation (3.5)). This is not a strong assumption on SARSA, as finite horizon problems (as usually considered also in ACO) do not require a discounted form for convergence [24, p. 72 (2nd ed.)]. The complement of pheromone update rate $1-\rho$ (3.3) plays the role of the learning rate ε in SARSA (3.5). ACO's probability rule represents a particular choice for the policy in SARSA. In RL, mostly single agents are considered, so we should take a single ant in ACO, too, when making this relationship more explicit.

3.4.3 Q-learning

Q-learning [31] is formally very similar to SARSA although it is an off-policy algorithm, i. e., belongs to a different class of RL algorithms. The update rule for the value function

$$Q'(s,a) = (1-\varepsilon)Q(s,a) + \varepsilon\Big(r + \gamma \max_{a'} Q(s',a')\Big), \tag{3.7}$$

is strikingly similar to (3.5), but the value function now has a different meaning: Whereas (3.5) measures the expected reward for a fixed policy, such as (3.6), the Q-leaning update rule (3.7) collects information for any actions that are then attributed to the best policy according to the present estimate of the value function [24]. The two algorithms differ also in the sense that the policy used in SARSA is not necessarily the best policy, while the estimate of the value obtained in Q-learning (3.7) is not necessarily the best estimate. Upon convergence to the optimal solution both will coincide, although convergence in SARSA means that the policy converges to a greedy policy after having started with a soft stochastic policy that becomes more and more restricted to rewarded actions. In Q-learning in contrast that convergence means that the actual policy follows the best-known policy.

3.4.4 Parameter interpretation

The parameter ρ plays a different role than both the discount factor γ and the eligibility trace parameter λ in reinforcement learning [24]. Namely, ρ has an effect only for the next trial, while γ controls the time horizon $(1-\gamma)^{-1}$ within a trial and λ controls the distribution of value estimates to previous states. Note that we exclude the possibility that states can be revisited in the same trial, but as the pheromone update is usually performed after the current solution is completed, this would not matter. The parameter γ (λ) determines the effect of the value of the next state on the previous state(s) such that the value estimate can take advantage of intermediate rewards. ACO, on the other hand is clearly a Monte Carlo algorithm that increases the likelihood estimate only after the path (or rather the solution) is completed. In reinforcement learning, Monte Carlo algorithms are considered as a limit of infinitely long eligibility traces, i. e., $\lambda \to 1$, and it is straightforward to define an ACO algorithm with $\lambda < 1$ by performing the pheromone update after each step along the path traversed so far, but exponentially decaying (with a scale of $(1-\lambda)^{-1}$) in the backward direction. In a TSP, relevant information would be available in each step, but also if less often information can be obtained as a more gradual and partial increase of the pheromone level is possible in this way. It can also be provided by the currently best ant only, if an RL-like version of the earlier mentioned Min-Max Ant System is to be realized.

3.5 Fisher–Eigen ACO

This section will reformulate the ACO algorithm in such a way that it becomes a representation of the Fisher–Eigen (FE) equation [32, 33] which has been used to describe the evolution of species ("quasispecies") and to provide a quantitative model of the survival of the fittest. It has been known for a long time that also RL algorithms can be represented in this form which gives us the opportunity to propose a Fisher–Eigen inspired version of ACO.

3.5.1 Update rule for probabilities

We consider the TS problem, and simplify the notation by considering only allowed links starting at a fixed city. Based on the probabilities (3.4) to move from this city to city j,

$$p_j = \frac{\tau_j \eta_j}{\sum_k \tau_k \eta_k}, \tag{3.8}$$

we can ask how the probabilities change from trial to trial. This will enable us to arrive at an update rule for the probabilities. We consider the case of a large number of ants such that we can replace the sampled results for ants by appropriate averages.

The new probability p_j' of link j depends on whether link j was actually chosen in which case the respective pheromone level has the expected value

$$\tau_j' = \rho\tau_j + (1-\rho)\lambda_j, \tag{3.9}$$

where λ_j is the expected fitness conditioned on link j. If the ant took any other link, we have simply

$$\tau_j' = \rho\tau_j. \tag{3.10}$$

Case (3.9) occurs with probability p_j and case (3.10) with probability $1 - p_j$, such that we expect the (unnormalized) probabilities for the next step to be

$$\begin{aligned} p_j' &\sim (p_j(\rho\tau_j + (1-\rho)\lambda_j) + (1-p_j)\rho\tau_j)\eta_j \\ &\sim p_j(1-\rho)\lambda_j\eta_j + \rho\tau_j\eta_j \end{aligned}$$

Consider for a moment the greedy case ($\rho = 0$),

$$p_j' = \frac{p_j \lambda_j \eta_j}{\sum_k p_k \lambda_k \eta_k} \tag{3.11}$$

which we want to reformulate using subtractive normalization

$$p_j' = p_j(\lambda_j\eta_j - X),$$

where we require

$$\sum_j p_j' = \sum_j p_j(\lambda_j \eta_j - X) \stackrel{!}{=} 1,$$

i. e., $X = -1 + \sum_j p_j \lambda_j \eta_j$. Thus

$$p_j' = p_i(1 + f_j - \bar{f}), \tag{3.12}$$

with

$$f_j = \lambda_j \eta_j$$

and the weighted average

$$\bar{f} = \sum_j p_j \lambda_j \eta_j.$$

For general ρ, we notice that $p_j \sim \tau_j \eta_j$, and use again subtractive normalization:

$$\begin{aligned} p_j' &= p_j((1-\rho)\lambda_j \eta_j - X) + \rho \tau_j \eta_j \\ &= p_j(\rho + (1-\rho)\lambda_j \eta_j - X) \end{aligned}$$

i. e., we require

$$\sum_j p_j' = \sum_j p_j(\rho + (1-\rho)\lambda_j \eta_j - X) \stackrel{!}{=} 1.$$

This leads to

$$X = -1 + \sum_j p_j(\rho + (1-\rho)\lambda_j \eta_j),$$

i. e., we can use again (3.12) but now with

$$f_j = \rho + (1-\rho)\lambda_j \eta_j$$

and

$$\bar{f} = \sum_j p_j(\rho + (1-\rho)\lambda_j \eta_j).$$

We have used here subtractive normalization instead of divisive normalization as in the original ant algorithms, in order to relate the ant colony dynamics to the Fisher–Eigen equation (see next section). Subtractive normalization can lead to negative values for the probabilities, which, however, does not occur in continuous time (see below). The different effects of subtractive and divisive normalization is also an important question in computational neuroscience [34].

3.5.2 The Fisher–Eigen equation

Equation (3.12) is the discrete-time version of the Fisher–Eigen equation [32, 33]. The interpretation is straightforward: If $\lambda_j\eta_j$ is larger than the average, p_j increases exponentially. As it increases, it causes also the average to increase, such that even if $\lambda_j\eta_j$ is maximal the increase will come to a halt for $p_j = 1$ which implies $\lambda_j\eta_j = \overline{\lambda\eta}$ such that p_j saturates at this value. On the other hand, the nonmaximal links will eventually fall short of the average that shifts in favor of the maximal link, such that all but the probability of maximal link approach zero. For a related representation of a reinforcement learning algorithm by the Fisher–Eigen equation, see [35].

Although equation (3.12) is (by construction) self-normalizing, it does not necessarily guarantee that p_j remains in the unit interval. Note that the continuous version of the Fisher–Eigen equation

$$\dot{p}_j = p_j(f_j - \bar{f}) \tag{3.13}$$

does assure that $p_j \in [0, 1]$. In the discrete case, violations can occur, if the deviation of $\lambda_j\eta_j$ from the average is larger than unity. This can be avoided by appropriate scaling, although smaller scales imply also a smaller convergence rate. In practical computations, a sanity check whether $p_j \in [0, 1]$ is thus advisable.

The Fisher–Eigen equation has some remarkable properties. Because of the form of the fitness average, it is self-normalizing. Interestingly, it is also possible to solve the equation if the conditional fitness is assumed to be constant.

$$p_j(t) = \frac{\exp(f_j t)p_j(0)}{\sum_k \exp(f_k t)p_k(0)} \tag{3.14}$$

In the limit $t \to \infty$ and in the generic case that all f_j are different,

$$p_j(t) \to \begin{cases} 1 & \text{for } j = \arg\max f_j \\ 0 & \text{otherwise} \end{cases}$$

If, e. g., two f's are identical and maximal, then the result will depend on the respective initial values as obvious from equation (3.14). It should be noted that in this case the stochasticity of the actual underlying process will lead to substantial deviations which in the nonlinear system may not simply average out over many trials. Nevertheless, any longer paths will converge to a zero probability and the optimal paths will clearly stick out eventually. Although (3.14) indicates that this can be expected already in logarithmically short time, the settling time can be large if the difference between the best solution(s) and the next best solution(s) is very small. Also, if the initial probability is small, then the settling time will be larger, which is typical in larger problems. The exponential convergence of (3.14) does not preclude good solutions, as has been shown in the application to RL [35] because the dynamics is realized by a stochastic gradient the precision of which depends on the complexity of the fitness landscape of the problem. Similar arguments apply for the discussion [24] of the difference between SARSA and Q-learning, which also depend on the sampling time in SARSA that is implied by the problem and the acceptable risk.

3.5.3 FE-RL algorithm

As discussed above, we will consider the action of moving to the next city and the city as synonymous, i. e., we consider a deterministic discrete environment. The policy can now be updated as implied by the FE equation based on the currently known Q values

$$\Delta\pi(a|s) = \varepsilon(Q(s,a) - \bar{Q}(s))\pi(a|s), \tag{3.15}$$

where τ is a time scale and

$$\bar{Q}(s) = \sum_a \pi(a|s)$$

is an average value of state s. $\bar{Q}$ refers to the same state as the action, but we can define it to include also information about the subsequent state in order to obtain a representation of the Q-learning or SARSA, namely

$$Q(s,a) = r_s + \gamma Q(s',a')$$

which is, as in SARSA an average over the current $\pi(a|s)$ that evolves by (3.15), i. e.,

$$Q(s,a) = r_s + \gamma \sum_a \pi(a|s) Q(s',a')$$

with a' sampled from $\pi(a'|s')$ for the FE-SARSA variant or

$$Q(s,a) = r_s + \gamma \sum_a \pi(a|s) \max_{a'} Q(s',a')$$

for the FE-Q-learning variant. The relation of these cases and the variants (3.1) and (3.2) is particularly interesting.

3.5.4 FE-ACO algorithm

Instead of equations (3.11), we can as well use a sliding (exponential) average, i. e., the standard (nongreedy) pheromone update, such we have in place of equation (3.12)

$$p_j' = p_j(1 + \tau_j\eta_j - \overline{\tau\eta}), \tag{3.16}$$

with

$$\overline{\tau\eta} = \sum_j p_j \tau_j \eta_j.$$

Note that equation (3.16) is not equivalent to equation (3.8) nor is it more efficient. We should also be aware that, in contrast to the standard assumptions of the Fisher–Eigen equation, the pheromone trail τ_j does not remain constant and depends on the balance of link probabilities at other nodes, i. e., we have to consider the system

$$p'_{ij} = p_{ij}(1 + \tau_{ij}\eta_{ij} - (\overline{\tau\eta})_i), \tag{3.17}$$

where

$$(\overline{\tau\eta})_i = \sum_j p_{ij}\tau_{ij}\eta_{ij}.$$

The variable in equation (3.17) is dependent in various ways: Any changes at node i can affect the fitness of all nodes. Changes of p_{ij} can affect the path of the ant(s), and thus whether or when ants arrive at other nodes, or how many neighbors are reachable from other nodes. This forms a complication as compared to the original formulation of the Fisher–Eigen equation, but is similar to typical models in evolution theory.

3.5.5 Discussion

In contrast to the assumption of the Fisher–Eigen equation, the fitness is not constant but depends on the reward in other actions (RL) or on the taboo lists (ACO). It is possible to deal with these dependencies and retain the original statements of the FE system, namely if the dynamics is considered on a global level. In this way, the ACO can be related easily to other algorithms that do not assume a stepwise transition, while practically the convenience of a serial procedure can be kept.

For this purpose, let $p_{ij}(a)$ be the probability of reaching state j as a consequence of action a being applied in state i. For a determinist environment, $p_{ij}(a)$ would be degenerate, but the state transitions may still be stochastic for a stochastic choice of actions by the policy $\pi(a|i)$. The probabilities of the state transitions are then given by

$$M_{ij} = \sum_a p_{ij}(a)\pi(a|i) \tag{3.18}$$

i. e., the value function or likewise the pheromone level will obey the iteration relation

$$V = r + \gamma MV$$

which is a linear equation with r begin the vector of state-dependent (average) rewards. The equation can be solved easily for any specific problem where (3.18) can be computed, and it can provide an estimate of the time for convergence of the value function a. k. a. pheromone trail.

3.6 Conclusions

We have presented a new vista on ACO and its relation to RL. The recent advances in RL, in particular in combination with deep neural networks, can stimulate the development of improved variants of ACO although, where admittedly the metaphorical ants are less and less suited as an inspiration for improvement. Instead the combination with powerful algorithms from machine learning appears to be a promising option for the future of this approach. We have provided here an intuitive explanation of some features of ACO, such as namely the advantage of using a collective of ants rather than single agents. Using more than one agent is not typical in RL. On the other hand, the effective and flexible reduction of the search space by introducing a time horizon is a tool from RL which as not been fully utilized in ACO, although it was discussed already in the Ant-Q algorithm [1]. By the composition of a joint framework of RL and ACO that is rooted within evolutionary dynamics, based on the Fisher–Eigen equations, we have outlined a framework that can be useful for the further cross-fertilization between the two classes of algorithms.

Bibliography

[1] Gambardella LM, Dorigo M. Ant-Q: a reinforcement learning approach to the traveling salesman problem. In: Machine learning proceedings 1995. Elsevier; 1995. p. 252–260.

[2] Dorigo M, Stützle T. Ant colony optimization: overview and recent advances. In: Handbook of metaheuristics. Springer; 2019. p. 331–351.

[3] Wilson DG, Cussat-Blanc S, Luga H, Miller JF. Evolving simple programs for playing Atari games. In: Proc. of the Genetic and Evolutionary Computation Conference. ACM; 2018. p. 229–236.

[4] Stützle T, Hoos H. Max–min ant system. Future Gener Comput Syst. 2000;16(8):889–914.

[5] Joyce T, Herrmann JM. A review of no free lunch theorems, and their implications for metaheuristic optimisation. In: Nature-inspired algorithms and applied optimization. Springer; 2018. p. 27–51.

[6] Dorigo M, Gambardella LM. Ant colony system: a cooperative learning approach to the traveling salesman problem. IEEE Trans Evol Comput. 1997;1(1):53–66.

[7] Stützle T, Dorigo M. Ant colony optimization and stochastic gradient descent. IEEE Trans Evol Comput. 2002;6(4):358–365.

[8] Dayan P. The convergence of $td(\lambda)$ for general λ. Mach Learn. 1992;8(3-4):341–362.

[9] Stützle T, Hoos H. Improvements on the ant system: introducing the max–min ant system. In: Albrecht R, Smith G, Steele N, editors. Proceedings of artificial neural nets and genetic algorithms 1997. Norwich, UK. Springer-Verlag; 1998. p. 245–249.

[10] Pellegrini P, Favaretto D, Moretti E. On max–min ant system's parameters. In: International workshop on ant colony optimization and swarm intelligence. Springer, 2006; p. 203–214.

[11] Favaretto D, Moretti E, Pellegrini P. On the explorative behavior of max–min ant system. In: International workshop on engineering stochastic local search algorithms. Springer, 2009; p. 115–119.

[12] Gaertner D, Clark KL. On optimal parameters for ant colony optimization algorithms. In: IC-AI. 2005; p. 83–89.

[13] Siemiński A. Ant colony optimization parameter evaluation. In: Multimedia and internet systems: theory and practice. Springer, 2013; p. 143–153.
[14] Stützle T, López-Ibánez M, Pellegrini P, Maur M, Montes De Oca M, Birattari M, Dorigo M. Parameter adaptation in ant colony optimization. In: Autonomous search. Springer, 2011; p. 191–215.
[15] Pellegrini P, Stützle T, Birattari M. A critical analysis of parameter adaptation in ant colony optimization. Swarm Intell. 2012;6:23–48.
[16] Dorigo M, Blum C. Ant colony optimization theory: a survey. Theor Comput Sci. 2005;344:243–278.
[17] Dorigo M, Stützle T. Ant colony optimization: overview and recent advances. In: Handbook of metaheuristics. Springer, 2010; p. 227–263.
[18] Gutjahr WJ. Mathematical runtime analysis of ACO algorithms: survey on an emerging issue. Swarm Intell. 2007;1(1):59–79.
[19] Meuleau N, Dorigo M. Ant colony optimization and stochastic gradient descent. Artif Life. 2002;8(2):103–121.
[20] Blum C, Dorigo M. Search bias in ant colony optimization: on the role of competition-balanced systems. IEEE Trans Evol Comput. 2005;9(2):159–174.
[21] de Campos LM, Fernández-Luna JM, Gámez JA, Puerta JM. Ant colony optimization for learning Bayesian networks. Int J Approx Reason. 2002;31(3):291–311.
[22] López-Ibáñez M, Stützle T, Dorigo M. Ant colony optimization: a component-wise overview. In: Handbook of heuristics. Springer, 2017; p. 1–37.
[23] Dorigo M, Gambardella LM. A study of some properties of ant-Q. In: International conference on parallel problem solving from nature. Springer, 1996; p. 656–665.
[24] Sutton RS, Barto AG. Reinforcement learning: an introduction. Cambridge: MIT Press; 1998.
[25] Monekosso N, Remagnino P, Szarowicz A. An improved Q-learning algorithm using synthetic pheromones. In: International workshop of Central and Eastern Europe on multi-agent systems. Springer, 2001; p. 197–206.
[26] Monekosso N, Remagnino P. The analysis and performance evaluation of the pheromone-Q-learning algorithm. Expert Syst. 2004;21(2):80–91.
[27] GhasemAghaei R, Rahman MdA, Gueaieb W, El Saddik A. Ant colony-based reinforcement learning algorithm for routing in wireless sensor networks. In: Instrumentation and measurement technology conference proceedings, 2007. IMTC 2007. IEEE; 2007. p. 1–6.
[28] Birattari M, Di Caro G, Dorigo M. Toward the formal foundation of ant programming. In Dorigo M, Di Caro G, Stützle T, editors, International workshop on ant algorithms. Springer; 2002. p. 188–201.
[29] Bellman RE. Dynamic programming. Princeton, NJ: Princeton University Press; 1957.
[30] Stefan P. Combined use of reinforcement learning and simulated annealing: algorithms and applications. VDM Publishing; 2009.
[31] Watkins CJCH. Learning from delayed rewards. PhD thesis, Cambridge University, 1989.
[32] Fisher R. Population genetics (The Croonian lecture). Proc R Soc B. 1953;141:510–523.
[33] Eigen M. Selforganization of matter and the evolution of biological macromolecules. Naturwissenschaften. 1971;58:465–523.
[34] Doiron B, Longtin A, Berman N, Maler L. Subtractive and divisive inhibition: Effect of voltage-dependent inhibitory conductances and noise. Neural Comput. 2001;13(1):227–248.
[35] Der R, Herrmann M. Self-adjusting reinforcement learning. In: Proc. 1996 int symp on nonlinear theory and its applications (NOLTA). Katsurahama-so, Kochi, Japan. Oct. 7–9, 1996. Research Society of NOLTA, IEICE; 1996. p. 441–444.

Tanmoy Som, Tuli Bakshi, and Arindam Sinharay

4 Fuzzy informative evidence theory and application in the project selection problem

Abstract: The decision theoretic problem is formed to address the quality and availability of the information. In this context, a new algorithm has been presented to help the decision support system (DSS) under imperfect and fuzzy information acquired from an unreliable and conflicting source of information. Dempster–Shafer theory (DSTE) of evidence is an important mathematical process to resolve uncertainty in the decision making procedure. A new idea, termed as Systematic Derivation of Mass Function (SDMF) is applied to use quantification evidence preference from different multiple criteria. A numerical example of the project selection problem with fuzzy information has been presented to demonstrate the concept.

Keywords: Mass function, Dempster–Shafer theory of evidence, fuzzy information, belief function, uncertainty

4.1 Introduction

The project selection problem is a strategic decision making problem. The project selection problem can be classified by multiple, conflicting and uncertain criteria. The problem of project selection can be framed as a multicriteria decision making (MCDM) problem under uncertainty. However, in reality, inevitably incorporate evidence, based on multiple criteria, instead of preferred unary criterion, is available. Therefore, it is rational to formulate the problem by the Dempster–Shafer Theory (DST) of evidence. DST of evidence [8] is preferred in such uncertain conflicting criterion. DST needs weaker conditions than the Bayesian theoretic method. While using DST of evidence, usually three problems have to be resolved. Primarily, a mass function needs to be defined to describe the uncertainties in the problem. Then a belief and plausibility function needs to be declared and defined on the basis of mass function (MF). Finally, one has to optimally carry out the DST rule of combination [21]. The choice of application of DSTE, by and large, depends upon the appropriate construction of MF [2].

Tanmoy Som, Department of Mathematical Sciences, Indian Institute of Technology (BHU), Varanasi, India, e-mail: tsom.apm@iitbhu.ac.in
Tuli Bakshi, Deptartment of Computer Science, Calcutta Institute of Technology, Howrah, India, e-mail: tuli.bakshi@gmail.com
Arindam Sinharay, Department of Information Technology, Future Institute of English and Management, Sonarpur, Kolkata - 700150, India, e-mail: arindam.sinharay@gmail.com

https://doi.org/10.1515/9783110671353-004

Presently, we have introduced a new hybrid MCDM procedure for project selection. The introduced method has some distinct properties:

First, the introduced method uses a systematically derived mass function from multivariate data space to apply the Dempster–Shafer theory of evidence to solve the problem.

Second, the introduced method incorporates linguistic data modeled as fuzzy numbers.

Third, distinguishable property uses quantitative data and qualitative representation in the decision making process.

Fourth, the real life MCDM problem can be solved in systematic manner by implementing the introduced method.

Fifth, it has been shown that the introduced method can efficiently compute large scale data in polynomial time.

4.2 Related works

Several study efforts have been made to the field of decision science. There is a good volume of studies regarding the project selection problem. The paper [17] has provided different quantitative and qualitative models of project selection. The study in [6] the selection problem using fuzzy theory is discussed. The explained project (in [4]) selection problem uses different MCDM tools and techniques such as ARAS and MOORA. Xiaoyi Dai and Mian [18], Dey and Gupta [11] applied AHP for project selection. Authors of [5] solved project selection using AHP and goal programming. Some researchers applied MCDM based fuzzy set theory for the decision making problem [17, 10]. DSTE was introduced by Shafer [20] in 1976 for representing and reasoning with uncertain, imprecise and incomplete information. Then [9] has shown that DSTE can differentiate between uncertainty and ignorance by using the belief function that satisfy axioms weaker than the probability function. In [15], authors said that DSTE has the ability to form information in a flexible way without requiring a probability to be assigned at each element in a set. D. Boloi [2] establishes the fact that DST can allow an expression of partial knowledge. The presenter [23] has successfully combined Dempster's rule of evidence with an overall Basic Probability Assignment (BPA). In artificial intelligence literature, there are many instances where DSTE has been combined with expert systems, risk assessment, information fusion, pattern recognition, multiple attribute decision analysis, etc. [14, 12, 3]. Meredith and Mantel [17], Seraji and Serrano [19] claimed that DSTE is vividly applied in network security, information fusion, object tracking and many more domains of software engineering.

4.3 Preliminaries

In this section, we simply introduce some relative mathematical tools and techniques in the order of their usage in our newly developed method.

4.3.1 Fuzzy set theory

Let U be a universal discourse. A fuzzy set $\tilde{A}$ in the universal discourse U is defined as a set of ordered pairs and is expressed as $\tilde{A} = \{(x, \mu_{\tilde{A}}(x)) : x \in U\}$, where $\mu_{\tilde{A}}(x)$ is the degree of membership function of x in $\tilde{A}$. The degree of membership $\mu_{\tilde{A}}(x)$ varies in the range from 0 and 1, i. e., $\mu_{\tilde{A}}(x) \in [0,1]$. The fuzzy number is a wider inclusion of the interval concept. The fuzzy number may be a normal and convex membership function on the real line R. The Membership Function (MF) of a fuzzy number is piecewise continuous. The scalability of a fuzzy numbers means $\exists x \in R, \forall x \mu_{\tilde{A}}(x) = 1$ and convex $\forall x_1 \in X, x_2 \in X, \forall \alpha \in [0,1], \mu_{\tilde{A}}(\alpha x_1 + (1-\alpha)x_2) \geq \min(\mu_{\tilde{A}}(x_1), \mu_{\tilde{A}}(x_2))$. A triangular fuzzy number (TFN) $\overline{A}$ is termed as a triplet (a, b, c). We used the concept of TFN as we have tested our technique on several other fuzzy number systems, but in this domain, TFN system has given us optimal results.

4.3.2 Linguistic variable

Linguistic variables usually described in "words" or "sentences" or some "artificial languages". Every linguistic variable is transformed into a fuzzy set. In the current paper, the importance of weights of different category and criteria are presented in terms of linguistic variable. These variables are expressed by TFN which is shown in Table 4.2.

4.3.3 Defuzzification

Defuzzification can be considered as an important step in a fuzzy Multicriteria Decision Making (MSDM) model. Defuzzification is used in converting fuzzy value into crisp value. There are different techniques available in defuzzification. In the present problem, the method of defuzzification finds the best nonperformance (BNP value). For finding the BNP, the center of area (COA) procedure has been used [25]. BNP is a simple and practical way without the necessity of experts' opinions. The BNP value of a fuzzy number $\overline{A} = (a, b, c)$ is defined as follows:

$$BNP = \frac{(c-a) + (b-a)}{3} + a.$$

4.3.4 A brief review of Dempster–Shafer theory of evidence (DSTE)

We briefly review the basic concepts and definitions of evidence theory.

Let Ω be a finite set, Ω is defined as a set of hypothesis called frame of discernment and in mathematical notation $\Omega = \{w_1, w_2, \ldots, w_n\}$. Clearly, it has to be composed of an exhaustive and exclusive hypothesis. From the frame of discernment Ω, let us construct the power set, 2^Ω. The power set is composed with 2^n proposition P of Ω.

$2^\Omega = \{\phi, \{w_1\}, \{w_2\}, \ldots, \{w_1 \cup w_2\}, \{w_1 \cup w_3\}, \ldots, \Omega\}$, where ϕ denotes the empty set. The main point in evidence theory is the basic probability assignment (BPA) method. Here, an element similar to probability distribution is considered and named as the mass of belief. Mass of belief differs by the fact that the unit mass is distributed among the elements of 2^Ω. This distribution of mass is applicable on the singletons in $w_n \in \Omega$ as well as composite hypothesis, too.

Definition 4.1 (Mass function). A Mass Function (MF) is a mapping from $m : 2^\Omega \to [0, 1]$, where Ω is a frame of discernment. The mass function is also termed as the BPA (Basic Probability of Assignment), which satisfies, $m(\phi) = 0$ and $\sum_{\alpha \in 2^\Omega} m(\alpha) = 1$, where ϕ is an empty set and α is an element of 2^Ω. The mass $m(\alpha)$ measures the value of belief that is exactly committed to α. $\alpha \in 2^\Omega$ and is called a focal element.

Definition 4.2 (Focal element). When the value of the mass function is greater than zero, if $m(\alpha) > 0$, α is called a focal element (FE). The union of all focal elements (FE) constructs the core of the mass function. With every Mass Function (MF), there is association with belief, plausibility and commonality function [21, 22].

Definition 4.3 (Belief function). The belief function associated with m is $bel : 2^\Omega \to [0, 1]$ such that for any $x \in 2^\Omega$, $bel(x) = \sum_{\alpha \in 2^\Omega, \alpha \subseteq x} m(\alpha)$. It is written as bel_m. Every Mass Function (MF) over Ω, there corresponds a unique belief function, and conversely. The relationship between the mass function m and its associated belief function bel [1, 7, 16, 24] is for $\alpha \subseteq 2^\Omega$, $m(\alpha) = \sum_{\beta \subseteq \alpha} (-1)^{|\alpha - \beta|} bel(\beta)$.

In the above expression [21], the belief functions are represented as the subjective degrees of belief.

Definition 4.4 (Plausibility function). The plausibility function is expressed as $plx : 2^\Omega \to [0, 1]$, which is associated with the mass function (MF) m as $\forall \alpha \in 2^\Omega$, $plx(x) = \sum_{\alpha \cap x \neq \Phi} m(\alpha)$. The interval $[bel(x), pls(x)]$ contain the precise probability of $x \subseteq \Omega$ in the classical sense. That is, $bel(x) \leq p(x) \leq pls(x)$. It is known that $pls(x) = 1 - bel(\bar{x})$.

4.3.5 Multivariate data space

We define the probability space $\langle \Delta, F, P \rangle$ where Δ is the set of data space, F is a σ-field on Δ is a collection of subsets and P is a probability function, $P : F \to [0, 1]$. If Δ is finite, then we can define the power set 2^Δ and it is σ field on Δ. Let us consider the set of attributes $R = \{r_1, r_2, \ldots, r_n\}$, which is a data space under discussion. $Dom(r_i)$ is termed as the domain of attribute r_i. Simplistically, we will use r_i instead of $Dom(r_i)$ in the equation without any ambiguity. The multivariate data space is then the Cartesian product of the domain $\Delta = \prod_{i=1}^{n} Dom(r_i)$.

A data set S is a sample of Δ, i. e., $S \subseteq \Delta$. An element $e \in \Delta$ is a simple tuple $\langle e_1, e_2, \ldots, e_n \rangle$ where $e_i \in dom(r_i)$. Again we consider a generalized tuple $\langle F, \leq, \cap, \cup \rangle$ such that $g_i \subseteq dom(r_i)$. Such a generalized tuple is called hypertuple over R. Clearly, a simple tuple is a special hypertuple. Therefore, Δ is embedded in F. Therefore, we can have $\Delta \subseteq F$. We can consider two hypertuples $\langle g_{11}, g_{12}, \ldots, g_{1n} \rangle$ and $\langle g_{21}, g_{22}, \ldots, g_{2n} \rangle$. If the first tuple is covered by the second tuple, then we can write $\langle g_{11}, g_{12}, \ldots, g_{1n} \rangle \leq \langle g_{21}, g_{22}, \ldots, g_{2n} \rangle$. It can further be shown that $\langle F, \leq, \cap, \cup \rangle$ is a lattice [13].

4.4 Description of introduced fuzzy evidence theoretic algorithm

In this section, we briefly describe our DSS model for project selection in a precise format.

- Step 1: Linguistic variables are termed in positive TFN (Table 4.2).
- Step 2: Different classes of category have been used for different alternatives w. r. t. the considered DM problem and the domain of category is defined.
- Step 3: Different classes of criteria has been described for different alternatives w. r. t. considered DM problem and the domain of criteria is established.
- Step 4: Categorywise linguistic matrix for different projects/alternatives has been expressed as per DM's opinion.
- Step 5: Criteriawise linguistic matrix for different projects/alternatives has been given as per DM's opinion.
- Step 6: Fuzzified the category matrix (from Step 4).
- Step 7: Fuzzified criteria matrix (from Step 5).
- Step 8: Defuzzified category matrix (from Step 6) to form the crisp data value oriented category matrix.
- Step 9: Defuzzified criteria matrix (from Step 7) to form the crisp data value oriented criteria matrix.
- Step 10: Pairwise combination of category value set and criteria value set has been done.

The resulting value set is the set of focal elements and any element of this set is a focal element.

- Step 11: Compute table depicting all set of values of alternatives related to the focal element by a define relation.
- Step 12: Computation of MF systematically using multivariate data for different FE.
- Step 13: The value of belief function and plausibility function have been computed w. r. t. different alternative using MF.

- Step 14: Uncertainty has been quantized using belief and plausibility and shortened in ascending order. Thereafter, rank is assigned accordingly.

4.5 Illustrative example

A synthetic example has been developed by which the entire technique has been illustrated. Here, a project selection problem is considered as typical MCDM problems. The initial considerations such as linguistic variables have been expressed in Table 4.1. In Table 4.2, different category values have been given w. r. t. the project selection problem. They are Highly Effective (HE), Effective (E), Mean (M) and Nonbeneficial (NB). Similarly, criteria values have been presented w. r. t. project selection problem as the Probability of Success (PS), Project Risk (PR), Present Value (PV) and Internal Rate of Return (IRR). It is not unusual that a company or executive using all these criteria for project selection.

4.5.1 Preliminary assumption

Let us consider a sample space $S = \{1, 2, 3, \text{and } 4\}$; Category space $A1 = \{HE, E, M, \text{and } NB\}$; Criteria space $A2 = \{PS, PR, PV, \text{and } IRR\}$. We associate the values of sample space with the Category space A1 and Criteria space A2 in Table 4.1.

Table 4.1: Category-criteria space valued table.

Category	HB	B	AVG	P
	1	2	3	4
Criteria	PS	PR	PV	IRR
	1	2	3	4

For the set of alternatives, we consider four different projects: Project 1 (P1), Project 2 (P2), Project 3 (P3) and Project 4 (P4) and we consider the linguistic value of the variable in Table 4.2.

Table 4.2: Linguistic value.

Linguistic Scale	Corresponding TFN
Very Poor (VP)	(0.0, 0.12, 0.32)
Poor (P)	(0.12, 0.32, 0.52)
Medium (M)	(0.32, 0.52, 0.72)
High (H)	(0.52, 0.72, 0.92)
Very high (VH)	(0.72, 0.92, 1.0)

The category value of each alternative is taken in linguistic term as per three decision maker's opinion shown in Table 4.3.

Table 4.3: Performance value of project (category value).

	Highly Effective			Effective			Mean			Nonbeneficial		
	DM1	DM2	DM3	DM1	DM2	DM3	DM1	DM2	DM3	DM1	DM2	DM3
Project 1	VH	H	M	H	H	M	VH	M	VH	VP	P	M
Project 2	H	VH	H	M	VH	H	VH	M	VH	M	M	H
Project 3	M	M	H	P	M	H	P	P	M	H	H	VH
Project 4	VH	M	VH	M	M	H	H	H	VH	P	P	M

How the different projects run with performance measures criteria such as PS, PR, PV, IRR are linked with each decision maker and is provided in Table 4.4.

Table 4.4: Performance value of project (criteria value).

	PS			PR			PV			IRR		
	DM1	DM2	DM3	DM1	DM2	DM3	DM1	DM2	DM3	DM1	DM2	DM3
Project 1	H	H	M	H	H	VH	VH	M	VH	M	VH	VH
Project 2	P	M	H	VH	M	H	H	H	VH	VH	M	VH
Project 3	M	P	H	M	VH	VH	VP	P	M	H	H	M
Project 4	M	H	VH	VP	M	H	P	P	M	P	M	M

4.5.2 Fuzzification of category matrix

From the category value taken from three decision makers (in terms of linguistic value) corresponding to each alternative are fuzzified using TFN (presented in Table 4.2). Accordingly, we get all values which are shown in Table 4.5.

Table 4.5: Overall fuzzy decision matrix (category value).

	Highly Effective	Effective	Mean	Nonbeneficial
Project 1	(0.52, 0.72, 0.88)	(0.45, 0.65, 0.85)	(0.58, 0.78, 0.90)	(0.14, 0.32, 0.52)
Project 2	(0.58, 0.78, 0.94)	(0.52, 0.72, 0.88)	(0.58, 0.78, 0.90)	(0.38, 0.58, 0.78)
Project 3	(0.38, 0.58, 0.78)	(0.32, 0.52, 0.71)	(0.18, 0.38, 0.58)	(0.58, 0.78, 0.94)
Project 4	(0.58, 0.78, 0.90)	(0.38, 0.58, 0.78)	(0.58, 0.78, 0.94)	(0.18, 0.38, 0.58)

Thereafter fuzzified, the criteria matrix shown in Table 4.6.

Table 4.6: Overall fuzzy decision matrix (criteria value).

	PS	PR	PV	IRR
Project 1	(0.45, 0.65, 0.85)	(0.58, 0.78, 0.94)	(0.58, 0.78, 0.90)	(0.58, 0.78, 0.90)
Project 2	(0.32, 0.52, 0.72)	(0.52, 0.72, 0.88)	(0.58, 0.78, 0.94)	(0.58, 0.78, 0.90)
Project 3	(0.32, 0.52, 0.72)	(0.58, 0.78, 0.90)	(0.14, 0.32, 0.52)	(0.45, 0.65, 0.85)
Project 4	(0.52, 0.72, 0.88)	(0.28, 0.45, 0.65)	(0.18, 0.38, 0.58)	(0.25, 0.45, 0.65)

4.5.3 Defuzzification of category matrix

The next step is how to carry out defuzzification using BNP (Best nonfuzzy performance) technique. In this method, a TFN number is denoted as (a, b, c) and the BNP method is applied as

$$BNP = \frac{(c - a) + (b - a)}{3} + a.$$

In addition, in a similar manner, we get all values which are shown in Table 4.7.

Table 4.7: Overall defuzzyfied decision matrix (category value).

	Highly Effective	Effective	Mean	Nonbeneficial
Project 1	0.70	0.65	0.75	0.32
Project 2	0.76	0.70	0.75	0.58
Project 3	0.58	0.51	0.38	0.76
Project 4	0.75	0.58	0.76	0.38

4.5.4 Defuzzification of criteria matrix

Similarly, we defuzzified the criteria matrix shown in Table 4.8.

Table 4.8: Overall defuzzified decision matrix (criteria value).

	PS	PR	PV	IRR
Project 1	0.65	0.76	0.75	0.75
Project 2	0.52	0.70	0.76	0.75
Project 3	0.52	0.75	0.32	0.65
Project 4	0.70	0.46	0.38	0.45

4.5.5 Construction of focal element

Now we are going to reveal how the focal element construction can be carried out. The permissible focal element for the given problem is of number 16 coming out from the exhaustive pairing of the category set and criteria set as shown in Table 4.9. Figure 4.1 clearly describes the coupling made between the category set, criteria set and shows the resulting focal element.

Table 4.9: Combination of category and criteria sets.

Focal elements	Category	Criteria
F_1	HE	PS
F_2	HE	PR
F_3	HE	PV
F_4	HE	IRR
F_5	E	PS
F_6	E	PR
F_7	E	PV
F_8	E	IRR
F_9	M	PS
F_{10}	M	PR
F_{11}	M	PV
F_{12}	M	IRR
F_{13}	NB	PS
F_{14}	NB	PR
F_{15}	NB	PV
F_{16}	NB	IRR

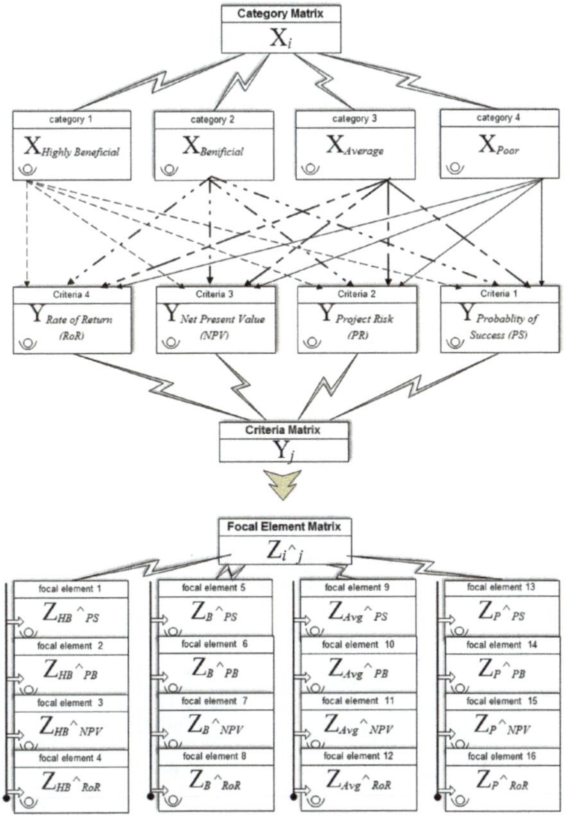

Figure 4.1: Construction of focal elements.

4.5.6 Determination of the values of focal elements

The focal element given in Table 4.9 has been populated by finite tuple having numerical values ranging between 0 and 1. Each of the alternatives for a particular focal element is populated by the minimum value taken into consideration from the category value of the respective alternative from defuzzified decision matrix (Table 4.7) and criteria value with respect to the same alternative from defuzzified decision matrix (Table 4.8). The focal point is calculated and is shown in Table 4.10.

Table 4.10: Set of possible combinations for all projects.

Set of Possible Combinations	Values of the Category			
	Project 1	Project 2	Project 3	Project 4
F_1	0.65	0.52	0.52	0.70
F_2	0.70	0.70	0.58	0.46
F_3	0.70	0.76	0.32	0.38
F_4	0.70	0.75	0.58	0.45
F_5	0.65	0.52	0.51	0.58
F_6	0.65	0.70	0.51	0.46
F_7	0.65	0.70	0.32	0.38
F_8	0.65	0.70	0.51	0.45
F_9	0.65	0.52	0.38	0.70
F_{10}	0.75	0.70	0.38	0.46
F_{11}	0.75	0.75	0.32	0.38
F_{12}	0.75	0.75	0.38	0.45
F_{13}	0.32	0.52	0.52	0.38
F_{14}	0.32	0.58	0.75	0.38
F_{15}	0.32	0.58	0.32	0.38
F_{16}	0.32	0.58	0.65	0.38

4.5.7 Calculation of mass function

Corresponds to the sample space S considered. We assume (category and criteria) values are numerical. The data of the corresponding alternatives can be shown in Table 4.11.

Table 4.11: Data samples distribution in the sample spaces.

Alternatives	Category	Criteria
Project 1	1	3
Project 2	2	4
Project 3	3	1
Project 4	4	2

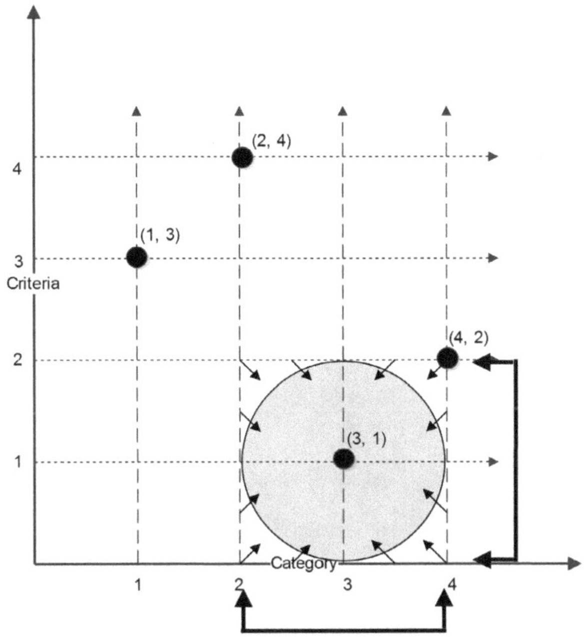

Figure 4.2: Simple representation of hypertuple.

Each intersection point is a simple representation of union of four quadrants of a circle, which we define as four hypertuples as shown in Fig. 4.2.

Therefore, it is not difficult to note that 16 permissible hypertuples coming out from four alternatives.

We construct a function $f(x, y) = (5y - xy)$, where x represents the category values and y represents the criteria values.

This function satisfies the following conditions:

$$\text{(i)} \quad f(0,0) = 0 \quad \text{(Proof is obvious)}$$

$$\text{(ii)} \quad \sum\sum f(x,y) = 100 = M = \text{total mass}$$

So we consider a mass function,

$$m(\alpha) = \frac{f(x,y)}{M}.$$

We can further prove that $m(\alpha)$ is a valid mass function as

$$\text{i)} \quad m(\phi) = \frac{f(0,0)}{M} = 0,$$

$$\text{ii)} \quad \sum m(\alpha) = \sum_{\substack{x=1\\y=1}}^{4} \frac{f(x,y)}{M} = 1.$$

Accordingly, we assigned mass value for all focal elements as shown in Table 4.12.

Table 4.12: Mass value of focal element.

Focal Element	Mass Function	Mass Value
F_1	$Z_{HB,PS}$	0.04
F_2	$Z_{HB,PR}$	0.08
F_3	$Z_{HB,PV}$	0.12
F_4	$Z_{HB,IRR}$	0.16
F_5	$Z_{B,PS}$	0.03
F_6	$Z_{B,PR}$	0.06
F_7	$Z_{B,PV}$	0.09
F_8	$Z_{B,IRR}$	0.12
F_9	$Z_{AVG,PS}$	0.02
F_{10}	$Z_{AVG,PR}$	0.04
F_{11}	$Z_{AVG,PV}$	0.06
F_{12}	$Z_{AVG,IRR}$	0.08
F_{13}	$Z_{P,PS}$	0.01
F_{14}	$Z_{P,PR}$	0.02
F_{15}	$Z_{P,PV}$	0.03
F_{16}	$Z_{P,IRR}$	0.04

4.5.8 Calculation of the belief function

After deriving all mass function value for all focal elements, the belief function is calculated for each alternative as we described $Bel(Alt_i) = \sum_{\alpha=1}^{\text{focal element}}(1-Max_{Z\neq Alt_i}F_\alpha(Z)*m(F_\alpha))$, where F_α are focal element & $m(F_\alpha)$ corresponding mass value. We have implemented the formula for calculation.

4.5.9 Calculation of plausibility function

After deriving all mass function value for all focal elements, the plausibility function is calculated for each alternative as we describe $Pls(Alt_i) = \sum_{\alpha=1}^{\text{focal element}} F_\alpha(Alt_i) * m(F_\alpha)$, where F_α are focal elements & $m(F_\alpha)$ corresponding mass value. We have calculated the plausibility of all the alternatives, which is shown in Table 4.13.

Table 4.13: Final ranking table.

Alternatives	Belief Value	Plausibility Value	Uncertainty Value	Final Rank
Project 1	0.2878	0.6530	0.3652	3
Project 2	0.3184	0.6934	0.3750	4
Project 3	0.2889	0.4625	0.1736	2
Project 4	0.2867	0.4448	0.1581	1

4.5.10 Calculation of uncertainty function

Uncertainty is calculated by the difference of plausibility and belief of the corresponding alternative and the alternative containing the lowest uncertainty is the best according to our introduced method. So,

$$\text{Uncertainty}(Alt_i) = Pls(Alt_i) - Bel(Alt_i).$$

Consequently, we have calculated the uncertainty of all of the alternatives, which is shown in Table 4.13. Finally, sorting the alternative according to the lowest to highest uncertainty values. According to Table 4.13, Project 4 is the best project.

4.6 Conclusion

In this problem, a new fuzzy number based DSTE is introduced. Throughout this paper, we set our aim toward the target on DSTE and introduced a systematically derived method for assigning mass values to the focal elements. Our method has developed the computation technique for belief, plausibility and commonality function in multivariate data space, which we have validated theoretically and computationally. The introduced mass function is composed by basic probability assignment. The previous studiers [9, 22] of DSS fall short of inferring implicit decisions from this imprecise and incomplete data set. We think that using the scalability factor the process can be extended to edit distance for strings to multidimensional attribute functions. Finally, we describe our introduced method in multivariate structured data space.

Bibliography

[1] Ashtiani B, Haghighirad F, Makui A et al.. Extension of fuzzy TOPSIS method based on interval-valued fuzzy sets. Appl Soft Comput. 2009;9(2):457–61.

[2] Baloi D, Price ADF. Modelling global risk factors affecting construction cost performance. Int J Proj Manag. 2003;21:261–9.

[3] Beynon MJ, Curry B, Morgan PH. The Dempster Shafer theory of evidence: an alternative approach to multi criteria decision modeling. Omega. 2000;28(1):37–50.

[4] Bakshi T, Sarkar B. MCA based performance evolution of project selection. Int J Softw Eng Appl. 2011;2:14–22.

[5] Bakshi T, Sarkar B, Sanyal SK. An optimal soft computing based AHP-QFD model using goal programming for decision support system. Int J Sci Eng Res. 2012;3(6). ISSN 2229-5518.

[6] Chu P, Hsu Y, Fehling M. A decision support system for project portfolio selection. Comput Ind. 1996;32(2):141–9.

[7] Chu TC, Lin Y-C. An extension to fuzzy MCDM. Comput Math Appl. 2009;57(3):445–54.

[8] Dempster AP. Upper and lower probabilities induced by a multi-valued mapping. Ann Math Stat. 1967;38:325–39.

[9] Denoeux T. Reasoning with imprecise belief structures. Int J Approx Reason. 1999;20(1):79–111.
[10] Deng Y. Plant location selection based on fuzzy TOPSIS. Int J Adv Manuf Technol. 2006;28:839–44.
[11] Dey PK, Gupta SS. Feasibility analysis of cross-country pipeline projects: A quantitative approach. Proj Manag J. 2001;32(4):50–8.
[12] Enea M, Salemi G. Fuzzy approach to the environmental impact evaluation. Ecol Model. 2001;135:131–47.
[13] Gratzer G. General Lattice Theory. Basel, Switzerland: Birkhauser; 1978.
[14] Jones RW, Lowe A, Harrison MJ. A framework for intelligent medical diagnosis using the theory of evidence. Knowl-Based Syst. 2002;15:77–84.
[15] Liu J, Yang J-B, Wang J, Sii HS. Review of uncertainty reasoning approaches as guidance for maritime and offshore safety-based assessment. Saf Reliab. 2002;23(1):63–80.
[16] Mahoor M, Abdel-Mottaleb M. A multimodal approach for face modeling and recognition. IEEE Trans Inf Forensics Secur. 2008;3(3):431–40.
[17] Meredith J, Mantel S. Project Management: A Managerial Approach. 4th ed. New York: Wiley; 2000.
[18] Mian SA, Xiaoyi Dai C. Decision-making over the project life cycle: An analytical hierarchy approach. Proj Manag J. 1999;30(1):40–52.
[19] Seraji H, Serrano N. A multisensor decision fusion system for terrain safety assessment. IEEE Trans Robot. 2009;25(1):99–108.
[20] Shafer G. Mathematical Theory of Evidence. Princeton, NJ: Princeton University Press; 1976.
[21] Shafer G. Dempster-Shafer Theory. 2002.
[22] Smets P, Kennes R. The transferable belief model. Artif Intell. 1994;66(2):191–234.
[23] Wang YM, Elhag TMS. A comparison of neural network, evidential reasoning and multiple regression analysis in modelling bridge risks. Expert Syst Appl. 2007;32:336–48.
[24] Zhu YM, Bentabet L, Dupuis O, Kaftandjian V, Babot D, Rombaut M. Automatic determination of mass functions in Dempster–Shafer theory using fuzzy c-means and spatial neighborhood information for image segmentation. Opt Eng. 2002;41(4):760–70.
[25] Zimmermann HJ. Fuzzy Set Theory and Its Applications. Boston: Kluwer Academic Publishers; 1991.

Mališa R. Žižović and Dragan Pamučar

5 The fuzzy model for determining criteria weights based on level construction

Abstract: In this paper is given one new subjective methodology for determining weighted coefficients in multicriteria decision making analysis based on fuzzy theory. The new method for calculation of criteria weights based on level construction allows including experts, lawyers or dispute parties opinion about significance ratio of the attributes in the process of rational decision determination. The method could be applied in practical implementation of specialized decision support systems and alternative dispute resolution in virtual environment. Starting with principles and established approaches, a problem-structuring methodology was developed which would condition the problem to allow a more thoughtful application of existing decision-making analytic methodologies.

Keywords: Multicriteria decision making analysis, weighted coefficients, subjective approach for determining criteria weights

5.1 Introduction

Determining weights of criteria is one of the key problems arising in multicriteria decision-making models. In addition to the fact that there is no unique definition of the notion of the weight of criteria, the problem of determining criteria weights becomes more complex by insufficient knowledge of mathematical logic of the existing methods for determining criteria weights and their suitability for application in particular decision-making situation. Having in mind the fact that weights of criteria can significantly influence results in a decision-making process, it is clear that particular attention must be paid to the objectivity of criteria weights. However, unfortunately that is not always present when solving practical problems.

Models for determining criteria weights have been the subject of research and scientific debate for many years. In the literature, it is possible to find many developed approaches for determining criteria weights. Traditional methods for determining weights of criteria include among others: Tradeoff method [1], proportional (ratio) method, SWING method [2], conjoint method [3], Analytic Hierarchy Process model (AHP) [4], SMART method (the Simple Multi Attribute Rating Technique) [5], MACBETH method (Measuring Attractiveness by Categorical Based Evaluation Technique)

Mališa R. Žižović, Faculty of Technical Sciences in Čačak, University of Kragujevac, Serbia, e-mail: zizovic@gmail.com
Dragan Pamučar, Department of Logistics, Military Academy, University of Defence, Belgrade, Serbia, e-mail: dpamucar@gmail.com

https://doi.org/10.1515/9783110671353-005

[6], direct point allocation method [7], ratio or direct significance weighting method [8], resistance to change method [9], AHP method [4], WLS method (Weighted Lest Square) [10] and FPP method (the Fuzzy Preference Programming method) [11]. Recent subjective methods include multipurpose linear programming [12], linear programming [13], DEMATEL (DEcision MAking Trial and Evaluation Laboratory) method [14, 15, 16], SWARA (Step-wise Weight Assessment Ratio Analysis) method [17, 18], BWM (Best Worst Method) [19], FUCOM (FUll COnsistency Method) [20] and LBWA (Level Based Weight Assessment model) [21].

Methods have also been developed in which criteria weights are calculated based on the information contained in the decision-making matrix, such as: Entropy method [22] and FANMA method whose name was derived from the names of the authors of the method [23]. Multiple experts or interested parties may be involved in the process of determining criteria weights, which requires the application of group decision-making methods including mathematical or social aggregation of individual weights [20].

This paper presents modification of the LBWA model with the purpose of determining weight coefficients of criteria in fuzzy environment. The LBWA model is selected because of the advantages it has over other subjective models for defining criteria weights, among which there are small number of comparisons of criteria ($n - 1$ only) and simple and rational mathematical algorithm that is not made more complex with increasing the number of criteria in multicriteria model. There are several goals in this paper. The first goal of the paper is to modify the LBWA model to solve complex MCDM models in uncertain environment. The second goal is to define fuzzy model which enables the calculation of objective values of the weight coefficients of criteria, taking into account uncertainties in group decision making. The third goal of the paper is to develop a model that can be easily implemented in solving practical problems.

The remaining part of the paper is organized as follows. The second section of the paper presents basic definitions of fuzzy approach, as well as arithmetic operations with fuzzy numbers. The third section presents mathematical algorithm of the fuzzy LBWA model (F-LBWA). In the fourth section of the paper the LBWA model is tested. The fifth section presents concluding considerations.

5.2 Fuzzy sets

A fuzzy set is extension and generalization of a discrete set [24]. It represents a set of elements with similar properties. The degree of membership of an element to a fuzzy set may be any real number from interval $[0, 1]$. Formally, fuzzy set A is defined as a set of regulated pairs

$$A = \{(x, \mu_A(x)) \mid x \in X, 0 \le \mu_A(x) \le 1\} \tag{5.1}$$

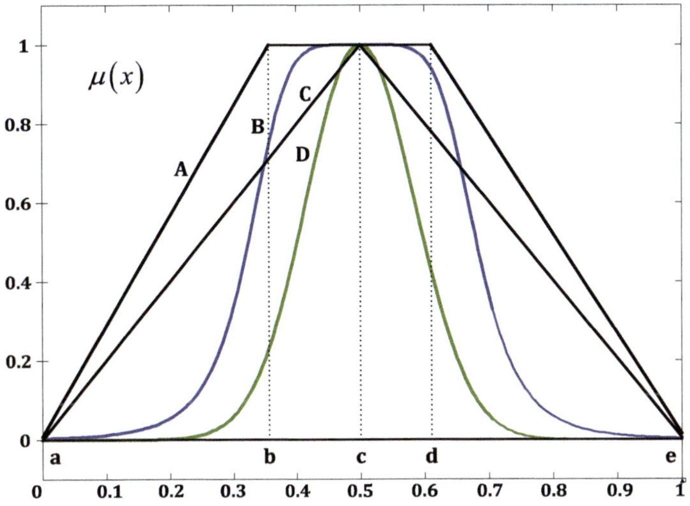

Figure 5.1: Most frequent forms of membership function.

If we define the reference set as $V = \{o, p, r, s, t\}$. One fuzzy set can be like this: $B = \{(0.3, o), (0.1, p), (0, r), (0, s), (0.9, t)\}$. This means that element o is included in set B with 0.3 degree, p with 0.1 degree, t with 0.9 degree and r and s are not included in set B.

Each fuzzy set can be represented by its membership function. If a reference set is discrete, the membership function is a set of discrete values from interval [0,1], as in the previous example. If a reference set is continual, we can express it by means of membership function.

The most frequently used forms of membership functions are:

- triangular function, Figure 5.1C
- trapezoid function, Figure 5.1A
- the Gauss curves, Figure 5.1D
- Bell-shaped curves, Figure 5.1B

In Figure 5.1, the ordinate represents the membership degree. Fuzzy variable x is shown on the abscissa. Mathematical expressions describing membership functions shown in Figure 5.1 have the following form:

$$\mu_C(x) = \left\{ \begin{array}{cc} 0, & 0 < x < a \\ (x-a)/(c-a), & a \le x \le c \\ (e-x)/(e-c), & c \le x \le e \\ 0, & x > e \end{array} \right\} \tag{5.2}$$

$$\mu_A(x) = \begin{Bmatrix} 0, & 0 < x < a \\ (x-a)/(b-a), & a \le x \le b \\ 1, & b \le x \le d \\ (e-x)/(e-c), & d \le x \le e \\ 0, & x > e \end{Bmatrix} \tag{5.3}$$

$$\mu_D(x) = e^{-\frac{1}{2}\left(\frac{x-c}{e}\right)^2} \tag{5.4}$$

$$\mu_B(x) = \frac{1}{1+(x-c)^2} \tag{5.5}$$

Most tools for fuzzy system design enable the user to define different random membership functions.

5.2.1 Arithmetic operations

Arithmetic operations on fuzzy numbers are defined by application of the interval arithmetic rules and the extension principle. α – intersections A_α and B_α can be defined for two fuzzy numbers A and B, where α may be interpreted as the degree of confidence that the number falls within the interval defined by the corresponding intersection. By rules of interval arithmetic, for sets, that is to say intervals $A_\alpha = [a_1^\alpha, a_2^\alpha]$ and $B_\alpha = [b_1^\alpha, b_2^\alpha]$ and arithmetic operator $\circ$ set $C_\alpha = [c_1^\alpha, c_2^\alpha]$ is defined representing the interval in which, with the same degree of confidence, the result of operation $\circ$ exists. For the basic arithmetic operation, it is:

$$[a_1^\alpha, a_2^\alpha] + [b_1^\alpha, b_2^\alpha] = [a_1^\alpha + b_1^\alpha, a_2^\alpha + b_2^\alpha] \tag{5.6}$$

$$[a_1^\alpha, a_2^\alpha] - [b_1^\alpha, b_2^\alpha] = [a_1^\alpha - b_2^\alpha, a_2^\alpha - b_1^\alpha] \tag{5.7}$$

$$[a_1^\alpha, a_2^\alpha] \times [b_1^\alpha, b_2^\alpha] = [a_1^\alpha \times b_1^\alpha, a_2^\alpha \times b_2^\alpha] \tag{5.8}$$

$$[a_1^\alpha, a_2^\alpha] : [b_1^\alpha, b_2^\alpha] = [a_1^\alpha : b_2^\alpha, a_2^\alpha : b_1^\alpha] \tag{5.9}$$

Based on these rules and the extension principle for two fuzzy numbers A and B and arithmetic operator $\circ$ so that $A \circ B = C$, fuzzy set C is defined as

$$C = \left\{(z, \mu_C(z)) \mid z = x \circ y,\ \mu_C(z) = \max_{x \circ y = z} \min[\mu_A(x), \mu_B(y)], \quad \forall x \in A, \forall y \in B\right\} \tag{5.10}$$

Using this general rule, arithmetic operations with triangular fuzzy numbers $A = (a_1, a_2, a_3)$ and $B = (b_1, b_2, b_3)$ takes the following forms.

Addition

The result of adding $A + B$ is triangular fuzzy number $C = (c_1, c_2, c_3)$,

$$C = (a_1 + b_1, a_2 + b_2, a_3 + b_3) \tag{5.11}$$

For example, adding fuzzy numbers $A = \tilde{4} = (3, 4, 5)$ and $B = \widetilde{2.5} = (2, 2.5, 3)$ gives fuzzy number $C = (5, 6.5, 8)$ as the result. It should be noted that the indefiniteness of results of arithmetic operations on two indefinite numbers is bigger than the indefiniteness of individual numbers [25].

Subtraction
The result of subtraction $A - B$ is triangular fuzzy number $C = (c_1, c_2, c_3)$

$$C = (a_1 - b_3, a_2 - b_2, a_3 - b_1) \tag{5.12}$$

Multiplication by constant
The result of multiplication of fuzzy number $A = (a_1, a_2, a_3)$ by constant k is a triangular number (ka_1, ka_2, ka_3) [26, 27].

Multiplication
Multiplication of two triangular fuzzy numbers $A \cdot B$ results in curved triangular fuzzy number $C = (c_1, c_2, c_3)$ [28].

$$C = (a_1 \cdot b_1, a_2 \cdot b_2, a_3 \cdot b_3) \tag{5.13}$$

Division
Assuming that 0 (zero) in not a member to fuzzy set B, the division of two triangular numbers $A : B$ results in curved triangular fuzzy number $C = (c_1, c_2, c_3)$,

$$C = (a_1/b_3, a_2/b_2, a_3/b_1) \tag{5.14}$$

5.3 Fuzzy Level Based Weight Assessment (F-LBWA) model

The following part of the paper presents fuzzy extension of the LBWA model [21]. Let us consider a multicriteria model with n criteria $S = \{C_1, C_2, \ldots, C_n\}$. Suppose that weight coefficients associated to these criteria are to be determined, i. e., they are not given in advance. In the following part, it is presented the process of obtaining the weight coefficients of criteria by applying the F-LBWA model:

Step 1. Determining the most important criterion from the set of criteria $S = \{C_1, C_2, \ldots, C_n\}$. A group of m experts are used in this step. Based on the consensus of decision makers (DM), they choose the most important criterion from a set of criteria $S = \{C_1, C_2, \ldots, C_n\}$ and classify the criteria by levels. Let the decision maker determine the most important criterion, i. e., let the criterion C_1 be the criterion in the set of criteria $S = \{C_1, C_2, \dots, C_n\}$ that is the most significant for the decision-making process.

Step 2. Grouping criteria by levels of significance. Let the DM's establish subsets of criteria in the following way:

Level S_1: At the level S_1 group, the criteria from the set S whose significance is equal to the significance of the criterion C_1 or up to twice as less as the significance of the criterion C_1.

Level S_2: At the level S_2 group, the criteria from the set S whose significance is exactly twice as less as the significance of the criterion C_1 or up to three times as less as the significance of the criterion C_1.

Level S_3: At the level S_3 group, the criteria from the set S whose significance is exactly three times as less as the significance of the criterion C_1 or up to four times as less as the significance of the criterion C_1.

...

Level S_k: At the level S_k group, the criteria from the set S whose significance is exactly k times as less as the significance of the criterion C_1 or up to $k + 1$ as less as the significance of the criterion C_1.

By applying the rules presented above, each DM establishes rough classification of the observed criteria, i. e., groups the criteria according to the levels of significance. If the significance of the criterion C_j is denoted by $s(C_j)$, where $j \in \{1, 2, \dots, n\}$, then we have $S = S_1 \cup S_2 \cup \cdots \cup S_k$, where for every level $i \in \{1, 2, \dots, k\}$. The following applies:

$$S_i = \{C_{i_1}, C_{i_2}, \dots, C_{i_s}\} = \{C_j \in S : i \le s(C_j) < i + 1\} \tag{5.15}$$

Also, for each $p, q \in \{1, 2, \dots, k\}$ such that $p \neq q$ holds $S_p \cap S_q = \emptyset$. Thus, in this way is well-defined partition of the set of criteria S.

Step 3. Within the formed subsets (levels) of the influence of the criteria each DM performs the comparison of criteria by their significance. Each criterion $C_{i_p} \in S_i$ in the subset $S_i = \{C_{i_1}, C_{i_2}, \dots, C_{i_s}\}$ is assigned with an integer $I_{i_p} \in \{0, 1, \dots, r\}$ such that the most important criterion C_1 is assigned with $I_1 = 0$, and if C_{i_p} is more significant than C_{i_q} then $I_p < I_q$, and if C_{i_p} is equivalent to C_{i_q} then $I_p = I_q$. Where the maximum value on the scale for the comparison of criteria is defined by applying the expression (5.16),

$$r = \max\{|S_1|, |S_2|, \dots, |S_k|\} \tag{5.16}$$

Based on the comparisons of criteria performed within every observed level for every DM are obtained the values $I_{i_p} \in \{0, 1, \dots, r\}$, depending on the preferences expressed. Considering that every DM independently compares the criteria within subsets (levels), deviations are expected in comparison of criteria that will be represented by fuzzy numbers. Applying the expression (5.17), crisp values of expert comparisons of criteria within subsets (levels) are transformed into fuzzy values. The expression (5.17) is also used to aggregate decision makers' preferences:

$$\tilde{I}_{i_p} = (a, b, c) \quad \Rightarrow \quad a = \min_m \{I_{i_p}^m\}; \quad b = \frac{1}{m}\sum_{e=1}^{m} I_{i_p}^e; \quad c = \max_m \{I_{i_p}^m\} \tag{5.17}$$

where m presents total number of decision makers participating in the research.

Step 4. Based on the defined maximum value of the scale for the comparison of criteria (r), the expression (5.16), it is defined the elasticity coefficient $r_0 \in N$ (where N presents the set of real numbers) which should meet the requirement where $r_0 > r$, $r = \max\{|S_1|, |S_2|, \ldots, |S_k|\}$.

Step 5. Calculation of the influence function of the criteria. The influence function $f : S \to R$ is defined in the following way. For every criterion $C_{i_p} \in S_i$ can be defined the influence function of the criterion

$$\tilde{f}(C_{i_p}) = \frac{r_0}{i \cdot r_0 + \tilde{I}_{i_p}} \tag{5.18}$$

where i presents the number of the level/subset in which is classified the criterion, r_0 presents the elasticity coefficient, while $\tilde{I}_{i_p}$ presents the fuzzy value assigned to the criterion C_{i_p} within the observed level.

Step 6. Calculation of the optimum fuzzy values of the weight coefficients of criteria. By applying the expression (5.19), it is calculated the fuzzy weight coefficient of the most significant criterion:

$$\tilde{w}_1 = \frac{1}{1 + \tilde{f}(C_2) + \cdots + \tilde{f}(C_n)} \tag{5.19}$$

The values of the weight coefficients of the remaining criteria are obtained by applying the expression (5.20),

$$\tilde{w}_j = \tilde{f}(C_j) \cdot \tilde{w}_1 \tag{5.20}$$

where $j = 2, 3, \ldots, n$ and n present total number of criteria.

Defuzzification was carried out using equation (5.21)

$$w_j = (w_j^{(l)} + 4w_j^{(m)} + w_j^{(r)}) \cdot 6^{-1} \tag{5.21}$$

where $w_j^{(l)}$ and $w_j^{(r)}$ are the left and right distribution trust intervals of the triangular fuzzy number, respectively, $w_j^{(m)}$ is the value in which the triangular function reaches its maximum value.

Final crisp values of the weight coefficients are obtained by applying the expression (5.22)

$$w_j = \frac{W_j}{\sum_{j=1}^{n} W_j} \tag{5.22}$$

5.4 Application of F-LBWA model

In the following section, it is presented the application of the F-LBWA model in determining weight coefficients of criteria in the multicriteria problem discussed in the literature. The problem of determining the weight coefficients of the criteria in the evaluation of the work of advisers in the transport of dangerous goods has been considered [29]. Nine criteria were identified for the evaluation of the work of advisers in the transport of dangerous goods: C_1 – Knowledge of regulations and professional development, C_2 – Analytic processing of established requirements, C_3 – Quality of proposed measures, C_4 – Level of realization of the proposed measures, C_5 – Quality of professional training of employees, C_6 – Response method in emergency situations, C_7 – Document preparation, C_8 – Method of solving professional questions and C_9 – Activity in professional institutions. Four experts participated in the research, who grouped the criteria by levels and compared them in pairs of criteria based on personal preferences. The weight coefficients of the criteria for evaluating the work of advisors in the transport of dangerous goods are defined using the F-LBWA model.

Step 1. Determining the most important criterion from the set of criteria $S = \{C_1, C_2, \ldots, C_9\}$. As the most significant/influential criterion, DMs defined the criterion C_5.

Step 2. DMs grouping criteria by levels of significance. In accordance with the preferences of the DMs, the criteria are grouped in the following subsets/levels:

Level S_1: The criteria C_1, C_8 and C_9 are up to twice as less important as the criterion C_5.
Level S_2: The criteria C_3 and C_4 are between twice or three times as less important as the criterion C_5.
Level S_4: The criterion C_6 is between four or five times as less important as the criterion C_5.
Level S_7: The criteria C_2 and C_7 are between seven or eight times as less important as the criterion C_5.

Then, based on the mentioned preferences of the decision makers, the criteria can be grouped in the following subsets/levels:

$$
\begin{aligned}
S_1 &= \{C_1, C_5, C_8, C_9\}, \\
S_2 &= \{C_3, C_4\}, \\
S_3 &= \emptyset, \\
S_4 &= \{C_6\}, \\
S_5 &= S_6 = \emptyset, \\
S_7 &= \{C_2, C_7\}.
\end{aligned}
$$

Step 3. Based on the expression (5.16), it defines the maximum value of the scale for the comparison of criteria:

$$\left.\begin{array}{l} S_1 = \{C_1, C_5, C_8, C_9\}, \\ S_2 = \{C_3, C_4\}, \\ S_3 = \emptyset, \\ S_4 = \{C_6\}, \\ S_5 = S_6 = \emptyset, \\ S_7 = \{C_2, C_7\}. \end{array}\right\} \Rightarrow \quad r = \max\{|S_1|, |S_2|, |S_4|, |S_7|\} = 4.$$

Based on the maximum value of the scale for comparison, it can be concluded that the scale for comparing the criteria ranges in the interval $I_{i_p} \in \{0, 1, \ldots, 4\}$. Based on the scale and the predefined set of criteria, the comparison of criteria can be performed within each individual set.

Level S_1: Based on the preferences of the DMs, the following relations are defined:

$$\begin{array}{ll} DM1: & I_5 = 0, \quad I_8 = 1, \quad I_9 = 2 \quad \text{and} \quad I_1 = 4; \\ DM2: & I_5 = 1, \quad I_8 = 1.5, \quad I_9 = 3.5 \quad \text{and} \quad I_1 = 5; \\ DM3: & I_5 = 0.5, \quad I_8 = 2.5, \quad I_9 = 2.5 \quad \text{and} \quad I_1 = 3.5; \\ DM4: & I_5 = 1.5, \quad I_8 = 2, \quad I_9 = 2.5 \quad \text{and} \quad I_1 = 4. \end{array}$$

Level S_2: Within the set S_2, the following relations are defined:

$$\begin{array}{ll} DM1: & I_3 = 0 \quad \text{and} \quad I_4 = 2; \\ DM2: & I_3 = 1 \quad \text{and} \quad I_4 = 2; \\ DM3: & I_3 = 1.5 \quad \text{and} \quad I_4 = 2.5; \\ DM4: & I_3 = 0.5 \quad \text{and} \quad I_4 = 2. \end{array}$$

Level S_4: Within the set S_4, the following relations are defined:

$$\begin{array}{ll} DM1: & I_6 = 1.5; \\ DM2: & I_6 = 2; \\ DM3: & I_6 = 2; \\ DM4: & I_6 = 3.5. \end{array}$$

Level S_7: Within the set S_7, the following relations are defined:

$$\begin{array}{ll} DM1: & I_2 = 1 \quad \text{and} \quad I_7 = 3; \\ DM2: & I_2 = 0.5 \quad \text{and} \quad I_7 = 2; \\ DM3: & I_2 = 1 \quad \text{and} \quad I_7 = 3.5; \\ DM4: & I_2 = 2 \quad \text{and} \quad I_7 = 3. \end{array}$$

Table 5.1: Fuzzy preferences.

Preferences	Triangular fuzzy number	Level/subset
I_5	(0.00, 0.75, 1.50)	Level 1
I_8	(1.00, 1.75, 2.50)	
I_9	(2.00, 2.63, 3.50)	
I_1	(3.50, 4.13, 5.00)	
I_3	(0.00, 0.75, 1.50)	Level 2
I_4	(2.00, 2.13, 2.50)	
I_6	(1.50, 2.25, 3.50)	Level 4
I_2	(0.50, 1.13, 2.00)	Level 7
I_7	(2.00, 2.88, 3.50)	

Table 5.2: The fuzzy influence functions of the criteria.

Functions	Triangular fuzzy number
$\tilde{f}(C_5)$	(0.769, 0.870, 1.000)
$\tilde{f}(C_8)$	(0.667, 0.741, 0.833)
$\tilde{f}(C_9)$	(0.588, 0.656, 0.714)
$\tilde{f}(C_1)$	(0.500, 0.548, 0.588)
$\tilde{f}(C_3)$	(0.435, 0.465, 0.500)
$\tilde{f}(C_4)$	(0.400, 0.412, 0.417)
$\tilde{f}(C_6)$	(0.213, 0.225, 0.233)
$\tilde{f}(C_2)$	(0.135, 0.138, 0.141)
$\tilde{f}(C_7)$	(0.130, 0.132, 0.135)

Based on the comparisons of criteria performed, using the expression (5.17) crisp values within the subsets (levels) are transformed into fuzzy values; Thus the following fuzzy are obtained preferences by levels, as in the Table 5.1.

Step 4. Based on the defined maximum value of the scale for the comparison of criteria $r = 4$, it is defined the elasticity coefficient such that $r_0 > r$, respectively, $r_0 > 4$.

Step 5. Defining the fuzzy influence function of the criteria. If it is known that $r_0 > 4$, arbitrarily is determined the value $r_0 = 5$. By applying the expression (5.18), the fuzzy influence functions of the criteria are calculated (Table 5.2).

Step 6. Calculation of the optimum fuzzy values of the weight coefficients of criteria. By applying the expression (5.19), it is calculated the fuzzy value of the weight coefficient of the most influential criterion

$$\tilde{w}_5 = \left(w_5^{(l)}, w_5^{(m)}, w_5^{(r)}\right) = \begin{cases} w_5^{(l)} = \dfrac{1}{1+1.00+0.833+0.714+\cdots+0.135} = 0.180 \\ w_5^{(m)} = \dfrac{1}{1+0.870+0.741+0.656...+0.132} = 0.193 \\ w_5^{(r)} = \dfrac{1}{1+0.769+0.667+0.588+\cdots+0.130} = 0.207 \end{cases}$$

Table 5.3: Fuzzy values of the weight coefficients of the criteria.

Criteria	Triangular fuzzy number
$\tilde{w}_5$	(0.180, 0.193, 0.207)
$\tilde{w}_8$	(0.120, 0.143, 0.172)
$\tilde{w}_9$	(0.106, 0.126, 0.148)
$\tilde{w}_1$	(0.090, 0.106, 0.122)
$\tilde{w}_3$	(0.078, 0.090, 0.103)
$\tilde{w}_4$	(0.072, 0.080, 0.086)
$\tilde{w}_6$	(0.038, 0.043, 0.048)
$\tilde{w}_2$	(0.024, 0.027, 0.029)
$\tilde{w}_7$	(0.023, 0.025, 0.028)

By applying the expression, (5.20) are obtained the fuzzy values of the weight coefficients of the remaining criteria, Table 5.3.

Using the expressions (5.21) and (5.22), fuzzy values of the weight coefficients are defuzzified and final vector of the weight coefficients is obtained:

$$w_j = (0.231, 0.173, 0.152, 0.127, 0.108, 0.095, 0.052, 0.032, 0.031)^T.$$

5.5 Conclusion

This paper presents fuzzy modification of the LBWA model [21], which is characterized by a simple and rational mathematical algorithm. The results of this research showed that the F-LBWA model enabled obtaining credible values of the weight coefficients under the conditions of uncertainty prevailing in group decision making.

Based on the results presented can be distinguished the following advantages of the F-LBWA model: (1) The F-LBWA model enables the calculation of weight coefficients with small number of criteria comparisons; (2) The algorithms of the F-LBWA model is not made more complex by the increase in the number of criteria, which makes it suitable for the application in complex MCDM models with more evaluation criteria; (3) The LBWA model enables decision makers to present their preferences through logical algorithm when prioritizing criteria. Using the F-LBWA model, fuzzy values of weight coefficients are obtained with simple mathematical apparatus that eliminates inconsistencies in expert preferences, which are tolerated in certain subjective models.

In addition to the mentioned advantages, it is also necessary to emphasize the flexibility of the F-LBWA model with respect to additional corrections of fuzzy weight coefficients using the elasticity coefficient (r_0). The elasticity coefficient allows decision makers to further adjust the values of the weight coefficients according to their own preferences. Besides, the elasticity coefficient enables the robustness of the

MCDM model to be analyzed by defining the influence of the change of the weights of criteria to the final decision.

Bibliography

[1] Keeney RL, Raiffa H. Decisions with multiple objectives. New York: Wiley; 1976.
[2] Weber M, Eisenfuhr F, von Winterfeldt D. The effects of splitting attributes on weights in multiattribute utility measurement. Manag Sci. 1988;34:431–45.
[3] Green PE, Srinivasan V. Conjoint analysis in marketing: new developments with implications for research and practice. J Mark. 1990;54:3–19.
[4] Saaty TL. Analytic hierarchy process. New York: McGraw-Hill; 1980.
[5] Edwards W, Barron FH. SMARTS and SMARTER: improved simple methods for multiattribute utility measurement. Organ Behav Hum Decis Process. 1994;60:306–25.
[6] Bana Costa C, Vansnick JC. MACBETH: an interactive path towards the construction of cardinal value functions. Int J Oper Res. 1994;1(4):489–500.
[7] Poyhonen M, Hamalainen R. On the convergence of multiattribute weighting methods. Eur J Oper Res. 2001;129:569–85.
[8] Weber M, Borcherding K. Behavioral influences on weight judgments in multiattribute decision making. Eur J Oper Res. 1993;67:1–12.
[9] Rogers M, Bruen M. A new system for weighting environmental criteria for use within ELECTRE III. Eur J Oper Res. 1998;107(3):552–63.
[10] Graham A. Nonnegative matrices and applicable topics in linear algebra. Chichester, UK: Ellis Horwood; 1987.
[11] Mikhailov L. A fuzzy programming method for deriving priorities in the analytic hierarchy process. J Oper Res Soc. 2000;51:341–9.
[12] Costa JP, Climaco JC. Relating reference points and weights in MOLP. J Multi-Criteria Decis Anal. 1999;8:281–90.
[13] Mousseau V, Slowinski R, Zielniewicz P. A user-oriented implementation of the ELECTRE-TRI method integrating preference elicitation support. Comput Oper Res. 2000;27:757–77.
[14] Chatterjee K, Pamučar D, Zavadskas EK. Evaluating the performance of suppliers based on using the R'AMATEL-MAIRCA method for green supply chain implementation in electronics industry. J Clean Prod. 2018;184:101–29.
[15] Roy J, Adhikary K, Kar S, Pamucar D. A rough strength relational DEMATEL model for analysing the key success factors of hospital service quality. Decis Mak Appl Manag Eng. 2018;1(1):121–42.
[16] Liu F, Aiwu G, Lukovac V, Vukic M. A multicriteria model for the selection of the transport service provider: a single valued neutrosophic DEMATEL multicriteria model. Decis Mak Appl Manag Eng. 2018;1(2):121–30.
[17] Valipour A, Yahaya N, Md Noor N, Antuchevičienė J, Tamošaitienė J. Hybrid SWARA-COPRAS method for risk assessment in deep foundation excavation project: an Iranian case study. J Civ Eng Manag. 2017;23(4):524–32.
[18] Veskovic S, Stevic Z, Stojic G, Vasiljevic M, Milinkovic S. Evaluation of the railway management model by using a new integrated model DELPHI-SWARA-MABAC. Decis Mak Appl Manag Eng. 2018;1(2):34–50.
[19] Rezaei J. Best-worst multi-criteria decision-making method. Omega. 2015;53:49–57.
[20] Pamučar D, Stević Ž, Sremac S. A new model for determining weight coefficients of criteria in MCDM models: Full Consistency Method (FUCOM). Symmetry. 2018;10(9):393.

[21] Žižović M, Pamučar D. New model for determining criteria weights: Level Based Weight Assessment (LBWA) model. Decis Mak Appl Manag Eng. 2019;2(2):126–37.
[22] Shannon CE, Weaver W. The mathematical theory of communication. Urbana: The University of Illinois Press; 1947.
[23] Srđević B, Medeiros YDP, Faria AS, Schaer M. Objektivno vrednovanje kriterijuma performanse sistema akumulacija. Vodoprivreda. 2003;35:163–76. (Only in Serbian).
[24] Jantzen J. Design of fuzzy controllers. Tech report No. 98-864. Technical. University of Denmark, Department of Automation; 1998.
[25] Pamučar D, Božanić D, Đorović B, Milić A. Modelling of the fuzzy logical system for offering support in making decisions within the engineering units of the Serbian army. Int J Phys Sci. 2011;3:592–609.
[26] Zadeh LA. Fuzzy sets. Inf Control. 1965;8:338–53.
[27] Teodorović D, Kikuchi S. Fuzzy sets and applications in traffic and transport. Serbia: Faculty of Transport and Traffic Engineering, University of Belgrade; 1994.
[28] Ćirović G, Pamučar D. Decision support model for prioritizing railway level crossings for safety improvements: application of the adaptive neuro-fuzzy system. Expert Syst Appl. 2013;40(6):2208–23.
[29] Pamučar D, Sremac S, Stević Ž, Ćirović G, Tomić D. New multi-criteria LNN WASPAS model for evaluating the work of advisors in the transport of hazardous goods. Neural Comput Appl. 2019;31:5045–68. https://doi.org/10.1007/s00521-018-03997-7.

Divya Chhibber, Dinesh C. S. Bisht, and Pankaj Kumar Srivastava

6 Fuzzy transportation problems and variants

Abstract: The given chapter focuses on the notion of fuzzy theory used to optimize transportation problem. The emergence of transportation problem and its types have been explained along with the concept of linear programming problem. The implementation of fuzzy theory into transportation problem has been proved to be a boon in the area of decision making. Researchers have worked substantially in this field and have evolved imperative techniques to solve it. Some of the ranking techniques established by esteemed scholars have been elucidated here. Types of the fuzzy transportation problem inclusive of their characteristics have been explained. Also, several variants of fuzzy transportation based upon membership function as well as classical approach have been described.

Keywords: Fuzzy transportation, multiobjective transportation, fuzzy set, fuzzy number, ranking of fuzzy set, intuitionistic fuzzy set, rough set

6.1 Introduction

Everyone must have witnessed the chaos happening because of time constraints. Be it preparing for exams which are going to happen back-to-back or finishing various assigned tasks before the deadline, allocation of limited time is an obstacle. Considering the objective and other constraints (such as grades, skill levels, available time, etc.), a work plan has to be fabricated. Likewise, in an organization, decisions have to be taken to make the best use of limited resources to attain organizational goals. There, the problem is based on how to make a decision regarding allocation to obtain the best result. These issues can be characterized by a linear programming problem (LPP). Ages back, George B. Dantzig [1], established a notion of LPP.

An LPP is to establish an optimum value of a linear function taking into account the constraints involved which are represented by linear inequalities or equations.

Its mathematical representation is

$$\begin{aligned}&\text{Maximize } \beta^T\theta\\&\text{subject to } \alpha\theta \le d\\&\text{and } \theta \ge 0\end{aligned}$$

where α is a coefficient matrix, β and d are coefficient vectors, θ represents the vector of unknown variables. The function which has to be optimized is called the objective

* **Pankaj Kumar Srivastava,** Department of Mathematics, Jaypee Institute of Information Technology, 201304 Noida, India, e-mail: pankaj.srivastava@jiit.ac.in
Divya Chhibber, Dinesh C. S. Bisht, Department of Mathematics, Jaypee Institute of Information Technology, 201304 Noida, India, e-mails: divyachhibber@gmail.com, drbisht.math@gmail.com

https://doi.org/10.1515/9783110671353-006

function and the inequalities $\alpha\theta \leq d$ and $\theta \geq 0$ based upon which the objective function is to be optimized, are called constraints.

Dantzig [2] also introduced the simplex method to solve it. His intense work benefitted various industries and fields. For example, it helped the airline industry to schedule crews and make fleet assignments, the oil company got the idea for refinery planning, etc. The pragmatic approach of linear programming problem (LPP) features its extensive applications in the field of operations research. It is used in company management, manufacturing, planning, production, transportation, advertising, etc. [3, 4, 5].

6.2 Transportation problem

The most crucial application of LPP is the transportation problem (TP) which was initiated by Hitchcock [6]. This standard form of LPP intends to curtail the transportation cost while fetching the commodities from distinct locations. Imagine that a person owns a network of retail shops of organic products in different areas. In order to run a successful business, he also has a warehouse to store his goods ready to be dispatched whenever there is a requirement in his stores. All of the products were being transported from his warehouse to all of his stores. Now, he has expanded his business and opened two more warehouses for which important decisions need to be taken in order to reduce cost and time. For example, from which warehouse to which store commodities need to be transported, source of transportation, time taken, etc. This problem describes the application of the transportation problem.

Let us deal with p origins φ_m $(m = 1, \ldots, p)$ and q destinations ψ_n $(n = 1, \ldots, q)$. Let α_m be the quantity to be transported to q destinations ψ_n to fulfill the demand of β_n units and ω_{mn} be the amount of commodity to be transported. Let Ω_{mn} be the penalty cost corresponding to fetching a unit of commodity form p to q.

The mathematical representation of a transportation problem is

$$\text{Minimize } \eta = \sum_{m=1}^{p} \sum_{n=1}^{q} \Omega_{mn}\omega_{mn}$$

subject to

$$\begin{aligned} \sum_{n=1}^{q} \omega_{mn} &= \alpha_m, \quad m = 1, \ldots, p, \\ \sum_{m=1}^{p} \omega_{mn} &= \beta_n, \quad n = 1, \ldots, q, \\ \omega_{mn} &\geq 0, \quad m = 1, \ldots, p,\ n = 1, \ldots, q. \end{aligned}$$

After the development of this useful technique, study was done on various types of transportation problems [6, 7, 8]. In 1947, Dantzig [9] introduced the concept of a simplex algorithm which was later modified by Charnes et al. [10]. Dantzig [11] was the man behind the innovative evolvement of Northwest corner rule as well. It is the technique to obtain initial basic feasible solution of the transportation problem. Afterwards, various methods like row minima, column minima, matrix minima, row column method, Vogel approximation method (VAM) were introduced by scholars to obtain an initial basic feasible solution. Thereafter, methods like modified distribution (MODI) technique, stepping stones method [12], were introduced in the literature. Thereupon, many new algorithms were introduced by scholars. Palanivel and Suganya [13] used the harmonic mean technique to solve a TP. Dinagar and Kaliyaperumal [14] gave a new technique of finding optimal solution by categorizing it under four different environments.

6.3 Multiobjective transportation problem

In practicality, transportation problem with multiple objectives, i. e., multiobjective transportation problem (MOTP), is of much more importance than that with single objective. Multiple objectives may include minimization of transportation cost along with delivery time or loss during transportation, etc.

Mathematical representation of MOTP:

$$\text{Minimize } \eta_t = \sum_{m=1}^{p} \sum_{n=1}^{q} \Omega_{mn}^t \omega_{mn}, \quad t = 1, \dots, T$$

subject to

$$\sum_{n=1}^{q} \omega_{mn} = \alpha_m, \quad m = 1, \dots, p,$$

$$\sum_{m=1}^{p} \omega_{mn} = \beta_n, \quad n = 1, \dots, q,$$

$$\omega_{mn} \geq 0, \quad m = 1, \dots, p, \ n = 1, \dots, q,$$

where the penalty criterion is symbolized by the subscript on η_t and superscript on Ω_{mn}^t. $\alpha_m > 0$ for all m, $\beta_n > 0$ for every n, $\Omega_{mn}^t \geq 0$ for every (m, n).

Lee et al. [15] primarily deliberated the concept of optimization of MOTP. Further, algorithms for the identification of nondominated solutions were developed by Diaz [16] and Isserman [17]. Analysis of multiobjective design of transportation networks was performed by Current et al. [18]. For the accomplishment of all the nondominated solutions of the MOTP, an alternative procedure was presented by Diaz [19]. The use of additive multiattribute utility function had been advocated by Edwards [20] and the two interactive algorithms for the linear MOTP was developed by Ringuest et al. [21].

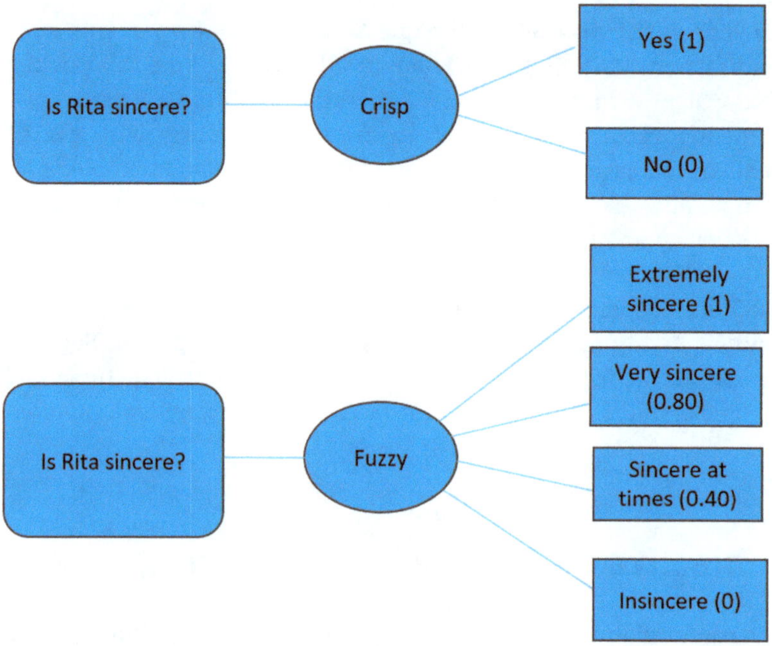

Figure 6.1: Example of fuzziness.

6.4 Balanced and unbalanced TP

Two types of TP exist; balanced TP and unbalanced TP. Balanced TP is when the total supply equals the total demand and unbalanced TP is when the total supply is not equal to the total demand. In the latter, a dummy destination is added to convert it into balanced TP.

6.5 Concept of fuzzy theory

Zadeh [22] introduced the ingenious notion of fuzzy theory. The word fuzzy means vague. Human intelligence comprehends fuzzy knowledge. Earlier also the perception of vagueness was there but Zadeh's strong description made it possible to realize the concept of fuzziness.

A fuzzy set Ê on a set χ is a function Ê: $\chi \to [0,1]$. It is characterized by its membership function $\mu_{\hat{E}} : \chi \to [0,1]$ which associates a real number $\mu_{\hat{E}} \in [0,1]$ with each $x \in \chi$. The value of $\mu_{\hat{E}}$ defines the degree to which $x \in \chi$. It is shown with the help of an example in Figure 6.1.

Fuzzy theory is used in various fields like psychology, philosophy, political science, etc.

6.6 Basics of fuzzy transportation problem

It is understood until now that TP is of utmost importance in real life. But there always exist some inexactness while estimating the transportation cost. Several factors like petrol/diesel prices, climatic conditions and traffic contribute to this impreciseness. Thus Belman and Zadeh [23] were influenced to expand the idea of fuzziness to transportation problem giving rise to the concept of Fuzzy Transportation Problem (FTP). As the membership value is considered in fuzzy sets, the dubiety of human decisions decreases. Henceforth, the theory of fuzzy sets had been enacted and generalized to FTP by several researchers [24, 25, 26, 27, 28].

6.7 Preliminaries

Definition 6.1 (Fuzzy set). Let Φ be the universe set. A fuzzy set $\tilde{B}$ in Φ is described as $\tilde{B} = \{(\lambda, \omega_{\tilde{B}}(\lambda)) : \lambda \in \Phi\}$, where $\omega_{\tilde{B}}(\lambda) \in [0, 1]$ indicates the membership degree of λ in $\tilde{B}$ and is called the membership function of $\tilde{B}$.

Definition 6.2 (Triangular fuzzy number). A real triangular fuzzy number (TFN) $\tilde{\alpha} = (\alpha_1, \alpha_2, \alpha_3)$ is a fuzzy subset of the set of real numbers $\Re$ with the membership function $\omega_{\tilde{B}}(\lambda)$ satisfying the conditions given below:
(i) $\omega_{\tilde{B}}(\lambda)$ defines a continuous mapping from $\Re$ to $[0, 1]$.
(ii) $\omega_{\tilde{B}}(\lambda) = 0\ \forall \alpha \in (-\infty, \alpha_1]$.
(iii) $\omega_{\tilde{B}}(\lambda)$ is strictly increasing and continuous on $[\alpha_1, \alpha_2]$.
(iv) $\omega_{\tilde{B}}(\lambda) = 1\ \forall \alpha = \alpha_2$.
(v) $\omega_{\tilde{B}}(\lambda)$ is strictly decreasing and continuous on $[\alpha_2, \alpha_3]$.
(vi) $\omega_{\tilde{B}}(\lambda) = 0\ \forall \alpha \in [\alpha_3, +\infty)$.

Membership function of a TFN is defined as

$$\omega_{\tilde{B}}(\lambda) = \begin{cases} \frac{z-\alpha_1}{\alpha_2-\alpha_1}; & \alpha_1 \le z \le \alpha_2 \\ \frac{\alpha_3-z}{\alpha_3-\alpha_2}; & \alpha_2 \le z \le \alpha_3 \\ 0, & \text{otherwise} \end{cases}$$

Diagrammatic representation of a TFN is depicted in Figure 6.2.

Definition 6.3 (Trapezoidal fuzzy number). A real trapezoidal fuzzy number $\tilde{\alpha} = (\alpha_1, \alpha_2, \alpha_3, \alpha_4)$ is a fuzzy subset of the set of real numbers $\Re$ with the membership function $\omega_{\tilde{B}}(\lambda)$ satisfying the conditions given below:
(i) $\omega_{\tilde{B}}(\lambda)$ defines a continuous mapping from $\Re$ to $[0, 1]$.
(ii) $\omega_{\tilde{B}}(\lambda) = 0\ \forall \alpha \in (-\infty, \alpha_1]$.
(iii) $\omega_{\tilde{B}}(\lambda)$ is strictly increasing and continuous on $[\alpha_1, \alpha_2]$.

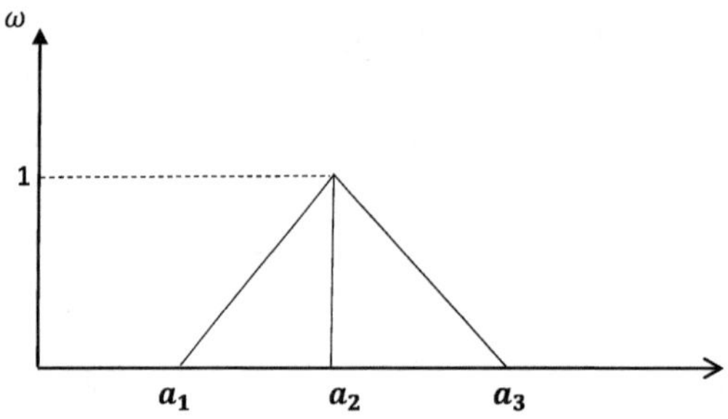

Figure 6.2: Triangular fuzzy number.

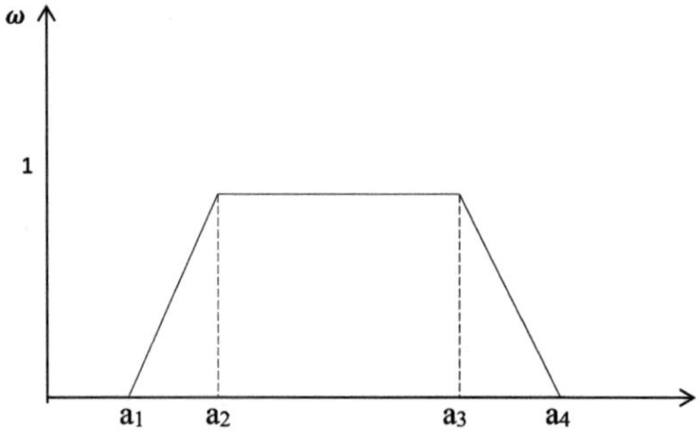

Figure 6.3: Trapezoidal fuzzy number.

(iv) $\omega_{\tilde{B}}(\lambda) = 1\ \forall \alpha \in [\alpha_2, \alpha_3]$.
(v) $\omega_{\tilde{B}}(\lambda)$ is strictly decreasing and continuous on $[\alpha_3, \alpha_4]$.
(vi) $\omega_{\tilde{B}}(\lambda) = 0\ \forall \alpha \in [\alpha_4, +\infty)$.

Membership function of a trapezoidal fuzzy number is defined as

$$\omega_{\tilde{B}}(\lambda) = \begin{cases} 0; & z \le \alpha_1 \\ \frac{z-\alpha_1}{\alpha_2-\alpha_1}; & \alpha_1 \le z \le \alpha_2 \\ 1; & \alpha_2 \le z \le \alpha_3 \\ \frac{\alpha_4-z}{\alpha_4-\alpha_3}; & \alpha_3 \le z \le \alpha_4 \\ 0; & z \ge \alpha_4 \end{cases}$$

Diagrammatic representation of a trapezoidal fuzzy number is depicted in Figure 6.3.

6.7.1 Arithmetic operations on triangular fuzzy numbers

Let $\tilde{M} = (m_1, m_2, m_3)$ and $\tilde{N} = (n_1, n_2, n_3)$ be two triangular fuzzy numbers. Then the arithmetic operations on them are defined as:

(i) *Addition*: $\tilde{M} \oplus \tilde{N} = (m_1 + n_1, m_2 + n_2, m_3 + n_3)$;
(ii) *Subtraction*: $\tilde{M} \ominus \tilde{N} = (m_1 - n_3, m_2 - n_2, m_3 - n_1)$;
(iii) *Multiplication*: $\tilde{M} \otimes \tilde{N} = (m_1 n_1, m_2 n_2, m_3 n_3)$;
(iv) *Scalar Multiplication*: $\lambda \times \tilde{M} = \begin{cases} (\lambda m_1, \lambda m_2, \lambda m_3), & \lambda \geq 0 \\ (\lambda m_3, \lambda m_2, \lambda m_1), & \lambda \leq 0. \end{cases}$

6.7.2 Arithmetic operations on trapezoidal fuzzy numbers

Let $\tilde{M} = (m_1, m_2, m_3, m_4)$ and $\tilde{N} = (n_1, n_2, n_3, n_4)$ be two triangular fuzzy numbers. Then the arithmetic operations on them are defined as:
(i) *Addition*: $\tilde{M} \oplus \tilde{N} = (m_1 + n_1, m_2 + n_2, m_3 + n_3, m_4 + n_4)$;
(ii) *Subtraction*: $\tilde{M} \ominus \tilde{N} = (m_1 - n_4, m_2 - n_3, m_3 - n_2, m_4 - n_1)$;
(iii) *Scalar Multiplication*: $\lambda \times \tilde{M} = \begin{cases} (\lambda m_1, \lambda m_2, \lambda m_3, \lambda m_4), & \lambda \geq 0 \\ (\lambda m_4, \lambda m_3, \lambda m_2, \lambda m_1), & \lambda < 0; \end{cases}$
(iv) *Multiplication*: $\tilde{M} \otimes \tilde{N} = (\theta_1, \theta_2, \theta_3, \theta_4)$;

where

$$\begin{aligned} \theta_1 &= \min\{m_1 n_1, m_1 n_4, m_4 n_1, m_4 n_4\} \\ \theta_2 &= \min\{m_2 n_2, m_2 n_3, m_3 n_2, m_3 n_3\} \\ \theta_3 &= \max\{m_2 n_2, m_2 n_3, m_3 n_2, m_3 n_3\} \\ \theta_4 &= \max\{m_1 n_1, m_1 n_4, m_4 n_1, m_4 n_4\} \end{aligned}$$

6.8 Ranking technique

Comparison of fuzzy numbers by ranking them is imperative. A function given by $\Psi : \Gamma(\mathfrak{R}) \to \mathfrak{R}$, where $\Gamma(\mathfrak{R})$ denotes the set of fuzzy numbers, is called a ranking function which maps each fuzzy number to the real line. Several approaches including coefficient of variation (CV index), distance between fuzzy sets, centroid point and original point, weighted mean value, etc., have been recommended by the scholars to rank the fuzzy numbers. Dubois and Prade [29] used maximizing sets to rank fuzzy numbers. Cheng [30] advocated the distance method for ranking of fuzzy numbers, i. e.,

$$\Gamma(z) = \sqrt{\hat{s}^2 + \hat{t}^2}, \quad \text{where } \hat{s} = \frac{\int_a^b s z_L ds + \int_b^c s ds + \int_c^d s z_R ds}{\int_a^b z_L ds + \int_b^c ds + \int_c^d z_R ds}, \quad \hat{t} = \frac{\int_0^1 r \underline{z} dr + \int_0^1 r \hat{z} dr}{\int_0^1 \underline{z} dr + \int_0^1 \hat{z} dr},$$

z_L, z_R are the left and right membership functions of fuzzy number z, and $(\underline{z}, \hat{z})$ is the parametric form. The obtained value of $\Gamma(z)$ is used to rank the fuzzy numbers. If $\Gamma(z_i) < \Gamma(z_j)$, then $z_i < z_j$. If $\Gamma(z_i) > \Gamma(z_j)$, then $z_i > z_j$ and if $\Gamma(z_i) = \Gamma(z_j)$, then $z_i \sim z_j$.

Choobineh [31] suggested the coefficient of variance (CV index), i. e., CV $= \sigma/|\mu|$, $\mu \neq 0$, $\sigma > 0$, where σ is the standard error and μ is the mean. In this technique, the fuzzy number having lesser CV index has been ranked higher. Later, Chu and Tsao [32] utilized the area between the centroid point and original point for ranking purpose whereas Yager [33] worked on weighted mean value for ranking. Comparison of fuzzy numbers based on the probability measure of fuzzy events had been stated by Lee and Li [34] while Wang [35] represented preference degree by introducing a fuzzy preference relation with membership function for the comparison of two fuzzy numbers. Henceforth, work on ranking techniques had been continued giving rise to numerous approaches for ranking triangular and trapezoidal fuzzy numbers of different types [36, 26, 25, 37]. Comparing fuzzy numbers is continued to be a foremost concern of fuzzy aspect. Today's highly competitive era demands better result-oriented techniques. As a result, the existing techniques are still being refined for generalized fuzzy numbers and new approaches are being worked upon.

6.9 Variants of fuzzy transportation

6.9.1 Type 1 Fuzzy transportation problem

At times, a decision maker might be unsure about the quantity of the commodity available at the distribution center or its requirement at the destination. For example, when a new product is introduced in the market, an ambiguity regarding supply and demand occurs due to consumer's behavior. This uncertainty gave rise to the concept of Type-1 FTP which is defined as,

> A TP in which the cost function is crisp valued and the demand and supply functions are fuzzy valued, is called Type-1 FTP.

Let us deal with a TP with p distribution centers and q stations. Let $\widetilde{\alpha_m}$ be the fuzzy valued quantity to be transported to q stations to fulfill the fuzzy valued demand of $\widetilde{\beta_n}$ units and ω_{mn} be the amount of commodity to be transported. Let Ω_{mn} be the penalty cost corresponding to fetching a unit of commodity form p to q.

Mathematical representation of type-1 FTP is

$$\text{Minimize } \eta = \sum_{m=1}^{p}\sum_{n=1}^{q} \Omega_{mn}\omega_{mn}$$

subject to

$$\sum_{n=1}^{q} \omega_{mn} = \widetilde{\alpha_m}, \quad m = 1, \dots, p,$$
$$\sum_{m=1}^{p} \omega_{mn} = \widetilde{\beta_n}, \quad n = 1, \dots, q,$$
$$\omega_{mn} \geq 0, \quad m = 1, \dots p, \; n = 1, \dots, q.$$

6.9.2 Type 2 Fuzzy transportation problem

Uncertainty regarding the cost function gave rise to the concept of Type 2 FTP. It is defined as

> A TP in which the cost function is fuzzy valued but the demand and supply functions are crisp valued, is called Type-2 FTP.

Mathematical representation of type-2 FTP is:

$$\text{Minimize } \eta = \sum_{m=1}^{p}\sum_{n=1}^{q} \widetilde{\Omega_{mn}}\omega_{mn}$$

subject to

$$\sum_{n=1}^{q} \omega_{mn} = \alpha_m, \quad m = 1, \ldots, p,$$
$$\sum_{m=1}^{p} \omega_{mn} = \beta_n, \quad n = 1, \ldots, q,$$
$$\omega_{mn} \geq 0, \quad m = 1, \ldots, p,\ n = 1, \ldots, q,$$

where $\widetilde{\Omega_{mn}}$ is the fuzzy valued penalty cost corresponding to fetching a unit of commodity from p to q.

6.9.3 Fully fuzzy transportation problem

> A TP in which the cost function, the demand and supply along with the amount of the commodity to be transported are in fuzzy form, it is called a fully FTP.

Mathematical representation of fully fuzzy FTP is

$$\text{Minimize } \eta = \sum_{m=1}^{p}\sum_{n=1}^{q} \widetilde{\Omega_{mn}}\widetilde{\omega_{mn}}$$

subject to

$$\sum_{n=1}^{q} \widetilde{\omega_{mn}} = \widetilde{\alpha_m}, \quad m = 1, \ldots, p,$$
$$\sum_{m=1}^{p} \widetilde{\omega_{mn}} = \widetilde{\beta_n}, \quad n = 1, \ldots, q,$$
$$\omega_{mn} \geq 0, \quad m = 1, \ldots, p, n = 1, \ldots, q.$$

6.9.4 Solid transportation problem

In real life TP, three dimensions need to be handled. Along with demand and supply, a third constraint, conveyance, is also considered. It is necessary to take a decision regarding the mode of transportation, through which the products need to be shipped from distinct warehouses to several stations, in order to reduce the total cost of transportation.

A TP which deals with three types of constraints viz., demand, supply and conveyance is called a solid transportation problem (STP).

The concept of STP had been flourished by Shell [38] and further its solution had been recommended by Haley [39]. Hereafter, many researchers [40, 41, 42, 43] investigated it and developed several techniques to solve it. Pramanik et al. [44] worked upon bicriterion STP, Tao and Xu [45] determined a rough programming model having multiple objectives to solve a fuzzy STP and Giri et al. [46] proposed a methodology to solve fully fuzzy STP.

Mathematical representation of STP is:

$$\text{Minimize } \eta = \sum_{m=1}^{p}\sum_{n=1}^{q}\sum_{r=1}^{t}\Omega_{mnr}\omega_{mnr}$$

subject to

$$\sum_{n=1}^{q}\sum_{r=1}^{t}\omega_{mnr} = \alpha_m, \quad m = 1,\dots,p,$$
$$\sum_{m=1}^{p}\sum_{r=1}^{t}\omega_{mnr} = \beta_n, \quad n = 1,\dots,q,$$
$$\sum_{m=1}^{p}\sum_{n=1}^{q}\omega_{mnr} = \varsigma_r,$$
$$\omega_{mnr} \geq 0, \quad m = 1,\dots,p,\ n = 1,\dots,q,\ r = 1,\dots t.$$

6.10 Based upon the membership function, FTP can further be classified as the following

6.10.1 Intuitionistic fuzzy transportation problem

The concept and importance of fuzzy sets have been discussed above. Fuzzy sets reduce the hesitancy of human decisions by considering the membership values of a function. In 1986, a Bulgarian mathematician, Krassimir Atanassov [47] took another

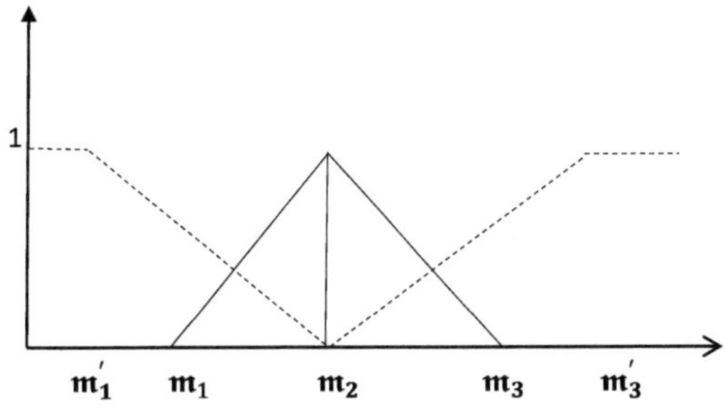

Figure 6.4: Triangular intuitionistic fuzzy number.

step to cut back the skepticism of human decisions by proposing the concept of intuitionistic fuzzy sets (IFS). Both the membership as well as nonmembership degree of a value, along with the hesitancy factor, are handled by IFSs. Posterior to this, Bharti and Singh [48] fixed up the theory of fuzzy sets into TP and brought up the idea of intuitionistic fuzzy transportation problem (IFTP). Many researchers [49, 50, 51, 52] worked upon this useful concept and further suggested techniques to solve it.

Definition 6.4 (Intuitionistic fuzzy set). An intuitionistic fuzzy set $\tilde{\mathfrak{I}}$, can be described as $\tilde{\mathfrak{I}} = \{\langle a, \mu_{\tilde{\mathfrak{I}}}(\mathfrak{a}), \nu_{\tilde{\mathfrak{I}}}(\mathfrak{a}) : \mathfrak{a} \in \tilde{\xi} \rangle\}$, where $\tilde{\xi}$ is a nonempty set, $\mu_{\tilde{\mathfrak{I}}}(\mathfrak{a}) : \tilde{\xi} \to [0,1]$ expresses the membership degree and $\nu_{\tilde{\mathfrak{I}}} : \tilde{\xi} \to [0,1]$, the nonmembership degree of $\mathfrak{a}$, satisfying $0 \le \mu_{\tilde{\mathfrak{I}}} + \nu_{\tilde{\mathfrak{I}}} \le 1$. The degree of hesitancy is represented as $\mathcal{H}_{\tilde{\mathfrak{I}}} = 1 - \mu_{\tilde{\mathfrak{I}}} - \nu_{\tilde{\mathfrak{I}}}$.

Definition 6.5 (Triangular intuitionistic fuzzy number). A fuzzy number of the form $\tilde{\mathfrak{N}} = \langle(\mathfrak{m}_1, \mathfrak{m}_2, \mathfrak{m}_3), (\mathfrak{m}_1', \mathfrak{m}_2, \mathfrak{m}_3')\rangle$, where $\mathfrak{m}_1' \le \mathfrak{m}_1 \le \mathfrak{m}_2 \le \mathfrak{m}_3 \le \mathfrak{m}_3'$ is called a triangular intuitionistic fuzzy number (TIFN). Let the membership function of a TIFN is represented by $\mu_{\tilde{\mathfrak{I}}}(\mathfrak{a})$ and the nonmembership function by $\nu_{\tilde{\mathfrak{I}}}(\mathfrak{a})$. These are defined as

$$\mu_{\tilde{\mathfrak{I}}}(\mathfrak{a}) = \begin{cases} 0, & \mathfrak{a} < \mathfrak{m}_1, \\ \frac{\mathfrak{a}-\mathfrak{m}_1}{\mathfrak{m}_2-\mathfrak{m}_1}, & \omega_1 \le \mathfrak{a} \le \mathfrak{m}_2, \\ 1, & \mathfrak{a} = \mathfrak{m}_2, \\ \frac{\mathfrak{m}_3-\mathfrak{a}}{\mathfrak{m}_3-\mathfrak{m}_2}, & \mathfrak{m}_2 \le \mathfrak{a} \le \mathfrak{m}_3, \\ 0, & \mathfrak{a} > \omega_3, \end{cases}$$

and

$$\nu_{\tilde{\mathfrak{I}}}(\mathfrak{a}) = \begin{cases} \frac{\mathfrak{m}_2-\mathfrak{a}}{\mathfrak{m}_2-\mathfrak{m}_1'}, & \mathfrak{m}_1' < \mathfrak{a} \le \mathfrak{m}_2 \\ 0, & \mathfrak{a} = \mathfrak{m}_2 \\ \frac{\mathfrak{a}-\mathfrak{m}_2}{\mathfrak{m}_3'-\mathfrak{m}_2}, & \mathfrak{m}_2 \le \mathfrak{a} < \mathfrak{m}_3' \\ 1, & \text{otherwise.} \end{cases}$$

Diagrammatic representation of TIFN is represented in Figure 6.4.

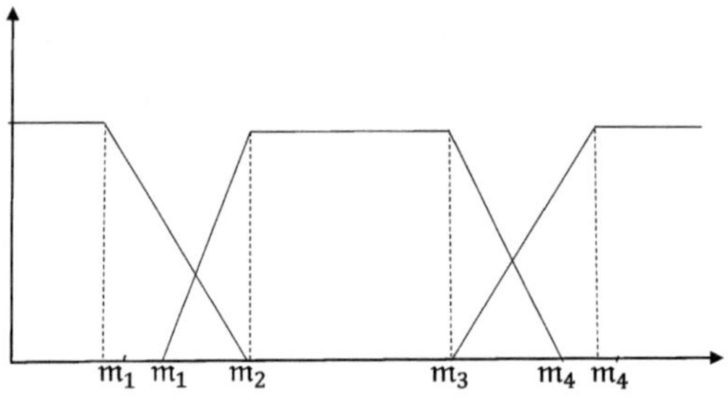

Figure 6.5: Trapezoidal intuitionistic fuzzy number.

Definition 6.6 (Trapezoidal intuitionistic fuzzy numbers). A fuzzy number of the form $\tilde{\mathfrak{N}} = \langle(\mathfrak{m}_1, \mathfrak{m}_2, \mathfrak{m}_3, \mathfrak{m}_4), (\mathfrak{m}'_1, \mathfrak{m}_2, \mathfrak{m}_3, \mathfrak{m}'_4)\rangle$, where $\mathfrak{m}'_1 \leq \mathfrak{m}_1 \leq \mathfrak{m}_2 \leq \mathfrak{m}_3 \leq \mathfrak{m}_4 \leq \mathfrak{m}'_4$ is called a trapezoidal intuitionistic fuzzy number (TrIFN). Let the membership function of a TrIFN is represented by $\mu_{\tilde{\mathfrak{J}}}(\mathfrak{a})$ and the nonmembership function by $\nu_{\tilde{\mathfrak{J}}}(\mathfrak{a})$. These are defined as

$$\mu_{\tilde{\mathfrak{J}}}(\mathfrak{a}) = \begin{cases} \frac{\mathfrak{a}-\mathfrak{m}_1}{\mathfrak{m}_2-\mathfrak{m}_1}, & \mathfrak{m}_1 \leq \mathfrak{a} \leq \mathfrak{m}_2 \\ 1, & \mathfrak{m}_2 \leq \mathfrak{a} \leq \mathfrak{m}_3 \\ \frac{\mathfrak{m}_4-\mathfrak{a}}{\mathfrak{m}_4-\mathfrak{m}_3}, & \mathfrak{m}_3 \leq \mathfrak{a} \leq \mathfrak{m}_4 \\ 0, & \text{otherwise} \end{cases}$$

and

$$\nu_{\tilde{\mathfrak{J}}}(\mathfrak{a}) = \begin{cases} \frac{\mathfrak{m}_2-\mathfrak{a}}{\mathfrak{m}_2-\mathfrak{m}_1}, & \mathfrak{m}'_1 \leq \mathfrak{a} \leq \mathfrak{m}_2 \\ 0, & \mathfrak{m}_2 \leq \mathfrak{a} \leq \mathfrak{m}_3 \\ \frac{\mathfrak{a}-\mathfrak{m}_3}{\mathfrak{m}'_4-\mathfrak{m}_3}, & \mathfrak{m}_3 \leq \mathfrak{a} \leq \mathfrak{m}'_4 \\ 1, & \text{otherwise} \end{cases}$$

Diagrammatic representation of TrIFN is depicted in Figure 6.5.

6.10.1.1 Algebraic operations on TIFN

Let $\tilde{\mathfrak{N}} = \langle(\mathfrak{m}_1, \mathfrak{m}_2, \mathfrak{m}_3), (\mathfrak{m}'_1, \mathfrak{m}_2, \mathfrak{m}'_3)\rangle$ and $\tilde{\mathfrak{P}} = \langle(\mathfrak{n}_1, \mathfrak{n}_2, \mathfrak{n}_3), (\mathfrak{n}'_1, \mathfrak{n}_2, \mathfrak{n}'_3)\rangle$ be two triangular intuitionistic fuzzy numbers and $\mathfrak{d}$ be any scalar. Then:

i. $\tilde{\mathfrak{N}} + \tilde{\mathfrak{P}} = \langle(\mathfrak{m}_1 + \mathfrak{n}_1, \mathfrak{m}_2 + \mathfrak{n}_2, \mathfrak{m}_3 + \mathfrak{n}_3), (\mathfrak{m}'_1 + \mathfrak{n}'_1, \mathfrak{m}_2 + \mathfrak{n}_2, \mathfrak{m}'_3 + \mathfrak{n}'_3)\rangle$;

ii. $\mathfrak{d}\tilde{\mathfrak{R}} = \langle(\mathfrak{d}\mathfrak{m}_1, \mathfrak{d}\mathfrak{m}_2, \mathfrak{d}\mathfrak{m}_3), (\mathfrak{d}\mathfrak{m}'_1, \mathfrak{d}\mathfrak{m}_2, \mathfrak{d}\mathfrak{m}'_3)\rangle$ if $\mathfrak{d} \geq 0$;

iii. $\mathfrak{d}\tilde{\mathfrak{N}} = \langle(\mathfrak{d}\mathfrak{m}_3, \mathfrak{d}\mathfrak{m}_2, \mathfrak{d}\mathfrak{m}_1), (\mathfrak{d}\mathfrak{m}'_3, \mathfrak{d}\mathfrak{m}_2, \mathfrak{d}\mathfrak{m}'_1)\rangle$ if $\mathfrak{d} < 0$;

iv. $\tilde{\mathfrak{N}} - \tilde{\mathfrak{P}} = \langle(\mathfrak{m}_1 - \mathfrak{n}_3, \mathfrak{m}_2 - \mathfrak{n}_2, \mathfrak{m}_3 - \mathfrak{n}_1), (\mathfrak{m}'_1 + \mathfrak{n}'_3, \mathfrak{m}_2 - \mathfrak{n}_2, \mathfrak{m}'_3 - \mathfrak{n}'_1)\rangle$;

v. $\tilde{\mathfrak{N}}.\tilde{\mathfrak{P}} = \langle(\mathfrak{m}_1\mathfrak{n}_1, \mathfrak{m}_2\mathfrak{n}_2, \mathfrak{m}_3\mathfrak{n}_3), (\mathfrak{m}'_1\mathfrak{n}'_1, \mathfrak{m}_2\mathfrak{n}_2, \mathfrak{m}'_3\mathfrak{n}'_3)\rangle$;

6.10.1.2 Mathematical representation of IFTP

Let us deal with an IFTP with p distribution centers and q stations. Let $\widetilde{\alpha_m^{\mathfrak{I}}}$ be the intuitionistic fuzzy valued quantity to be transported to q stations to fulfill the intuitionistic fuzzy valued demand of $\widetilde{\beta_n^{\mathfrak{I}}}$ units and $\widetilde{\omega_{mn}^{\mathfrak{I}}}$ be the amount of commodity to be transported. Let $\widetilde{\Omega_{mn}^{\mathfrak{I}}}$ be the intuitionistic fuzzy penalty cost corresponding to fetching a unit of commodity form p to q.

$$\text{Minimize } \eta = \sum_{m=1}^{p} \sum_{n=1}^{q} \widetilde{\Omega_{mn}^{I}} \widetilde{\omega_{mn}^{I}}$$

subject to

$$\sum_{n=1}^{q} \widetilde{\omega_{mn}^{\mathfrak{I}}} = \widetilde{\alpha_m^{\mathfrak{I}}}, \quad m = 1, \ldots, p,$$
$$\sum_{m=1}^{p} \widetilde{\omega_{mn}^{\mathfrak{I}}} = \widetilde{\beta_n^{\mathfrak{I}}}, \quad n = 1, \ldots, q,$$
$$\widetilde{\omega_{mn}^{\mathfrak{I}}} \geq 0, \quad m = 1, \ldots, p, \ n = 1, \ldots, q.$$

6.10.2 Rough set

The problem of insufficient awareness has been handled by many researchers in the past decades. The concept of fuzzy sets, which is one of the ways of handling this problem, has been discussed above. Another way of dealing with the unpredictability of human decisions has been emerged by Z. Pawlak [53] in 1982. It is analogous to fuzzy theory. The difference between the two is that rough sets deal with multiple membership functions while fuzzy sets consider partial membership function.

Uncertainty is specified by the concept of rough set by applying a boundary of a set rather than the membership functions. Empty boundary region shows that the set is deterministic while the nonempty boundary set shows that the set is rough, which implies insufficient knowledge to define a set. This theory has gained the attention of many researchers [54, 55, 56, 57, 58] who have contributed in developing techniques to solve it.

Let this case be discussed as following.

Suppose a universal set is defined by $\mathcal{T}$ and $\mathcal{P} \subseteq \mathcal{T} \times \mathcal{T}$ be an equivalence relation representing inadequate knowledge about the objects of $\mathcal{T}$. Our task is to characterize $\mathcal{Z} \subseteq \mathcal{T}$ w.r.t $\mathcal{P}$. For this, some basic concepts of rough set theory need to discussed:

(i) $\mathcal{P}$-lower approximation of $\mathcal{Z}$ is denoted by $\mathcal{P}_\star(z) = \bigcup_{z\in\mathcal{Z}}\{\mathcal{P}(z) \colon \mathcal{P}(z) \subseteq \mathcal{Z}\}$.

(ii) $\mathcal{P}$-upper approximation of $\mathcal{Z}$ is denoted by $\mathcal{P}^\star(z) = \bigcup_{z\in\mathcal{Z}}\{\mathcal{P}(z)\cap\mathcal{Z} \neq \phi\}$.

(iii) $\mathcal{P}$-boundary region of $\mathcal{Z}$ is denoted by $\mathcal{P}_{\mathcal{B}}(z) = \mathcal{P}^\star(z) - \mathcal{P}_\star(z)$

Then, if $\mathcal{P}_{\mathcal{B}}(z)$ is empty, the set is crisp and if $\mathcal{P}_{\mathcal{B}}(z)$ is nonempty, then the set is rough.

Thus, it can be stated that rough sets are defined by approximations which have the following properties:

1. $\mathcal{P}_\star(z) \subseteq \mathcal{Z} \subseteq \mathcal{P}^\star(z)$
2. $\mathcal{P}_\star(\emptyset) = \mathcal{P}^\star(\emptyset) = \emptyset$ and $\mathcal{P}^\star(\mathcal{T}) = \mathcal{P}_\star(\mathcal{T}) = \mathcal{T}$
3. $\mathcal{P}^\star(\mathcal{Z} \cup \mathcal{S}) = \mathcal{P}^\star(\mathcal{Z}) \cup \mathcal{P}^\star(\mathcal{S})$
4. $\mathcal{P}_\star(\mathcal{Z} \cap \mathcal{S}) = \mathcal{P}_\star(\mathcal{Z}) \cap \mathcal{P}_\star(\mathcal{S})$
5. $\mathcal{P}_\star(\mathcal{Z} \cup \mathcal{S}) \supseteq \mathcal{P}_\star(\mathcal{Z}) \cup \mathcal{P}_\star(\mathcal{S})$
6. $\mathcal{P}^\star(\mathcal{Z} \cap \mathcal{S}) \subseteq \mathcal{P}^\star(\mathcal{Z}) \cap \mathcal{P}^\star(\mathcal{S})$
7. $\mathcal{P}^\star(-\mathcal{Z}) = -\mathcal{P}_\star(\mathcal{Z})$
8. $\mathcal{P}_\star(-\mathcal{Z}) = -\mathcal{P}^\star(\mathcal{Z})$
9. $\mathcal{P}_\star\mathcal{P}_\star(\mathcal{Z}) = \mathcal{P}^\star\mathcal{P}_\star(\mathcal{Z}) = \mathcal{P}_\star(\mathcal{Z})$
10. $\mathcal{P}^\star\mathcal{P}^\star(\mathcal{Z}) = \mathcal{P}_\star\mathcal{P}^\star(\mathcal{Z}) = \mathcal{P}^\star(z)$

Instead of approximations, rough sets can also be explained using rough membership function [54], as given below:

$$\mu_{\mathcal{Z}}^{\mathcal{P}}: \mathcal{T} \to [0,1], \quad \text{where } \mu_{\mathcal{Z}}^{\mathcal{P}} = \frac{|\mathcal{Z} \cap \mathcal{P}(z)|}{|\mathcal{P}(z)|}$$

$|\mathcal{Z}|$ represents the cardinality of $\mathcal{Z}$.

Using rough membership function, the basic approximations and the boundary set can be defined as:

(i) $\mathcal{P}_\star(z) = \{z \in \mathcal{T} : \mu_{\mathcal{Z}}^{\mathcal{P}}(z) = 1\}$
(ii) $\mathcal{P}^\star(z) = \{z \in \mathcal{T} : \mu_{\mathcal{Z}}^{\mathcal{P}}(z) > 0\}$
(iii) $\mathcal{P}_{\mathcal{B}}(z) = \{z \in \mathcal{T} : 0 < \mu_{\mathcal{Z}}^{\mathcal{P}}(z) < 1\}$

The rough membership function holds the following properties:

1. $\mu_{\mathcal{Z}}^{\mathcal{P}}(z) = 1$ iff $z \in \mathcal{P}_\star(\mathcal{Z})$
2. $\mu_{\mathcal{Z}}^{\mathcal{P}}(z) = 0$ iff $z \in \mathcal{T} - \mathcal{P}^\star(\mathcal{Z})$
3. $0 < \mu_{\mathcal{Z}}^{\mathcal{P}}(z) < 1$ iff $z \in \mathcal{P}_{\mathcal{B}}(\mathcal{Z})$
4. $\mu_{\mathcal{T}-\mathcal{Z}}^{\mathcal{P}}(z) = 1 - \mu_{\mathcal{Z}}^{\mathcal{P}}$ for any $z \in \mathcal{T}$
5. $\mu_{\mathcal{Z}\cup\mathcal{W}}^{\mathcal{P}}(z) \geq \max(\mu_{\mathcal{Z}}^{\mathcal{P}}(z), \mu_{\mathcal{W}}^{\mathcal{P}}(z))$ for any $z \in \mathcal{T}$
6. $\mu_{\mathcal{Z}\cap\mathcal{W}}^{\mathcal{P}}(z) \leq \min(\mu_{\mathcal{Z}}^{\mathcal{P}}(z), \mu_{\mathcal{W}}^{\mathcal{P}}(z))$ for any $z \in \mathcal{T}$

Property 5th and 6th highlights the major difference between fuzzy sets and rough sets. In fuzzy sets, union and intersection of membership functions cannot be deduced from the fundamental membership functions, whereas in rough sets they can be found.

6.10.3 Rough fuzzy transportation problem

After the origination of the theory of rough sets, researchers implemented it into fuzzy transportation problem and succeeded in handling impreciseness in a better way. Tao and Xu [45] detected multiobjective STP by implementing rough sets into it, Pramanik et al. [44] devised this concept by converting bicriterion STP into a single objective problem, Kundu et al. [59] worked upon STP having transportation cost as rough valued, Kundu et al. [60] further equipped the concept of rough set theory in STP with product blending by considering all the parameters to be rough valued, etc.

6.11 Various approaches to solve variants of fuzzy transportation problems

FTP has proved to be of utmost importance in real life problems. Various researchers have contributed in its development. Several methods have been proposed by the distinct scholars to solve the FTP. Few of them are mentioned below.

Liu and Kao [61] advised to solve FTP using extension principle, developed by Zadeh. α-cuts have been found and used to derive membership functions. Demand and supply values are taken as triangular and trapezoidal numbers.

Ritha and Vinotha [62] conferred a FTP in two stages having multiple objectives in which an optimal solution has been obtained using the fuzzy geometric programming approach. Considering the demand and supply values as trapezoidal fuzzy numbers, a fuzzy programming approach has been defined as follow to find the compromise solution:

Step 1: Solve each of the objective functions individually with the same set of constraints and obtain the optimal solutions.

Step 2: Find the lower and upper bound of each of the objective function. Let $\mathcal{L}_r$ be the lower bound and $\mathcal{U}_r$ be the upper bound of the rth objective function, where $r = 1, 2, \ldots, \mathcal{R}$.

Step 3: Let the membership function be defined as

$$\mu_r \mathcal{M}^r(x) = \begin{cases} 1, & \mathcal{M}^r(x) \le \mathcal{L}_r \\ \frac{\mathcal{U}_r - \mathcal{M}^r(x)}{\mathcal{U}_r - \mathcal{L}_r}, & \mathcal{L}_r < \mathcal{M}^r(x) < \mathcal{U}_r \\ 0, & \text{otherwise} \end{cases}$$

Step 4: The fuzzy programming problem constructed by using the membership function defined above is

$$\operatorname{Max} \operatorname*{Min}_{r=1,2,\ldots,R} \mu_r \mathcal{M}^r(x)$$

subject to

$$\sum_{s=1}^{n} x_{st} = a_s, \quad s = 1, 2, \ldots, m$$

$$\sum_{t=1}^{n} x_{st} = b_t, \quad t = 1, 2, \ldots, n$$

$$x_{st} \geq 0.$$

Step 5: Form its equivalent LPP

$$\text{Max}\, y$$

subject to

$$y \leq \mu_r \mathcal{M}^{r}(x)$$

$$\sum_{s=1}^{t} x_{st} = a_s, \quad s = 1, 2 \ldots, m$$

$$\sum_{s=1}^{n} x_{st} = b_t, \quad t = 1, 2, \ldots, n$$

$$0 \leq y \leq 1, \quad x_{st} \geq 0.$$

Step 6: Using a suitable software package, obtain the solution of the problem defined above.

Likewise, many scholars have worked on FTPs. While Samuel and Venkatachalapathy [63] modified the classical Vogel's approximation method to solve FTP, Buckley and Jowers [64] proposed Monte Carlo method to solve it. Kiruthiga et al. [65] suggested the row minima method to solve the two-stage FTP while Kaur and Kumar [27] preferred the ranking function to get the solution of FTP, etc. The research is still continued to establish new methods [66, 67, 68].

6.12 Future scopes

A research on FTP has a wide scope. Making it more oriented in terms of practical situations and developing heuristic approaches to improve it motivate the researchers to keep thinking more about it.

Bibliography

[1] Dantzig GB. Linear programming: the story about how it began. In: History of Mathematical Programming: A Collection of Personal Reminiscences. North-Holland; 1991. p. 19–31.
[2] Nash JC. The (Dantzig) simplex method for linear programming. Comput Sci Eng. 2000;2:29–31.

[3] Huangfu Q, Hall JJ. Parallelizing the dual revised simplex method. Math Program Comput. 2018;10:119–42.

[4] Doi A, Chiba T. An automatic image registration method using downhill simplex method and its applications. In: Int. Conf. Netw.-Based Inf. Syst. Springer; 2017. p. 635–43.

[5] Yan C-H, Liu F, Hu CF, Cui H, Luo ZQ. Improved simplex method algorithm used in power market clearing. J Phys Conf Ser. 2019;1304:012025.

[6] Hitchcock FL. The distribution of a product from several sources to numerous localities. J Math Phys. 1941;20:224–30.

[7] Harrath Y, Kaabi J. New heuristic to generate an initial basic feasible solution for the balanced transportation problem. Int J Ind Syst Eng. 2018;30:193–204.

[8] Dinagar DS, Palanivel K. The transportation problem in fuzzy environment. Int J Algorithms Comput Math. 2009;2:65–71.

[9] Dantzig GB. Linear programming & extensions. Princeton Univ. Press; 1963.

[10] Cooper WW, Henderson A, Charnes A. An introduction to linear programming: an economic introduction to linear programming. Lectures on the mathematical theory of linear programming. Wiley; 1953.

[11] Mhlanga A, Nduna IS, Matarise F, Machisvo A. Innovative application of Dantzig's north-west corner rule to solve a transportation problem. Int J Educ Res. 2014;2:1–12.

[12] Charnes A, Cooper WW. The stepping stone method of explaining linear programming calculations in transportation problems. Manag Sci. 1954;1:49–69.

[13] Palanivel M, Suganya M. A new method to solve transportation problem – harmonic mean approach. Eng Technol Open Access J. 2018;2(3):555586.

[14] Dinagar DS, Kaliyaperumal P. The transportation problem in fuzzy environment. Int J Algorithms Comput Math. 2009;2:65–71.

[15] Lee SM, Moore LJ. Optimizing transportation problems with multiple objectives. AIIE Trans. 1973;5:333–8.

[16] Diaz JA. Finding a complete description of all efficient solutions to a multiobjective transportation problem. Ekon-Mat Obz. 1979;15:62–73.

[17] Isermann H. The enumeration of all efficient solutions for a linear multiple-objective transportation problem. Nav Res Logist Q. 1979;26(1):123–39.

[18] Current J, Min H. Multiobjective design of transportation networks: taxonomy and annotation. Eur J Oper Res. 1986;26:187–201.

[19] Diaz JA. Solving multiobjective transportation problem. Ekon-Mat Obz. 1978;14:267–74.

[20] Edwards W. How to use multiattribute utility measurement for social decisionmaking. IEEE Trans Syst Man Cybern. 1977;7:326–40.

[21] Riggs JL, Inoue MS. Introduction to operations research and management science. McGraw-Hill; 1975.

[22] Zadeh LA. Fuzzy sets. Inf Control. 1965;8:338–53.

[23] Bellman RE, Zadeh LA. Decision-making in a fuzzy environment. Manag Sci. 1970;17:B-141.

[24] Ebrahimnejad A. A simplified new approach for solving fuzzy transportation problems with generalized trapezoidal fuzzy numbers. Appl Soft Comput. 2014;19:171–6.

[25] Gupta G, Anupum K. An efficient method for solving intuitionistic fuzzy transportation problem of type-2. Int J Appl Comput Math. 2017;3:3795–804.

[26] Singh SK, Yadav SP. A new approach for solving intuitionistic fuzzy transportation problem of type-2. Ann Oper Res. 2016;243:349–63.

[27] Kaur A, Kumar A. A new method for solving fuzzy transportation problems using ranking function. Appl Math Model. 2011;35:5652–61.

[28] Gupta A, Kumar A. A new method for solving linear multi-objective transportation problems with fuzzy parameters. Appl Math Model. 2012;36:1421–30.

[29] Dubois D, Prade H. Operations on fuzzy numbers. Int J Syst Sci. 1978;9:613–26.
[30] Cheng C-H. A new approach for ranking fuzzy numbers by distance method. Fuzzy Sets Syst. 1998;95:307–17.
[31] Choobineh F, Li H. An index for ordering fuzzy numbers. Fuzzy Sets Syst. 1993;54:287–94.
[32] Chu T-C, Tsao C-T. Ranking fuzzy numbers with an area between the centroid point and original point. Comput Math Appl. 2002;43:111–7.
[33] Yager RR. A procedure for ordering fuzzy subsets of the unit interval. Inf Sci. 1981;24:143–61.
[34] Lee ES, Li R-J. Comparison of fuzzy numbers based on the probability measure of fuzzy events. Comput Math Appl. 1988;15:887–96.
[35] Wang Y-J. Ranking triangle and trapezoidal fuzzy numbers based on the relative preference relation. Appl Math Model. 2015;39:586–99.
[36] Kaur A, Kumar A. A new approach for solving fuzzy transportation problems using generalized trapezoidal fuzzy numbers. Appl Soft Comput. 2012;12:1201–13.
[37] Roy SK, Ebrahimnejad A, Verdegay JL, Das S. New approach for solving intuitionistic fuzzy multi-objective transportation problem. Sādhanā. 2018;43:3.
[38] Shell E. Distribution of a product by several properties. In: Directorate of Management Analysis. Proc. Second Symp. Linear Program. vol. 2. 1955. p. 615–42.
[39] Haley KB. New methods in mathematical programming – the solid transportation problem. Oper Res. 1962;10(4):448–63.
[40] Cui Q, Sheng Y. Uncertain programming model for solid transportation problem. 2012.
[41] Dalman H, Güzel N, Sivri M. A fuzzy set-based approach to multi-objective multi-item solid transportation problem under uncertainty. Int J Fuzzy Syst. 2016;18:716–29.
[42] Pramanik S, Jana DK, Maity K. A multi objective solid transportation problem in fuzzy, bi-fuzzy environment via genetic algorithm. Int J Adv Oper Manag. 2014;6:4–26.
[43] Dalman H. Uncertain programming model for multi-item solid transportation problem. Int J Mach Learn Cybern. 2018;9:559–67.
[44] Pramanik S, Jana DK, Maiti M. Bi-criteria solid transportation problem with substitutable and damageable items in disaster response operations on fuzzy rough environment. Socio-Econ Plan Sci. 2016;55:1–13.
[45] Tao Z, Xu J. A class of rough multiple objective programming and its application to solid transportation problem. Inf Sci. 2012;188:215–35.
[46] Giri PK, Maiti MK, Maiti M. Fully fuzzy fixed charge multi-item solid transportation problem. Appl Soft Comput. 2015;27:77–91.
[47] Atanassov KT. Intuitionistic fuzzy sets. Fuzzy Sets Syst. 1986;20:87–96.
[48] Bharati SK, Singh SR. Solving multi objective linear programming problems using intuitionistic fuzzy optimization method: a comparative study. Int J Model Optim. 2014;4:10.
[49] Gani AN, Abbas S. A new method on solving intuitionistic fuzzy transportation problem. Ann Pure Appl Math. 2017;15:163–71.
[50] Kumar PS, Hussain RJ. A systematic approach for solving mixed intuitionistic fuzzy transportation problems. Int J Pure Appl Math. 2014;92:181–90.
[51] Hussain RJ, Kumar PS. Algorithmic approach for solving intuitionistic fuzzy transportation problem. Appl Math Sci. 2012;6(80):3981–9.
[52] Pramila K, Uthra G. Optimal solution of an intuitionistic fuzzy transportation problem. Ann Pure Appl Math. 2014;8(2):67–73.
[53] Pawlak Z. Rough sets. Int J Comput Inf Sci. 1982;11:341–56.
[54] Pal SK, Skowron A. Rough-fuzzy hybridization: a new trend in decision making. Springer; 1999.
[55] Lin TY, Cercone N. Rough sets and data mining: analysis of imprecise data. Springer; 2012.
[56] Ziarko W. Introduction to the special issue on rough sets and knowledge discovery. Comput Intell. 1995;11(2):223–6.

[57] Lin TY. Introduction to the special issue on rough sets. Int J Approx Reason. 1996;15:287–9.
[58] Kondo M. Algebraic approach to generalized rough sets. In: Rough Sets, Fuzzy Sets, Data Mininig, and Granular Computing. Berlin, Heidelberg: Springer; 2005. p. 132–40.
[59] Kundu P, Kar S, Maiti M. Some solid transportation models with crisp and rough costs. Int J Math Comput Sci. 2013;7(1):14–21.
[60] Kundu P, Kar MB, Kar S, Pal T, Maiti M. A solid transportation model with product blending and parameters as rough variables. Soft Comput. 2017;21:2297–306.
[61] Liu S-T, Kao C. Solving fuzzy transportation problems based on extension principle. Eur J Oper Res. 2004;153:661–74.
[62] Ritha W, Vinotha JM. A priority based fuzzy goal programming approach for multi-objective solid transportation problem. J Compos Theory. 2012.
[63] Samuel AE, Modified VM. Vogel's approximation method for fuzzy transportation problems. Appl Math Sci. 2011;5(28):1367–72.
[64] Buckley JJ, Jowers LJ. Fuzzy transportation problem. In: Buckley JJ, Jowers LJ, editors. Monte Carlo Methods Fuzzy Optim. Berlin, Heidelberg: Springer; 2008. p. 217–21.
[65] Kiruthiga M, Lalitha M, Loganathan C. Solving two stage fuzzy transportation problem by row minima method. Int J Comput Appl. 2013;65(22):1–4.
[66] Chhibber D, Srivastava PK, Bisht DCS. Average duo triangle ranking technique to solve fully and type-2 intuitionistic fuzzy transportation problem. Nonlinear Stud. 2019;26(3):487–504.
[67] Srivastava PK, Bisht DCS. An efficient fuzzy minimum demand supply approach to solve fully fuzzy transportation problem. Math Eng Sci Aerosp. 2019;10(2):253–69.
[68] Bisht DCS, Srivastava PK. Trisectional fuzzy trapezoidal approach to optimize interval data based transportation problem. J King Saud Univ, Sci. 2020;32(1):195–9.

B. Farhadinia

7 Hesitant fuzzy sets: distance, similarity and entropy measures

Abstract: Throughout the present study, we are going to present a thorough and systematic review of hesitant fuzzy set (HFS) information measures including the distance, similarity and entropy measures.

Keywords: Hesitant fuzzy set, distance measure, similarity measure, entropy measure

7.1 Introduction

Fuzzy set is an initial and seminal concept which mainly models those types of uncertainty information arising in the case of imprecision and vagueness. However, such a concept is not able to deal with the situation in which the various vagueness sources appear simultaneously. Following this concept and in order to remove its limitations, some other interesting extensions have been introduced so far. A number of such extensions are: interval-valued fuzzy set [1] in which the membership degree of elements are in the form of closed subintervals of $[0, 1]$; type II fuzzy set [2] which describes the belongingness degree of each element by the use of again a fuzzy set on $[0, 1]$; fuzzy multiset [3] whose elements are described by multiple sets; and intuitionistic fuzzy set [4, 5] in which each element is described by the two notions: membership and non-membership degrees.

Sometimes an element should be described by the use of a number of membership degrees, but not like that performed in the definition of type II fuzzy set where the element is described by using possibly distribution. For modeling such a situation realistically, Torra [6, 7] proposed a fruitful concept as an extension of fuzzy set and he named it as the hesitant fuzzy set (HFS). HFS has become more and more popular and it has been applied in numerous studies [8, 9, 11, 12, 13, 14, 15, 16, 17, 19, 20, 21, 22, 23, 24, 25].

Generally, a HFS on the reference set X is defined as the functional $h : X \to \wp([0, 1])$ where $\wp([0, 1])$ denotes the set of all values belonging to the interval $[0, 1]$.

In what follows, we will denote all HFSs on the reference set X by $\mathbb{HFS}(X)$.

By the way, Xia and Xu [5] further provided a mathematical representation of a HFS as

$$A = \{\langle x, h_A(x)\rangle \mid x \in X\},$$

where the term $h_A(x)$ is known as the hesitant fuzzy element (HFE).

B. Farhadinia, Dept. Math., Quchan University of Technology, Quchan, Iran, e-mail: bfarhadinia@qiet.ac.ir

https://doi.org/10.1515/9783110671353-007

We organize this contribution into the following sections for dealing with three issues which are related together. Section 7.2 investigates the review of HFS distance measures, and moreover, HFS similarity measures are discussed in Section 7.3. There, by studying the relationship between HFS distance and similarity measures, we are able to generate interchangeably different desire measures. In Section 7.4, we provide the researcher with a class of other kind of information measures, called HFS entropy measures. Such a class of HFS information measures can be classified in two groups: HFS information-based and HFS axiomatic-based entropy measures.

7.2 HFS distance measures

HFS distance measure [26, 27, 28, 29] has been employed in different and numerous fields of pattern recognition, market prediction and decision making problems.

Here, a number of HFS distance measures are reviewed and their issues are discussed from different aspects.

Of the pioneer contributions on HFS distance measures are those proposed by Xu and Xia [30]. Then, by investigating Xu and Xia's [30] distance measures, Peng et al. [31] presented a class of HFS distance measures whose weights are taken into consideration, and then they introduced the generalized form of weighted distance measures, called generalized hesitant fuzzy synergetic weighted distance measures. In the sequel, Xu and Xia's [30] axiomatic definition of HFS distance measures are modified by Zhou and Li [32]. Farhadinia [10] asserted that Xu and Xia's [30] and Peng et al.'s [31] HFS distance measures satisfy less properties compared to the ideal HFS distance measures, and he then proposed some other axioms. In the computation process of all the above-mentioned HFS distance measures, it is only considered the difference between two length-unified HFSs, but the hesitancy concept of HFSs is not taken into account. These limitations are released by considering the two aspects which introduce the deviation-based and cardinal-based hesitancy indices. The former was suggested by Zhang and Xu [33] and the latter was proposed by Li et al. [34]. Zhang and Xu's [33] distance measures are computed on the basis of length-unification process which is priory done for any pair of HFSs, meanwhile, Li et al.'s [34] distance measures are calculated based on the length-unification process for all considered HFSs. However, the above-mentioned distance measures bear some drawbacks: (1) they should be length-unified in advance; (2) the appended value to a HFS is generally considered the maximum value or the minimum value of that HFS; (3) the appended value does strongly depend on the extremum value which is sensitive to the subjectivity of decision maker.

By removing these limitations, Hu et al. [35] defined a class of HFS distance measures, and they showed that these distance measures do not need to consider an appended value or the risk preferences of decision makers. Same to the existing fuzzy set, interval-valued fuzzy set, type II fuzzy set or intuitionistic fuzzy set distance measures, a HFS distance measures should satisfy a class of known properties. In this regard, the followings are the modified version of properties, those must be considered for defining a reasonable HFS distance measure.

Suppose that $A = \{\langle x, h_A(x)\rangle \mid x \in X\}$, $B = \{\langle x, h_B(x)\rangle \mid x \in X\}$ and $C = \{\langle x, h_C(x)\rangle \mid x \in X\}$ are three HFSs defined on the reference set X. A HFS distance measure generally satisfies those properties in the forms of

(D0) $0 \le d(A, B) \le 1$ (*Boundary axiom*);
(D1) $d(A, B) = d(B, A)$ (*Symmetry axiom*);
(D2) $d(A, A^C) = 1$ if and only if A is the empty HFS O^* or the full HFS I^* (*Complementarity axiom*);
(D3) $d(A, B) = 0$ if and only if $A = B$ (*Reflexivity axiom*);
(D4) If $A \le B \le C$, then $d(A, B) \le d(A, C)$ and $d(B, C) \le d(A, C)$ (*Inequality axiom*).

Here, A^C denotes the complement of A, and it is defined as $A^C = \{\langle x, h_{A^c}(x) = \bigcup_{\gamma \in h_A(x)} \{1 - \gamma\}\rangle \mid x \in X\}$.

7.3 HFS similarity measures

HFS similarity measure has been employed in different and numerous fields of pattern recognition, clustering analysis and medical diagnosis problems.

It is noticeable that there is a reciprocal relationship between HFS distance and similarity measures that enables the interested researcher to reach from one measure to another measure. Such a tied relationship allows the existing HFS distance measures to be transformed to those measures which are used for evaluating the degree of similarity between HFSs. Here, we should notice that this relationship is not the only way of constructing a HFS similarity measure, and there exist the other forms of HFS similarity measures, those are not resulted directly from HFS distance measures.

Suppose that $A = \{\langle x, h_A(x)\rangle \mid x \in X\}$, $B = \{\langle x, h_B(x)\rangle \mid x \in X\}$ and $C = \{\langle x, h_C(x)\rangle \mid x \in X\}$ are three HFSs defined on the reference set X. A HFS similarity measure generally satisfies those properties in the forms of

(S0) $0 \le S(A, B) \le 1$ (*Boundary axiom*);
(S1) $S(A, B) = S(B, A)$ (*Symmetry axiom*);
(S2) $S(A, A^C) = 0$ if A is the empty HFS O^* or the full HFS I^* (*Complementarity axiom*);
(S3) $S(A, B) = 1$ if and only if $A = B$ (*Reflexivity axiom*);
(S4) If $A \le B \le C$, then $S(A, C) \le S(A, B)$ and $S(A, C) \le S(B, C)$ (*Inequality axiom*).

However, if we take a monotone and decreasing function Z into account where $Z(1) = 0$ and $Z(0) = 1$, for example, $Z(t) = 1 - t$; $Z(t) = \frac{1-t}{1+t}$; $Z(t) = 1 - te^{t-1}$ and $Z(t) = 1 - t^2$, then we are able to obtain a class of different HFS similarity measures by the use of a verity of HFS distance measures. This rule is stated as $S_d(A,B) = \frac{Z(d(A,B))-Z(1)}{Z(0)-Z(1)}$.

Another construction method is that introduced by Zhang and Xu [33] in the form of $S(A,B) = \frac{d(A,B^c)}{d(A,B)+d(A,B^c)}$ where B^c stands for the complement of B, and moreover, $d(A,B)$ denotes any HFS distance measure between A and B. What is observable from Zhang and Xu's [33] definition of HFS similarity measure is that such a HFS similarity measure not only examines how the two HFSs are similar, but also it examines how the two HFSs are dissimilar.

Up to now, there exist a lot of application contexts, those are based on the HFS similarity measure such as clustering analysis [36], pattern recognition [37], image processing [38], approximate reasoning [39], decision making [40], medical diagnosis [41], etc.

7.4 HFS entropy measures

As the first attempt to introduce entropy measure for HFSs, we may refer to the work of Xu and Xia [42]. Then Farhadinia [18] showed the HFS entropy measures introduced by Xu and Xia's [42] are not able properly to distinguish HFSs in different situations, and they also cannot properly quantify the fuzziness degree of HFSs any time. Moreover, Farhadinia [18] by considering the axiomatic definitions of HFS information measures, presented a framework in order to transform HFS distance, similarity and entropy measures interchangeably. Following that contribution, Hu et al. [35] presented a number of HFS information measures for which it does not need to consider HFSs with the same length together with risk preference of decision makers like those considered by Farhadinia's [18] and Xu and Xia's [42]. Furthermore, by encountering the previous-mentioned shortcomings, Zhao et al. [43] redefined a number of axioms for HFS entropy measure, known as fuzziness and nonspecificity properties. In what follows, we are going to present the existing detailed forms of axioms related to the HFS entropy measures.

7.4.1 HFS information-based entropy measure

Suppose that $h_A(x) = \{h_A^{\sigma(j)}(x)\}_{j=1}^{l_x}$ and $h_B(x) = \{h_B^{\sigma(j)}(x)\}_{j=1}^{l_x}$ are two HFSs on X. A HFS information-based entropy measure E possesses the following properties given by Xu and Xia [42]:

(E0) $0 \leq E(h_A(x)) \leq 1$ (*Boundary axiom*);

(E1) $E(h_A(x)) = 0$ if and only if $h_A(x) = O^*$ or $h_A(x) = I^*$;

(E2) $E(h_A(x)) = 1$ if and only if $h_A^{\sigma(j)} + h_A^{\sigma(l_x-j+1)} = 1$ for $j = 1, \dots, l_x$ (*Reflexivity axiom*);

(E3) $E(h_A(x)) = E(h_{A^c}(x))$ (*Complementarity axiom*);

(E4) $E(h_A(x)) \leq E(h_B(x))$, if $h_A^{\sigma(j)} \leq h_B^{\sigma(j)}$ for $h_B^{\sigma(j)} + h_B^{\sigma(l_x-j+1)} \leq 1$ or $h_A^{\sigma(j)} \geq h_B^{\sigma(j)}$ for $h_B^{\sigma(j)} + h_B^{\sigma(l_x-j+1)} \geq 1$ where $j = 1, \ldots, l_x$ (*Inequality axiom*).

7.4.2 HFS distance-based entropy measure

Once again, suppose that $h_A(x) = \{h_A^{\sigma(j)}(x)\}_{j=1}^{l_x}$ and $h_B(x) = \{h_B^{\sigma(j)}(x)\}_{j=1}^{l_x}$ are two HFSs on X. A HFS distance-based entropy measure E_d possesses the following properties given by Farhadinia [18]:

(ED0) $0 \leq E_d(A) \leq 1$ (*Boundary axiom*);
(ED1) $E_d(A) = 0$ if and only if $A = O^*$ or $A = I^*$;
(ED2) $E_d(A) = 1$ if and only if $A = \overline{\{\frac{1}{2}\}}$ (*Reflexivity axiom*);
(ED3) $E_d(A) = E_d(A^c)$ (*Complementarity axiom*);
(ED4) If $d(A, \overline{\{\frac{1}{2}\}}) \geq d(B, \overline{\{\frac{1}{2}\}})$, then $E_d(A) \leq E_d(B)$ (*Inequality axiom*),

where $\overline{\{\frac{1}{2}\}}$ stands for $\overline{\{\frac{1}{2}\}} = \{\langle x, \frac{1}{2}\rangle \mid x \in X\}$.

Subsequently, Farhadinia [18] verified that HFS information measures may be transformed to each other. Among the theorems which are proved by Farhadinia [18], there exists a theorem which suggests the following HFS distance-based entropy measure:

$$E_d(A) = \frac{Z(2d(A, \overline{\{\frac{1}{2}\}})) - Z(1)}{Z(0) - Z(1)},$$

in which $Z : [0,1] \to [0,1]$ is a strictly monotone and decreasing function.

7.4.3 HFS similarity-based entropy measure

Taking a HFS distance-based similarity measure into consideration, Farhadinia [18] presented the following formula:

$$E_d(A) = \frac{Z(2Z^{-1}(S_d(A, \overline{\{\frac{1}{2}\}}))) - Z(1)}{Z(0) - Z(1)},$$

in which Z is considered like that in previous subsection.

7.4.4 Hesitant operation-based entropy measure

Similar to that presented by Shang and Jiang [44] for fuzzy sets, Hu et al. [35] proposed the hesitant operation-based entropy measure

$$E_h(h_A(x)) = \frac{1}{l_{h_A(x)}} \sum_{h_A^{\sigma(j)}(x) \in h_A(x)} \frac{\min\{h_A^{\sigma(j)}(x), 1 - h_A^{\sigma(j)}(x)\}}{\max\{h_A^{\sigma(j)}(x), 1 - h_A^{\sigma(j)}(x)\}},$$

in which the above mentioned measure returns the union and intersection of the HFS together with its complement.

7.4.5 Fuzziness and nonspecificity-based entropy measures

In a general expression, the concept of fuzziness describes the HFS departing from the nearest crisp set, meanwhile, the concept of nonspecificity relates to the imprecision knowledge of the HFS. By taking these two concepts into account, and supposing that $h_A(x) = \{h_A^{\sigma(j)}(x)\}_{j=1}^{l_x}$ and $h_B(x) = \{h_B^{\sigma(j)}(x)\}_{j=1}^{l_x}$ are two HFSs on X, a pair of HFS entropy measures (E_F, E_{NS}) possesses the following properties given by Zhao et al. [43]:

($\mathbf{E_F}0$) $0 \le E_F(h_A(x)) \le 1$ (*Boundary axiom*);
($\mathbf{E_F}1$) $E_F(h_A(x)) = 0$ if and only if $h_A(x) = O^*$ or $h_A(x) = I^*$;
($\mathbf{E_F}2$) $E_F(h_A(x)) = 1$ if and only if $h_A(x) = \{\overline{\tfrac{1}{2}}\}$ (*Reflexivity axiom*);
($\mathbf{E_F}3$) $E_F(h_A(x)) = E_F(h_{A^c}(x))$ (*Complementarity axiom*);
($\mathbf{E_F}4$) If $h_A^{\sigma(j)}(x) \le h_B^{\sigma(j)}(x) \le \frac{1}{2}$ or $h_A^{\sigma(j)}(x) \ge h_B^{\sigma(j)}(x) \ge \frac{1}{2}$, then $E_F(h_A(x)) \le E_F(h_B(x))$ (*Inequality axiom*);

and

($\mathbf{E_{NS}}0$) $0 \le E_{NS}(h_A(x)) \le 1$ (*Boundary axiom*);
($\mathbf{E_{NS}}1$) $E_{NS}(h_A(x)) = 0$ if and only if $h_A(x)$ is a singleton, that is, $h_A(x) = \{y\}$;
($\mathbf{E_{NS}}2$) $E_{NS}(h_A(x)) = 1$ if and only if $h_A(x) = \{0, 1\}$ (*Reflexivity axiom*);
($\mathbf{E_{NS}}3$) $E_{NS}(h_A(x)) = E_{NS}(h_{A^c}(x))$ (*Complementarity axiom*);
($\mathbf{E_{NS}}4$) If $|h_A^{\sigma(i)}(x) - h_A^{\sigma(j)}(x)| \le |h_B^{\sigma(i)}(x) - h_B^{\sigma(j)}(x)|$ for any $i, j = 1, 2, \ldots, l_x$, then $E_{NS}(h_A(x)) \le E_{NS}(h_B(x))$ (*Inequality axiom*).

7.5 Conclusions

In this contribution, the main aim was to give a thorough review of HFS information measures, known as the distance, similarity, and entropy measures from different viewpoints.

Bibliography

[1] Turksen IB. Interval valued fuzzy sets based on normal forms. Fuzzy Sets Syst. 1986;20:191–210.
[2] Dubois D, Prade H. Fuzzy sets and systems: theory and applications. New York: Academic Press; 1980.
[3] Miyamoto S. Multisets and fuzzy multisets. In: Liu ZQ, Miyamoto S, editors. Soft Computing and Human-Centered Machines. Berlin: Springer; 2000. p. 9–33.

[4] Atanassov K. Intuitionistic fuzzy sets, theory and applications. Heidelberg, New York: Physica-Verlag; 1999.
[5] Xu ZS, Xia MM. Distance and similarity measures for hesitant fuzzy sets. Inf Sci. 2011;181:2128–38.
[6] Torra V, Narukawa Y. On hesitant fuzzy sets and decision. In: The 18th IEEE International Conference on Fuzzy Systems. Korea: Jeju Island; 2009. p. 1378–82.
[7] Torra V. Hesitant fuzzy sets. Int J Intell Syst. 2010;25:529–39.
[8] Farhadinia B. A novel method of ranking hesitant fuzzy values for multiple attribute decision-making problems. Int J Intell Syst. 2013;28:752–67.
[9] Farhadinia B. Study on division and subtraction operations for hesitant fuzzy sets, interval-valued hesitant fuzzy sets and typical dual hesitant fuzzy sets. J Intell Fuzzy Syst. 2015;28:1393–402.
[10] Farhadinia B. Distance and similarity measures for higher order hesitant fuzzy sets. Knowl-Based Syst. 2014;55:43–8.
[11] Farhadinia B. Correlation for dual hesitant fuzzy sets and dual interval-valued hesitant fuzzy sets. Int J Intell Syst. 2014;29:184–205.
[12] Farhadinia B. A series of score functions for hesitant fuzzy sets. Inf Sci. 2014;277:102–10.
[13] Farhadinia B. Multiple criteria decision-making methods with completely unknown weights in hesitant fuzzy linguistic term setting. Knowl-Based Syst. 2016;93:135–44.
[14] Farhadinia B. Hesitant fuzzy set lexicographical ordering and its application to multi-attribute decision making. Inf Sci. 2016;327:233–45.
[15] Farhadinia B. Determination of entropy measures for the ordinal scale-based linguistic models. Inf Sci. 2016;369:63–79.
[16] Farhadinia B. A multiple criteria decision making model with entropy weight in an interval-transformed hesitant fuzzy environment. Cogn Comput. 2017;9:513–25.
[17] Farhadinia B. Improved correlation measures for hesitant fuzzy sets. In: 2018 6th Iranian Joint Congress on Fuzzy and Intelligent Systems (CFIS). 2018. https://doi.org/10.1109/CFIS.2018.8336664.
[18] Farhadinia B. Information measures for hesitant fuzzy sets and interval-valued hesitant fuzzy sets. Inf Sci. 2013;240:129–44.
[19] Farhadinia B, Herrera-Viedma E. Entropy measures for hesitant fuzzy linguistic term sets using the concept of interval-transformed hesitant fuzzy elements. Int J Fuzzy Syst. 2018;20:2122–34.
[20] Farhadinia B, Herrera-Viedma E. Sorting of decision-making methods based on their outcomes using dominance-vector hesitant fuzzy-based distance. Soft Comput. 2019;23:1109–21.
[21] Farhadinia B, Xu ZS. Novel hesitant fuzzy linguistic entropy and cross-entropy measures in multiple criteria decision making. Appl Intell. 2018;48:3915–27.
[22] Farhadinia B, Xu ZS. Ordered weighted hesitant fuzzy information fusion-based approach to multiple attribute decision making with probabilistic linguistic term sets. Fundam Inform. 2018;159:361–83.
[23] Farhadinia B, Xu ZS. Hesitant fuzzy information measures derived from T-norms and S-norms. Iran J Fuzzy Syst. 2018;15(5):157–75.
[24] Farhadinia B. An overview on hesitant fuzzy information measures. ICSES Trans Neural Fuzzy Comput. 2019;2:1–8.
[25] Farhadinia B, Xu ZS. Distance and aggregation-based methodologies for hesitant fuzzy decision making. Cogn Comput. 2017;9:81–94.
[26] Liao HC, Xu ZS, Xia MM. Multiplicative consistency of hesitant fuzzy preference relation and its application in group decision making. Int J Inf Technol Decis Mak. 2014;13:47–76.
[27] Liao H, Xu ZS, Zeng XJ, Merigo JM. Qualitative decision making with correlation coefficients of

hesitant fuzzy linguistic term sets. Knowl-Based Syst. 2015;76:127–38.
[28] Liao HC, Xu ZS, Zeng XJ. Distance and similarity measures for hesitant fuzzy linguistic term sets and their application in multi-criteria decision making. Inf Sci. 2014;271:125–42.
[29] Liao HC, Xu ZS. Subtraction and division operations over hesitant fuzzy sets. J Intell Fuzzy Syst. 2014;27:65–72.
[30] Xia MM, Xu ZS. Hesitant fuzzy information aggregation in decision making. Int J Approx Reason. 2011;52:395–407.
[31] Peng DH, Gao ChY, Gao ZhF. Generalized hesitant fuzzy synergetic weighted distance measures and their application to multiple criteria decision making. Appl Math Model. 2013;37:5837–50.
[32] Zhou XQ, Li QG. Some new similarity measures for hesitant fuzzy sets and their applications in multiple attribute decision making. Computing Research Repository. arXiv:1211.4125.
[33] Zhang XL, Xu ZS. Novel distance and similarity measures on hesitant fuzzy sets with applications to clustering analysis. J Intell Fuzzy Syst. 2015;28:2279–96.
[34] Li DQ, Zeng WY, Li JH. New distance and similarity measures on hesitant fuzzy sets and their applications in multiple criteria decision making. Eng Appl Artif Intell. 2015;40:11–6.
[35] Hu J, Zhang X, Chen X, Liu Y. Hesitant fuzzy information measures and their applications in multi-criteria decision making. Int J Syst Sci. 2016;47:62–76.
[36] Yang MS, Lin DC. On similarity and inclusion measures between type-2 fuzzy sets with an application to clustering. Comput Math Appl. 2009;57:896–907.
[37] Li DF, Cheng CT. New similarity measures of intuitionistic fuzzy sets and application to pattern recognitions. Pattern Recognit Lett. 2002;23:221–5.
[38] Pal SK, King RA. Image enhancement using smoothing with fuzzy sets. IEEE Trans Syst Man Cybern. 1981;11:495–501.
[39] Wang TJ, Lu ZD, Li F. Bidirectional approximate reasoning based on weighted similarity measures of vague sets. Comput Eng Sci. 2002;24:96–100.
[40] Xu ZS. A method based on distance measure for interval-valued intuitionistic fuzzy group decision making. Inf Sci. 2010;180:181–90.
[41] Szmidt E, Kacprzyk J. Intuitionistic fuzzy sets in intelligent data analysis for medical diagnosis. In: Alexandrov VN, Dongarra J, Juliano BA, Renner RS, Tan CJK, editors. ICCS 2001. LNCS. vol. 2074. Heidelberg: Springer; 2001. p. 263–71.
[42] Xu Z, Xia M. Hesitant fuzzy entropy and cross-entropy and their use in multiattribute decision-making. Int J Intell Syst. 2012;27:799–822.
[43] Zhao N, Xu ZS, Liu FJ. Uncertainty measures for hesitant fuzzy information. Int J Intell Syst. 2015;30:1–19.
[44] Shang XG, Jiang WS. A note on fuzzy information measures. Pattern Recognit Lett. 1997;18:425–32.

Manal Zettam, Jalal Laassiri, and Nourddine Enneya

8 A prediction MapReduce-based TLBO-FLANN model for angiographic disease status value

Abstract: Functional link artificial neural networks (FLANNs) are sensitive to weights' initialization and adopted learning algorithms. Using an efficient learning algorithm and randomly initialized weights leads to improve FLANNs efficiency and performance. The performance of the TLBO-FLANN model was proved in literature through simulation study and comparison studies involving GA-FLANN, PSO-FLANN and HS-FLANN. The current chapter presents a MapReduce-based TLBO-FLANN model to predict angiographic disease status value. The experimentations were carried out on data sets to prove the performance of the MapReduce-based TLBO-FLANN model.

Keywords: Classification, teaching–learning based optimization, Hadoop, functional link artificial neural network, data mining

8.1 Introduction

In data mining field, classification and forecasting are well known as the method of learning from data. Several applications of classification methods on different areas of science and engineering have been published in recent years [1]. G. P. Zhang [3] defined an artificial neural networks (ANN) model as an alternative to conventional classification methods [2]. Indeed, ANNs are capable of generating complex mapping between input and output space, thus ANNs are capable of creating arbitrarily complex nonlinear boundaries for decision making. In this chapter, an attempt is made to apply a prediction MapReduce-based TLBO-FLANN model to predict angiographic disease status value.

A great number of FLANN models and their applications in forecasting, classification and prediction were reported in the literature. Several attempts are also made to enhance the performance of Functional link artificial neural network (FLANN) models using different techniques of optimization. The current section reports previously published works of FLANN models employed in the data mining field especially in classification, prediction and forecasting methods. The reference [4] introduced FLANN as a classification method with a lower complex architecture than multilayer perceptron. The introduced FLANN model proves its efficiency in handling linearly nonseparable groups via increasing the input space dimension. In most cases, it is found that FLANN model's efficiency and processing time are higher than the other

Manal Zettam, Jalal Laassiri, Nourddine Enneya, Informatics, Systems and Optimization Laboratory, Department of Computer Science, Faculty of Science, Ibn Tofail University, Kenitra, Morocco, e-mails: zettammanal@gmail.com, laassiri@uit.ac.ma, ENNEYA.NOURDDINE@uit.ac.ma

https://doi.org/10.1515/9783110671353-008

models. Dehuri and Cho in [5] addressed the basic concept of FLANN, associated basic functions, learning schemes and FLANN development over time. Authors proposed a new hybrid FLANN model. By testing on the same benchmark data sets, the proposed model outperforms the classical FLANN. The reference [6] proposed an effective FLANN method for rendering stock price prediction. The trigonometric FLANN proposed by [6] surpasses MLP. The reference [7] compared MLP and supporting vector machines (SVM) and FLANN classifiers. The experiments show that FLANN surpasses SVM and MLP. Chakravarty and Dash [8] introduced a neural-fuzzy functional link (FLNF). Chakravarty and Dash [8] compared the proposed FLNF with FLANN. The results show that FLNF overcomes the FLANN model. Majhi et al. [9] demonstrates that recursive least square learning FLANN model has lower computational complexity compared to recursive least square learning FLANN model. Dehuri and Cho [10] proposed a compact and accurate hybrid FLANN classifier (HFLNN). The comparison study demonstrates that HFLANN performs better than FLANN and RBFN classifiers. Mishra et al. [11] performs the classification of biomedical data by means of FLANN, MLP and PSO-FLANN models. The traditional artificial neural network MLP shows better results compared to sophisticated FLANN and PSO-FLANN models. An improved PSO based FLANN classifier (IPSO-FLANN) has been proposed later by Dehuri et al. [12]. The classifier surpasses SVM, MLP and FLANN-GDL model. Naik et al. [13] introduced a classifier based on FLANN and PSO-GA. The parameters of FLANN are iteratively adjusted by means of GA, PSO and the gradient descent search algorithms. This leads to obtain better accuracy compared to other alternatives. The method proposed by Naik et al. [14] mimics the honey bee mating behavior. Naik et al. [14] also carried out a comparison study to prove the efficiency of the proposed honey bee mating optimization (HMBO) among GA based FLANN, FLANN, HS based FLANN and PSO based FLANN models. Naik et al. [15] introduced an HS-FLANN classifier able to classify data more efficiently than PSO-FLANN and GA-FLANN.

The FLANN models presented above implement learning methods of forecasting and classification. The main concept of those methods consists of learning from data's sets. The FLANNs' performance depends on the choice of the learning algorithm. The efficiency of FLANNs depends on weights' initialization. As pinpointed earlier in this section, the optimization algorithms such as GA, PSO, improved PSO, HS, HMBO and others [2] enhance both the efficiency and performance of FLANN models. Despite the fact that optimization techniques in draws benefices to FLANN models, the controlling parameters represent a major shortcoming of the use of those techniques. Indeed, the efficiency of GA based FLANN (GA-FLANN) [12], PSO based FLANN (PSO-FLANN) [12], IPSO based FLANN (IPSO-FLANN) [12], HBMO based FLANN (HBMO-FLANN) [14] and HS based FLANN (HS-FLANN) [15] depends on the following controlling parameters:

- Defining the type of crossover and mutation rate in both GA and GA-FLANN.
- Defining the adequate values of c_1, c_2 coefficients and inertia weight (k) in both PSO in IPSO-FLANN and PSO-FLANN.

- Defining drone and worker ratio and the types of crossover operator selection in both HBMO and HBMO-FLANN.
- And defining bandwidth, harmony memory, pitch adjustment rate consideration rate in both HS and HS-FLANN.

Any change in the controlling parameters may increase the complexity of used optimization technique and also the effort to develop the program. In addition, the efficiency of FLANN model can decrease drastically. To overcome this issue, Rumelhart et al. [16] introduced a TLBO-FLANN model with gradient descent learning (GDL) scheme for classification. The authors of [2] discuss the complexity of calibrating weights of the FLANN model by means the TLBO with gradient descent learning (GDL). TLBO has been applied in several problems such as the economic emission load dispatch problem [2], real-parameter optimization problems [2] and others. Readers can refer to Table 8.2 for more details.

Crepinsek et al. [17] has pinpointed issues of TLBO such as:
- Improper experiment settings.
- Evaluating incorrectly fitness functions by means of formula.
- Controlling parameters.

Waghmare [18] has examined, commented and tested TLBO. The experiments carried out in Waghmare [18] by considering a number of both constrained and unconstrained functions show that TLBO surpasses the existing evolutionary algorithms. Waghmare [18] proves also that the TLBO algorithm is free of the issues reported previously in [17]. Furthermore, Waghmare [18] claimed that the only parameters required by TLBO are those required for testing the functions and their results. In this chapter, we propose a MapReduce-based TLBO-FLANN based on a combination of TLBO-FLANN and MapReduce-based TLBO previously introduced in literature.

The second section of the current chapter introduced the Teaching learning based optimization algorithm. The third section of the current chapter presented the different forms of TLBO used in literature. The FLANN-TLBO based prediction model is explained in the fourth section. The fifth and sixth sections carry out the experimental results and present the concluding remarks.

8.2 Teaching learning based optimization

The Teaching Learning optimization (TLBO) algorithm was first introduced by [19] to solve the continuous optimization problems. TLBO is defined as a parameterless algorithm represented by the teacher and the student phases. The teacher phase consists of three steps. In the first step, the design variables mean is calculated. The best solution of the population is designated in the second. Finally, a randomly selected solution is modified via equation (8.1):

$$X_{\text{new}} = X_{\text{old}} + rand(X_{\text{teacher}} - (T_F)Mean) \tag{8.1}$$

where *rand* is a random generated number within the $[0, 1]$ interval, X_{teacher} denotes the teacher, X_{old} is the randomly selected solution, *Mean* designates a vector of design variables mean. T_F is a teaching factor equal to 1 or 2.

In the student phase, TLBO algorithm improves the fitness of one solution via equation (8.2):

$$X_{\text{new}} = X_i + r(X_i - X_j), \tag{8.2}$$

where *rand* is a random number within the $[0, 1]$ interval, X_j, X_i, respectively, denote the solution with the lower and greater fitness value.

8.3 Different teaching learning based optimization algorithms

8.3.1 Improved teaching learning based optimization (iTLBO)

In the teacher phase, learners' level is enhanced by means of weighted differential vector. In the learner phase, the learners' level is enhanced via interactions [20]. The random weighted differential vector is updated randomly by means of the following:

$$0.5 * (1 + rand) \tag{8.3}$$

where *rand* is a random number within the $[0, 1]$ interval. So this weighted differential scale factor's mean value is 0.75. This enables for unpredictable differences in a differential vector development, and thus helps to maintain the reasonable likelihood of getting a better place on the multimodal functional surface. Consequently, the fitness of a population's best vector is much less likely to become static until a global optimum is achieved.

So the new set of enhanced learners is as follows:

$$X_{\text{new}^g(i)} = X_{(i)^g} + 0.5 * (1 + rand) * (X_{\text{Teacher}^g} - T_F M^g) \tag{8.4}$$

The set of learners can be updated as follows:

$$X_{\text{new}^g(i)} = \begin{cases} X^g_{(i)} + 0.5 * (1 + rand) * (X^g_{(i)} - X^g_{(r)}) & \text{iff } (X^g_{(i)}) \prec f(X^g_{(r)}) \\ X^g_{(i)} + 0.5 * (1 + rand) * (X^g_{(r)} - X^g_{(i)}) \end{cases} \tag{8.5}$$

8.3.2 Weighted teaching learning based optimization (wTLBO)

Generally, a teacher requests his/her student to acquire the same expertise as him/her in a short time. Teaching-learning process is an iterative process in which ongoing knowledge transfer interaction takes place [21]. Often learners might not remember all skills. Therefore, every learner has a parameter known as "weight." The part of its previous value is considered when determining the new learner value and that is determined by a weight factor w. In wTLBO, during the early stages of the search, each learner is sampled to specific search space zone. The movements of trial solutions are carefully tuned in the later stages so that they can explore the interior of a relatively small space in which the global optimum predicted lies. To achieve this goal, we reduce the weight factor value linearly with time from a maximum to a minimum value:

$$w = w_{\max} - \left(\frac{w_{\max} - w_{\min}}{maxiteration} \right) * i \tag{8.6}$$

where the minimum and maximum values of the weight factor w are min and max, the current iteration number is iteration and the maximum number of iterations permitted is maxiteration. $w_{\max}$ and $w_{\min}$ are selected to be 0.9 and 0.1, respectively. Hence, in the teacher phase the new set of improved learners can be

$$w = w_{\max} - \left(\frac{w_{\max} - w_{\min}}{maxiteration} \right) * i \tag{8.7}$$

and a set of improved learners in learner phase as

$$X_{new^g(i)} = \begin{cases} X^g_{(i)} + 0.5 * (1 + rand(0,1)) * (X^g_{(i)} - X^g_{(r)}) & \text{iff } (X^g_{(i)}) \prec f(X^g_{(r)}) \\ X^g_{(i)} + 0.5 * (1 + rand(0,1)) * (X^g_{(r)} - X^g_{(i)}) \end{cases} \tag{8.8}$$

8.3.3 Orthogonal teaching-learning-based optimization (OTLBO)

Considering an experiment involving several factors each with various values named levels. Let us assume that there are Q levels for each factor of a set of P factors and Q_p as the number of combinations [22]. Orthogonal design is developed by [22] to study multifactor and multilevel problems as a mathematical tool. The developed orthogonal design constructs an orthogonal array L with the three following proprieties:

- The array L is represented by a subset of M combinations.
- Each factor is represented by a factor.
- To ensure diversity of the search space uniformly scatters the selected subset.

The proposed method in [22] generates an orthogonal array L, where

$$M = Q * Q \quad \text{and} \quad P = Q + 1$$

The steps of the proposed method are as follows:

Procedure for generating an orthogonal array L

Input: The number of levels Q
Output: An orthogonal array L
Calculate $M = Q * Q$ and $P = Q + 1$
Initialize L with M rows and P columns

For $i = 1$ to M, do

$$L_{i,1} = \mathrm{mod}(\lfloor (i-1)/q \rfloor, q)$$
$$L_{i,2} = \mathrm{mod}(i-1, q)$$

For $j = 1$ to $P - 2$ do

$$L_{i,2+j} = \mathrm{mod}(L_{i,j} * j + L_{i,2}, q)$$

end
end

8.3.4 Modified teaching-learning-based optimization (mTLBO)

In contrary to the learner phase, the teacher phase of this version remains same as in original TLBO. Generally, learners' performance depends on their investment, on the teacher level and also on their interactions. In the case where one learner or more have more knowledge than the teacher, only k reminding learners learn from the teacher. Therefore, an additional term is added to the learner phase equation of TLBO [24]. This term is scaled by scale factor in a random manner in the range (0.5, 1) by using the following:

$$0.5 * (1 + rand(0,1)) \tag{8.9}$$

where $rand(0,1)$ is a uniformly distributed random number within the range [0, 1]. The mean value of the scale factor is 0.75. The added extra term allows stochastic variations within the population. As a result, it preserves learners' diversity

$$X_{\mathrm{new}^g(i)} = \begin{cases} X^g_{(i)} + rand * (X^g_{(i)} - X^g_{(r)}) + 0.5 * (1 + rand) * (X^g_{(\mathrm{Teacher})} - X^g_{(i)}) \\ \text{iff } (X^g_{(i)}) \prec f(X^g_{(i)}) \\ X^g_{(i)} + rand * (X^g_{(i)} - X^g_{(r)}) + 0.5 * (1 + rand) * (X^g_{(\mathrm{Teacher})} - X^g_{(i)}) \\ \text{otherwise} \end{cases} \tag{8.10}$$

The third term in the equation (8.10) represents interaction between the teacher and learner. The comparative study conducted by [24] reveals that mTLBO surpasses few variants of PSO, DE and TLBO.

Table 8.1: Parameters of TLBO and its variants.

TLBO Variant	Parameters
Teaching-Learning-Based Optimization (TLBO)	n denotes the students number d denotes the design variables number X_i denotes the classroom's initial students X_{new} denotes the new student's value $X_{teacher}$ denotes the teacher's value T_f denotes the factor of teaching
Improved Teaching-Learning-Based Optimization (iTLBO)	$X^g_{(i)}$ denotes the added weight to performance of a given learner M^g denotes the mean weight of learners population
Weighted Teaching-Learning-Based Optimization (wTLBO)	w denotes the weight factor of a given learner w_{max} denotes the maximum weight factor w_{min} denotes the minimum weight factor
Modified Teaching-Learning-Based Optimization (mTLBO)	$X^g_{(teacher)}$ denotes the weight factor for teacher $X_g(i)$ denotes the added weight to performance of a given learner

8.3.5 TLBO and its variants

The current subsection presents a comparison of TLBO and some of its variants. Table 8.1 contains parameters involved in TLBO and some of its variants. Readers interested in applications of TLBO and its variants can refer to references: [20, 21, 22, 23, 24, 25, 26, 27] and [28].

8.4 TLBO-FLANN based prediction model

FLANN is defined as a single layer neural network and TLBO-FLANN as a hybrid prediction model (see Figure 8.1). Let us assume we have a population of P individuals. Each individual represents an eventual weight vector of TLBO-FLANN model. Each weight vector is constituted by D weight values. Population is updated via TLBO in a manner to minimize the mean square error (MSE).

The stages of the prediction model are as follows:

- Stage 1. Data preparation and normalization
- Stage 2. Initialization of the population
- Stage 3. Determine (K-L) testing and L training feature sets
- Stage 4. Apply algorithm to L training sets. Each set of parameter correspond to an individual of the population. The mean square error (MSE) of ith individual is determined as follows:

$$MSE(i) = \frac{\sum_{l=1}^{L} e_l^2}{L} \tag{8.11}$$

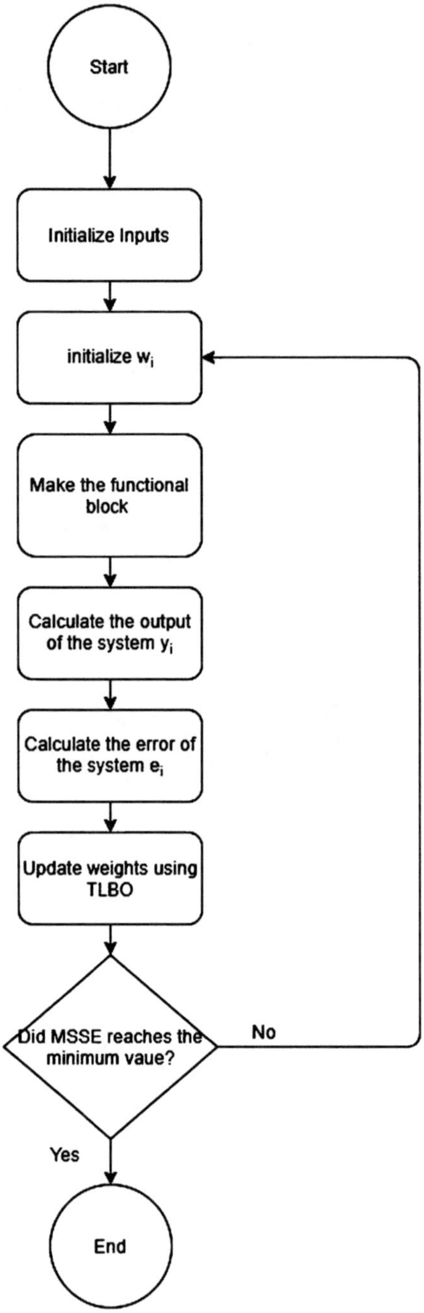

Figure 8.1: TLBO-FLANN flowchart.

where e_i is defined as follows:

$$e_i = d_i - y_i \tag{8.12}$$

- The ith set of features output is estimated as follows:

$$y_i = \sum_{d=1}^{D} w_{i,d} x_{i,d} \tag{8.13}$$

- Stage 5.
  - The weight (an individual of the population) is updated in two phases:
    - in the teacher phase, the weight is updated using (1).
    - in the learning phase using (2).
  - Only individuals with lower fitness values are selected in both teacher and learner phase.
  - The entire process is repeated until the stopping criteria are met.
- Stage 6. The learning process is stopped when the minimum MSE (MMSE) reaches the minimum level.

8.5 A MapReduce-based TLBO-FLANN based prediction model

Google set up MapReduce in 2004 to tackle problems faced by SQL database users. Indeed, SQL databases can neither capture unstructured and large data sets into a schema nor scale it and analyzed it by means of SQL queries as is the case of small structured data sets. In order to process unstructured large sets Google used the "map" and "reduce" functions of the functional programming paradigm as the main bases of MapReduce. Up now, Hadoop is the most widely known implementation of the MapReduce paradigm. For more details, readers can refer to [28]. MapReduce is also renowned for the efficiency and the performance it gives to distributed evolutionary algorithms. Some distributed evolutionary algorithms based on MapReduce are introduced [29, 30, 31, 32]. In this subsection, the implementation in MapReduce of algorithms used in this chapter to perform the study are described in detail.

The data structure used to store an individual is shown in Figure 8.2. This key-value structure is used as the input and output data of MapReduce. The key part contains an ID assigned to each individual of the population. The value part contains the mean square error (MSE), the generation number and the chromosome. The weights are randomly initialized within the interval $[0.0, 0.1]$.

The main TLBO-FLANN algorithm is implemented in the MapReduce Framework where each generation of the TLBO-FLANN is performed by a MapReduce job.

Key		Value		
ID		Fitness	Generation	Chromosome
int		double	int	Vector of doubles

Figure 8.2: The key-value structure to store an individual.

Map phase at each iteration of TLBO-FLANN
Function map (key, value)
if population not initialized
population ← INITIALIZATION (populationSize)
else
The mean of weights X_{mean} in population X is calculated.
The fitness of weights in X is calculated.
The weight with the maximum fitness is defined as teacher (X_{teacher}).

$$X_{\text{new}} = X_{\text{old}} + r(X_{\text{teacher}} - (T_F)Mean)$$

Determine the new population X_{next} using old weights X_{mean}, X, T_F and X_{teacher}:
For $i = 1$ to size of X

$$X_{\text{new}} = X_i + r(X_i - X_j)$$

Endfor
Endif

Reduce phase at each iteration of TLBO-FLANN
Function reduce (key, value)
Update the old population of X
For $i = 1$ to size of X
if $(X(i) \prec X_{\text{new}^g(i)})$

$$X(i) = X_{\text{new}^g(i)}$$

endif
endfor

8.6 Simulation study and result analysis

The simulation study is carried out to estimate the prediction performance of the proposed model (TLBO-FLANN). The data set has been taken from https://archive.ics.uci.edu/ml/datasets/Heart+Disease. Simulation has been done for the prediction of the presence of heart disease by using java code (see https://github.com/

Table 8.2: Overall error of TLBO-FLANN on the 4 data sets.

Data set	Min overall error	Overall Error	Number of epoch	Learning rate
Switzerland	0.001	0.001	10	0.3
Hungarian	0.001	0.001	10	0.3
Cleveland	0.001	0.001	10	0.3
VA Long Beach	0.001	0.001	10	0.3

PacktPublishing/Neural-Network-Programming-with-Java-SecondEdition) [33]. Table 8.2 shows the overall error of TLBO-FLANN on the 4 data sets. Moreover, the execution time was reduced by applying MapReduce-based TLBO-FLANN models. The results obtained in Table 8.2 confirm that the learning process was successful for all data sets with 10 epochs and 0.3 learning rate.

8.7 Conclusion

In this chapter, we presented a MapReduce-based TLBO-FLANN model. This method was inspired by two previously introduced methods in literature. The first one is the TLBO-FLANN that was promoted by authors due to accuracy of the model obtained. The second method is the MapReduce-based TLBO algorithm. We combined the two methods to gain on accuracy of the model and to reduce the computational time.

Bibliography

[1] Yang P, Gao W, Tan Q, Wong K. A link-bridged topic model for cross-domain document classification. Inf Process Manag. 2013;49(6):1181–93.
[2] Naik B, Nayak J, Behera HS. A TLBO based gradient descent learning-functional link higher order ANN: an efficient model for learning from non-linear data. J King Saud Univ, Comput Inf Sci. 2018;30(1):120–39.
[3] Zhang GP. Neural networks for classification: a survey. IEEE Trans Syst Man Cybern, Part C, Appl Rev. 2000;30(4):451–62.
[4] Mishra BB, Dehuri S. Functional link artificial neural network for classification task in data mining. J Comput Sci. 2007;3(12):948–55.
[5] Dehuri S, Cho S. A comprehensive survey on functional link neural networks and an adaptive PSO–BP learning for CFLNN. Neural Comput Appl. 2009;19(2):187–205.
[6] Patra JC, Lim W, Thanh N, Meher P. Computationally efficient FLANN-based intelligent stock price prediction system. In: IEEE proceedings of international joint conference on neural networks. Atlanta, Georgia, USA, June 14–19. 2009. p. 2431–8.
[7] Sun J, Patra J, Lim W, Li Y. Functional link artificial neural network-based disease gene prediction. In: IEEE proceedings of international joint conference on neural networks. Atlanta, Georgia, USA. June 14–19. 2009. p. 3003–10.

[8] Chakravarty S, Dash PK. Forecasting stock market indices using hybrid network. In: IEEE world congress on nature & biologically inspired computing. 2009. p. 1225–30.
[9] Majhi R, Panda G, Sahoo G. Development and performance evaluation of FLANN based model for forecasting of stock markets. Expert Syst Appl. 2009;36:6800–8.
[10] Dehuri S, Cho S-B. Evolutionarily optimized features in functional link neural network for classification. Expert Syst Appl. 2010;37:4379–91.
[11] Mishra S, Shaw K, Mishra D, Patnaik S. An enhanced classifier fusion model for classifying biomedical data. Int J Comput Vis Robot. 2012;3(1/2):129–37.
[12] Dehuri S, Roy R, Cho S, Ghosh A. An improved swarm optimized functional link artificial neural network (ISO-FLANN) for classification. J Syst Softw. 2012;85:1333–45.
[13] Naik B, Nayak J, Behera HS. A novel FLANN with a hybrid PSO and GA based gradient descent learning for classification. In: Proceedings of the 3rd international conference on frontiers of intelligent computing: theory and applications (FICTA). Advances in intelligent systems and computing. vol. 327. 2015. p. 745–54. https://doi.org/10.1007/978-3-319-11933-5_84.
[14] Naik B, Nayak J, Behera HS. A honey bee mating optimization based gradient descent learning – FLANN (HBMOGDL-FLANN) for classification. In: Proceedings of the 49th annual convention of the Computer Society of India CSI – emerging ICT for bridging the future. Advances in intelligent systems and computing. vol. 338. 2015. p. 211–20. https://doi.org/10.1007/978-3-319-13731-5_24.
[15] Naik B, Nayak J, Behera HS, Abraham A. A harmony search based gradient descent learning-FLANN (HS-GDLFLANN) for classification. In: Computational intelligence in data mining, vol. 2. Smart innovation, systems and technologies. vol. 32. 2015. p. 525–39.
[16] Rumelhart DE, Hinton GE, Williams RJ. Learning representations by back-propagating errors. Nature. 1986;323(9):533–6.
[17] Crepinsek M, Liu S-H, Mernik L. A note on teaching–learning-based optimization algorithm. Inf Sci. 2012;212:79–93.
[18] Waghmare G. Comments on "A note on teaching–learning-based optimization algorithm". Inf Sci. 2013;229:159–69.
[19] Rao RV, Savsani VJ, Vakharia DP. Teaching–learning-based optimization: a novel method for constrained mechanical design optimization problems. Comput Aided Des. 2011;43(3):303–15.
[20] Rao RV, Patel V. An improved teaching-learning-based optimization algorithm for solving unconstrained optimization problems. Sci Iran. 2013;20(3):710–20.
[21] Satapathy SC, Naik A, Parvathi K. Weighted teaching-learning-based optimization for global function optimization. Appl Math. 2013;4(3):429–39.
[22] Satapathy SC, Naik A, Parvathi K. A teaching learning based optimization based on orthogonal design for solving global optimization problems. SpringerPlus. 2013;2(1):130.
[23] Zheng H, Wang L, Wang S. A co-evolutionary teaching-learning-based optimization algorithm for stochastic RCPSP. In: 2014 IEEE congress on evolutionary computation (CEC). 2014. p. 587–94.
[24] Satapathy SC, Naik A. Modified teaching–learning-based optimization algorithm for global numerical optimization – a comparative study. Swarm Evol Comput. 2014;16:28–37.
[25] Rao RV, Savsani VJ, Vakharia DP. Teaching–learning-based optimization: a novel method for constrained mechanical design optimization problems. Comput Aided Des. 2011;43(3):303–15.
[26] Rao RV, Savsani VJ, Vakharia DP. Teaching–learning-based optimization: an optimization method for continuous non-linear large scale problems. Inf Sci. 2012;183(1):1–15.
[27] Wang KL, Wang HB, Yu LX, Ma XY, Xue YS. Teaching-learning-based optimization algorithm for dealing with real-parameter optimization problems. In: Vehicle, mechatronics and information technologies, Applied mechanics and materials. vol. 380. 2013. p. 1342–5.
[28] Perera S, Gunarathne T. Hadoop MapReduce cookbook. 2013. Packt Publishing.

[29] McNabb AW, Monson CK, Seppi KD. (MRPSO) MapReduce particle swarm optimization. In: Genetic and evolutionary computation conference (GECCO 2007). London, UK. 2007. p. 177.
[30] Tagawa K, Ishimizu T. Concurrent differential evolution based on MapReduce. Int J Comput. 2010;4(4):161–8.
[31] Wu B, Wu G, Yang M. A MapReduce based ant colony optimization approach to combinatorial optimization problems. In: 8th international conference on natural computation (ICNC 2012). 2012.
[32] Hans N, Mahajan S, Omkar S. Big data clustering using genetic algorithm on Hadoop MapReduce. Int J Sci Technol Res. 2015;4:58–62.
[33] Soares FM, Souza AMF. Neural network programming with Java. 2nd ed. Packt; 2017.

Part II: **Advanced applications**

Tanmoy Som, Pankhuri Jain, and Anoop Kumar Tiwari

9 Analysis of credit card fraud detection using fuzzy rough and intuitionistic fuzzy rough feature selection techniques

Abstract: With the emergence of advanced Internet technology, online banking has become a major channel for business and retail banking. In the last two decades, online banking fraud has been found to be a serious concern in financial crime management for all banking services. Credit card fraud has become a major problem in banking financial transactions and are responsible for the loss of billions of dollars every year. Credit card fraud detection is an interesting issue for the computational intelligence and machine-learning communities. In credit card fraud detection, three aspects namely: imbalanced data, feature selection and selection of appropriate learning algorithms, play the vital role in enhancing the prediction performance. Credit card fraud data sets are usually found to be imbalanced, which results in the classifier to be biased toward majority class. Feature selection is applied as a key factor of credit card fraud detection problem that aims to choose more relevant and nonredundant data features and produce more explicit and concise data descriptions. Furthermore, a suitable learning algorithm can enhance the prediction of fraud in credit card fraud data. In this chapter, SMOTE (Synthetic Minority Oversampling Technique) is employed as an oversampling technique to convert imbalanced data sets into optimally balanced data sets. Furthermore, fuzzy and intuitionistic fuzzy rough sets assisted feature selection approaches are implemented to choose relevant and nonredundant features from the credit card fraud data sets as the fuzzy and intuitionistic fuzzy rough set theories have been widely applied to cope with uncertainty in real-valued data or even in complex data. Moreover, various learning algorithms are applied on credit card fraud data sets and performances are analyzed. Finally, we observe that kernel logistic regression (KLR) is the best performing learning algorithm on reduced optimally balanced credit card fraud data sets for the prediction of fraud. From the experimental results, it can be inferred that the performance of different learning algorithms for the classification of fraud and nonfraud data sets can be easily improved by selecting optimally balanced reduced training data sets consisting of credit card fraud, which can be achieved by suitably modifying the class distribution followed by fuzzy and intuitionistic fuzzy rough set based feature selection techniques.

Tanmoy Som, Pankhuri Jain, Department of Mathematical Sciences, Indian Institute of Technology (Banaras Hindu University), Varanasi, 221005, India, e-mails: tsom.apm@itbhu.ac.in, pankhurijain20@gmail.com
Anoop Kumar Tiwari, Department of Computer Science, Banaras Hindu University, Varanasi, 221005, India, e-mail: anoop.phd2014@gmail.com

https://doi.org/10.1515/9783110671353-009

Keywords: Feature selection, imbalanced data set, SMOTE, fuzzy-rough set, intuitionistic fuzzy-rough set, credit card fraud

9.1 Introduction

Credit card fraud is an illegal act [1, 2]. It may lead to serious damage and huge loss to financial institutions and personals. The increasing approval of electronic payments is causing new viewpoints to fraudsters and requires for advance counter measures to their illegal activities. The financial and banking transactions are extremely essential parts in our modern life, where almost every individual has to cope with banks either physically or electronically. The major issue for modern credit card fraud detection systems is how to enhance fraud detection accuracy with an increasing number of transactions made by the user per second. The credit card fraud detection system is facing heavy workloads due to the increase in a number of users and online transactions [3]. Moreover, advancement of the banking information system has tremendously increased the productivity and profitability of both the private and public sectors. Nowadays, most of the e-commerce transactions are made through credit cards as well as online net banking. These systems are susceptible by modern attacks and techniques at an alarming rate. In 2007, a loss of $3.6 billion was estimated and it was recorded as $4 billion in 2008, which was an increment of 11 %. In 2014, worldwide fraud recorded for a loss of $16.31 billion and this number is rising day by day as fraudsters are employing new analytical approaches to change the regular operating activities of the credit card fraud detection system. It was reported during August 2014 that the Bank of America had to compensate 16.5 billion USD for settling down the financial fraud cases [4, 5]. In 2016, CyberSource (CyberSource, 2016) [2] reported that the 1.4 % of the total e-commerce credit card transactions in Latin America were fraudulent. In India, 17,504 fraud cases of government and private banks were reported between 2013 and 2017, where RBI declared a total loss of 10, 12, 89, 35, 216.40 billion US dollars (Rs. 66,066 crore) for the respective year [6, 7]. The performances of various machine learning algorithms are degrading due to the increment in volume, velocity and variability of modern data. Various predictive models for credit card fraud detection are frequently applied [8, 7, 9]. A plethora of data mining techniques and applications are available in the literature, however, very few data mining approaches are implemented for the credit card fraud detection [10, 11, 12]. Among these, most of the researchers presented neural networks based concepts [13, 14], however, it became popular in the 1990s. A summary of these techniques are presented in [15, 8, 7], which considers analytic techniques for universal fraud detections, comprising credit card fraud. In recent years, credit card fraud detection techniques included case based reasoning and hidden Markov models. The latest research [1] established the various techniques based on support vector machines and random forests to predict the credit card fraud [16, 15, 8].

The existence of large volume of credit card fraud data [9, 17, 18] from various banking sources makes the process of data analysis often inaccurate and challenging due to availability of irrelevant or redundant features in the data. Feature selection [19] is defined as a preprocessing phase that eliminates the irrelevant and redundant features and produces a final set of the most informative attributes or features, which leads to better performance in the task of data mining. Rough set (as proposed by Pawlak [20]) based feature selection is one of the widely used techniques due to its characteristics of acquiring information from the data set itself [21, 22]. The success of the rough set based approach can be categorized into three categories [23]. First, only the interesting facts in the data are analyzed. Second, it receives information from the data set itself and no additional information about the data is needed for data analysis such as expert knowledge or thresholds on a particular domain. Third, it discovers a minimal knowledge representation for data. However, this method cannot be applied to real-valued data sets directly. Discretization methods are applied in order to handle the real-valued data sets by rough set theory and this may lead to information loss. In order to deal with this problem, fuzzy and intuitionistic fuzzy rough sets based approaches were presented and successfully implemented for real-valued data sets [24, 25, 26, 27, 28, 29, 30, 31, 32, 33, 34, 35, 36, 37, 38].

In the last few years, many researchers presented different techniques for feature selection based on the fuzzy rough set theory as well as intuitionistic fuzzy rough set theory and presented their applications to solve various problems [39, 40, 41, 42, 43, 44, 45, 46, 47]. A brief description of these approaches can be given as follows.

In order to eliminate redundancy available in high-dimensional data sets, Jensen et al. [32] proposed a novel feature selection approach based on the fuzzy rough set model by combining the fuzzy [48, 49] and rough sets concepts in 2004. Moreover, this approach was applied on an arbitrary data set and promising results were presented. Finally, the proposed approach was applied to the problem of web categorization and it produced very interesting results as the reduced data sets caused the minimal information loss. A feature selection method was established by Shen et al. (2004) [50] using fuzzy-rough set theory to minimize the information loss due to quantization of the underlying features. This approach performed better than classical entropy, random-based and PCA methods by maintaining the semantics of the data sets. Finally, the proposed approach was applied to complex systems monitoring and the results from the conducted experiments proved the supremacy of this technique.

In 2005, Bhatt et al. [51] examined the termination criteria of the fuzzy rough set based feature selection algorithm and revealed that this technique is inadequate to handle many real-valued data sets. Furthermore, this study established a new concept of the fuzzy-rough set model for a compact computational domain by using natural characteristics of fuzzy t-norm and t-conorm. Thereafter, it was applied to improve the computational efficiency of the fuzzy rough attribute reduction algorithm. During 2005, another interesting method of the attribute reduction method based on fuzzy rough set model was proposed by Jensen et al. [33] to compute the optimal reduct sets. This technique was based on the concept of an ant colony optimization mechanism. Moreover, this approach was implemented for complex systems monitoring. Finally, a comparative study of the proposed approach by using support vector machine classifiers against entropy-based feature selector, PCA, and a transformation-based reduction was presented. Experimental results showed the superiority of the proposed approach over the existing methods until then.

In 2006, Hu et al. [52] introduced a unified data reduction method for hybrid data by presenting an information measure to compute discernibility power of fuzzy and crisp equivalence relations, which is the key notion of both fuzzy-rough set model and classical rough set model. Based on the information measure, a general definition for significance of numeric, nominal and fuzzy features were discussed. This study redefined the independence of reduct, relative reduct, and hybrid feature or attribute subset. Furthermore, this paper has given two greedy reduction algorithms for supervised and unsupervised dimensionality reduction. Finally, this article presents an experimental study to justify that the proposed concepts produce better results when compared with the traditional approaches based on rough set theory.

Various approaches for dimensionality reduction were established during 2007 by the researchers. An efficient hybrid attribute reduction technique based on a generalized fuzzy-rough model was given by Hu et al. [47]. In this article, a theoretic framework of the fuzzy-rough model based on fuzzy relations was introduced. Moreover, this model was applied to evaluate significant measures of attributes. The experimental results based on the proposed algorithm proved the importance of this study. Jensen et al. published two articles on fuzzy rough feature selection during 2007 [34, 53]. They discussed the way to handle the noisy data and the way to retain the semantics of the data sets. In the first article, they introduced the two extensions of rough sets, namely tolerance rough sets and fuzzy rough sets. In the second study, a novel fuzzy rough set based approach for feature selection was demonstrated, which retains data set semantics. These approaches were applied on various challenging domains namely: forensic glass fragment identification, web content classification and complex systems monitoring where a feature reducing step was important. These techniques were compared empirically with several dimensionality reducers based on the experimental results. In the conducted experiments, this technique produced the same or improved classification accuracy when compared to the results of the reduced data sets by other existing approaches as well as the results of unreduced data. These results revealed the fact that the proposed approaches were outperforming the others.

The key aspect of the feature selection techniques is that after elimination of the features, the decision systems should maintain the discernibility. Degree of dependency based approaches allows the decision systems to maintain the discernibility. In the standard attribute reduction approaches, attributes are estimated to be qualitative rather than considering quantitative.

In 2008, a more flexible methodology for dimensionality reduction by using a vaguely quantified rough set was proposed by Cornelis et al. [22]. The experimental results of this technique produced various interesting facts.

In 2009, Jensen et al. [54] presented a fuzzy rough feature selection based rule extraction technique, where the if-then rule was implemented on reduced data sets as produced by the proposed feature selection algorithm. The algorithm was evaluated against leading machine learning algorithms and was proved to be effective. Jensen et al. [55] discussed three approaches in 2009 to deal with the problems of imprecision and uncertainty. In the first technique, fuzzy T-transitive similarity relations were used to approximate the decision concepts, which was based on fuzzy lower approximations. Furthermore, a dependency function for the evaluation between decision and conditional features was introduced. The second technique utilized the information in the fuzzy boundary region to lead the feature search process. The third approach extended the methodology of the discernibility matrix to deal with the fuzzy case. Finally, all the three techniques were successfully applied for the real world data sets.

The degree of dependency based feature selection techniques had been a poor performer and lacked robustness to noisy information. In 2010, Hu et al. [56] proposed a novel feature selection method by combining fuzzy and rough sets to deal with this problem. This article established a novel fuzzy rough set model called soft fuzzy rough sets. In this article, a new dependency function was introduced and it was validated that this function is robust with respect to noisy information. This model was successfully carried out on real-valued data sets. The experimental results proved the effectiveness of the new model as it successfully reduced the influence of noise. Dimensionality reduction techniques based on the fuzzy rough set theory are the frequently and widely used techniques for feature selection of supervised data sets as it uses the information of decision class labels [57]. However, feature selection techniques for unsupervised data based on fuzzy rough set theory are rarely discussed in the literature. In 2010, Parthalain et al. [58] published an article for the fuzzy rough feature selection approach which successfully attempted to handle unsupervised data sets. This technique was applied to maintain the semantics of the data sets without any requirements of thresholding or domain knowledge.

Fuzzy rough sets assisted feature selection technique has become a popular method to deal with the real-valued features. Dimensionality reduction for decision systems with real-valued conditional features and symbolic decision attributes was successfully discussed and implemented by different researchers. However, feature selection based on fuzzy rough set theory for the decision systems with real-valued decision features was rarely discussed. Parthalain et al. [59] proposed a new concept in 2013 to reduce the data by removing both instance and feature by using fuzzy rough sets. They [60] presented a feature selection approach for the unsupervised data that preserved semantics of the data set while reducing the reduct size. Fuzzy rough prototype selection was introduced by Derrac et al. [61] (2013). They retrieved the quality of instances by fuzzy rough feature selection approach and then applied the wrapper method for instance pruning.

In 2014, Cornelis et al. [62] constructed feature selection framework by using multiadjoint fuzzy rough sets, where a family of adjoint fuzzy sets was utilized to calculate the lower approximations.

In 2015, Inuiguchi et al. [63] established the relationship between plausibility and belief functions in the Dempster–Shafer theory of evidence as well as between lower and upper approximations in fuzzy rough set theory and presented its practical applications in various machine learning domains. In order to eliminate overhead and complexity, feature grouping and neighborhood approximation were introduced for fuzzy rough feature selection by Jensen et al. [55, 64]. Onan (2015) [65] showed a new classification approach for breast cancer data by combining instance selection, feature selection with the machine learning algorithm. A weak gamma evaluator was applied to eliminate the useless instances. A novel consistency based feature selection was applied in conjunction with reranking for feature selection. Parthalain et al. [66] demonstrated two new feature selection techniques using different interpretations of the flock of starling algorithm to remove redundant and irrelevant features. Qian et al. [67] proposed an accelerator for fuzzy rough feature selection to speed up sample and dimensionality reduction. Vluymans et al. [68] developed a novel weighted kNN regression method followed by fuzzy rough distributed prototype selection technique to handle big data in 2015. Fuzzy rough incremental feature selection in hybrid information system was established by Zeng et al. (2015) [69] by combining new hybrid distance based on the value difference metric and Gaussian kernel.

Different fuzzy rough set based feature selection methods were presented in 2016 further. These methods were successfully applied to solve various real world problems. In the same year, a novel feature selection approach based on the fuzzy rough set concept was developed by ArunKumar and Ramakrishnan [70] by using a correlation coefficient and applied it for preprocessing of cancer microarray data. By addressing the classification algorithm as features and transforming ensemble predictions into training samples, the notion of feature selection to assist classifier ensemble reduction was developed by Diao et al. [71] (2016). Thereafter, a global heuristic harmony search is applied to choose feature subsets. Guo et al. [72] devised fuzzy rough feature selection on the basis of an invasive weed optimization for mammographic risk analysis and implemented it for early diagnosis of breast cancer. Lasisi et al. [73] (2016) established a fuzzy vaguely quantified rough set model and extended it for feature selection. Moreover, it was coupled with an artificial immune recognition system and clonal selection algorithm and applied for mining agriculture data. Based on information entropy, Zhang et al. [74] (2016) introduced a novel fuzzy rough feature selection to compute the reduct set of heterogeneous data.

During 2017, many researchers established various extensions of fuzzy rough set models and successfully implemented these for dimensionality reduction. By experimenting with genes microarray data, Kumar et al. [75] showed that fuzzy rough feature selection is faster and produces a highly reduced feature subset than the correlation based filter approaches. In this study, they demonstrated a comparative study of fuzzy rough set based feature selection against filter, wrapper approaches based on execution time, number of features and classifier accuracy. In 2017, the fuzzy rough feature selection method was successfully implemented for rule extraction and decision making. Moreover, Qu et al. [76] proposed a fuzzy rough assisted feature selection technique for multiclass data sets by constructing association rules implied in the class labels and taking into account each set of sublabels as a unique class. Furthermore, Su et al. [27] (2017) utilized ordered weighted averaging aggregation of fuzzy similarity relations by considering interaction between features to improve the performance of fuzzy rough feature selection. During 2017, Wang et al. [77] established a fitting model for dimensionality reduction to prohibit misclassification of instances by proposing a fuzzy decision of sample based on parameterized fuzzy relations and fuzzy neighborhoods to characterize information granules.

In 2018, various fuzzy rough feature selection approaches were presented in different articles. Based on dependency degree, Lin et al. [78] developed a novel fuzzy rough feature selection by combining binary shuffled frog leaping search strategy in 2018. ArunKumar and Ramakrishnan [79] (2018) presented a two-steps feature selection and applied it for lung cancer microarray genes expression data. In the first step, dimensionality reduction was introduced by using information gain based entropy. In the second step, fuzzy rough feature selection was presented with the help of customized similarity relation. Two fuzzy rough feature selection techniques were initiated by Dai et al. [80] based on weighted reduced maximal discernibility pair selection and reduced maximal discernibility pair selection and more efficient results were presented. Han et al. [81] and Hu et al. [82] showed two different approaches of fuzzy rough feature selection using a Laplace weighted summation operator and a multikernel concept to enhance the classification performances. Javidi et al. [83] proposed fuzzy rough sets assisted feature selection by utilizing an ant colony with information gain as an evaluation measure. This approach was implemented for gene selection. Li et al. [84] constructed a robust multilabel kernelized fuzzy rough set model and extended it for feature selection, in which two kernel functions were used, one to access the degree of overlap between labels and another to reveal similarity between samples. By defining the score vector to assess probability of different class' samples, Lin et al. [78] introduced a novel multilabel fuzzy rough feature selection approach [12]. In this study, distance between samples was calculated from local sampling and it was found robust to noise. Based on divergence measure, Sheeja et al. [85] presented a novel fuzzy rough feature selection method. In order to characterize attribute reduction better, Wang et al. [1] established fuzzy rough set models based on a fixed and variable parameter and extended these models for attribute reduction. Zhang et al. [86] presented a novel kernel based fuzzy rough feature selection and applied for an intrusion detection system during 2018. Zhang et al. [87] proposed a new approach in 2018 as an accelerator for fuzzy rough feature selection based on information entropy as an evaluation measure to select features. In the same year, another approach was presented by Zhang et al. [88] to select representative instances based on coverage ability of fuzzy granules. Thereafter, a heuristic algorithm for feature selection was developed by utilizing implication preserving reduction by maintaining discriminative information of selected instances.

In 2019, some of the interesting fuzzy rough feature selection techniques were presented. Wang et al. [89] established a novel fuzzy rough set model based on distance measure with a fixed and variable parameter. Furthermore, they introduced the feature selection concept by using these models. By combining the basic selection technique with membership function determination of fuzzy c-means and fuzzy equivalence, Zhao et al. [90] established a novel fuzzy rough feature selection. Consequently, this method took complete advantage of information regarding data sets.

The intuitionistic fuzzy rough set is another interesting concept for handling uncertainty available in the information systems, but rarely discussed by the researchers despite the fact that it is the combination of intuitionistic fuzzy [91, 92, 93] and rough sets and both capture specific aspects of the same notion-imprecision. Jena et al. [94], Chakrabarty et al. [95] and Nanda et al. [96] revealed the facts that lower and upper approximations of intuitionistic fuzzy rough sets once more showed the characteristics of intuitionistic fuzzy sets. During the last few years, by combining intuitionistic fuzzy and rough sets concepts, many research articles presented different intuitionistic fuzzy rough set models and showed its applications [97, 98, 35, 99]. Many researchers explored the wide applicability of intuitionistic fuzzy rough set theory and its applications to solve different real world problems. In the recent years, different extension of intuitionistic fuzzy rough set models were established and successfully implemented for attribute reduction [100, 101, 37, 38, 102, 103]. In the current study, we have used an intuitionistic fuzzy rough set based feature subset selection presented by Tan et al. [36]. This technique was composed of three steps. In the first step, based on intuitionistic fuzzy relations, fuzzy information granules was determined and further, it was applied for characterizing the hierarchical structures of the lower and upper approximations of the intuitionistic fuzzy rough set model within the structure of granular computing. The explored lower and upper approximations were employed for knowledge reduction in the second step. In the third step, significant measures were established to evaluate the approximate characteristics and classification properties of intuitionistic fuzzy relations based on the approximations of the intuitionistic fuzzy rough set. Moreover, this article developed a forward heuristic algorithm to obtain one optimal reduct for intuitionistic fuzzy information systems. Finally, this algorithm was applied on benchmark data sets to evaluate the effectiveness and efficiency on the basis of a number of selected attributes, classification accuracy and computational time.

The objective of this study is to improve the prediction performance of various machine learning algorithms for credit card fraud data. To achieve this goal, we apply

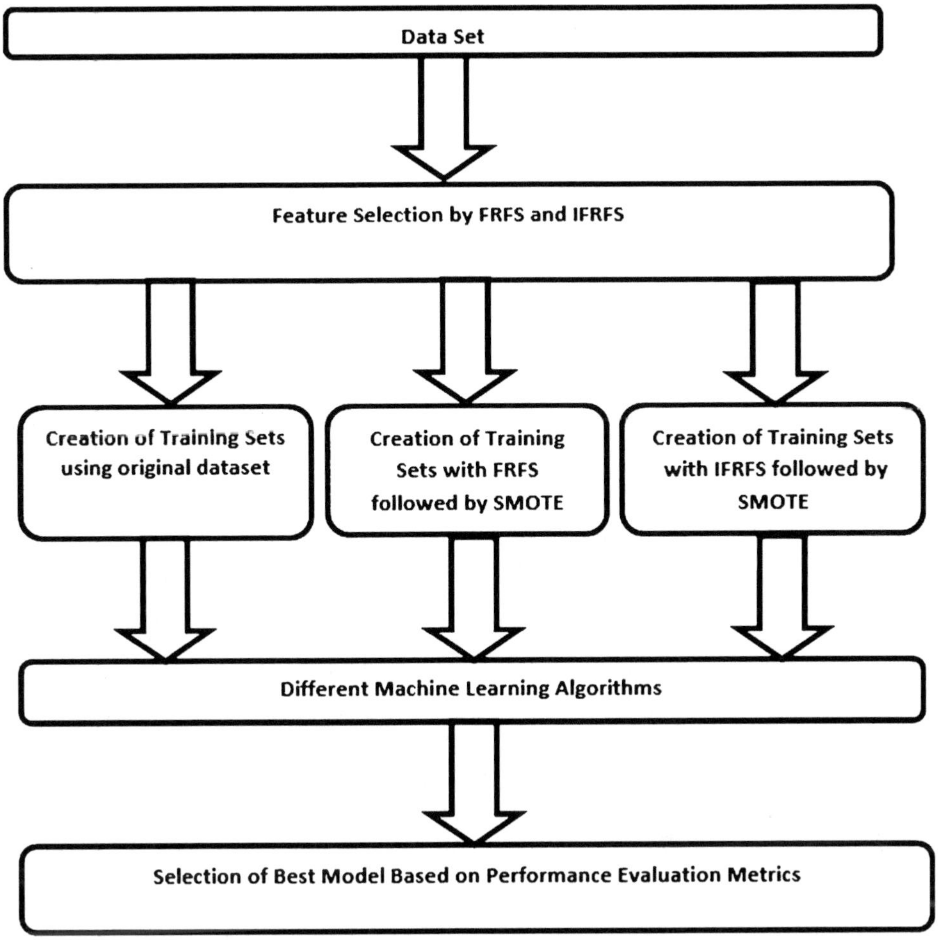

Figure 9.1: Schematic representation of proposed methodology.

fuzzy and intuitionistic fuzzy rough feature selection techniques to obtain the most informative features from credit card fraud data sets. Then we apply SMOTE (Synthetic Minority Oversampling Technique) on reduced data sets to achieve the optimal balancing ratio by suitably modifying the class distribution. Furthermore, various machine learning algorithms are applied on these optimally balanced reduced data sets and their performance is recorded based on a percentage split of 80:20. Moreover, we present our experimental results by using the ROC curve for the better visualization of classifiers performance. Finally, a comparative study for the results produced by our proposed methodology with the previously reported results is presented. From the experimental results, it can be observed that our proposed methodology produces the best results for credit card fraud prediction until the date . We have given a schematic representation of our proposed methodology in Figure 9.1.

9.2 Materials and methods

9.2.1 Data set

In this study, the entire experiments are conducted on Australian credit approval data sets. This data set was collected from UCI repository [104]. The Australian credit approval data set contains 690 instances and 14 attributes or features (with 8 numerical, 6 nominal), where each instance represents credit approval status, which is decided on the basis of corresponding entries available in different attributes or features and the status can be either accepted or rejected. It is an imbalanced data set as the ratio of positive and negative class is different from 1:1.

9.2.2 SMOTE

The performances of various machine learning techniques tend to be biased toward the majority class when the number of instances in the majority class highly exceeds the number of instances in the minority class [105, 106, 107, 15, 4, 108, 68], which is undesirable. This type of data set is defined as the imbalanced data set, which misleads the classification task and influences the results. The Synthetic Minority Oversampling Technique (SMOTE) [109, 110] is one of the most prominently applied sampling techniques to deal with imbalanced data. SMOTE is used to synthetically increase the instances of minority class on the basis of k-nearest neighbors to produce the balanced data set. The SMOTE randomly generates new examples or instances of the minority class from the k-nearest neighbors of the line joining the sample of the minority class to enlarge the number of instances. These new examples or instances become identical to the original instances of the minority class as they are generated based on the characteristics of the original data set. The SMOTE samples are characterized as the linear combinations of two identical samples corresponding to minority class (t and t^k) and are given by

$$z = t + j * (t^k - t) \tag{9.1}$$

where t^k is chosen randomly among the 5 minority class nearest neighbors of t and j changes from 0 to 1 and the default value of nearest neighbors for SMOTE is defined as 5 in WEKA [111].

9.2.3 Feature selection techniques

We have employed two approaches of feature selection. These are as follows:
1. Fuzzy rough set based feature selection.
2. Intuitionistic fuzzy rough set based feature selection.

In the last few years, the fuzzy rough set theory is a widely used concept for developing feature selection techniques. For a given information system $(U', C' \cup D')$, fuzzy set theory is applied to calculate the similarity between the samples using a fuzzy relation $S_r = \{\mu_{S_r}(x,y) \mid \forall x, y \in U'\}$ which satisfies:

Reflexivity: $\mu_{S_r}(x,x) = 1, \forall x \in U'$
Symmetry: $\mu_{S_r}(x,y) = \mu_{S_r}(y,x), \forall x, y \in U'$

That is, $S_r : U' \times U' \longrightarrow [0,1]$ is a mapping that assigns a degree of similarity to each distinct pair of objects. Let $U' \setminus D' = \{D'_1, D'_2, \ldots D'_k\}$ be a crisp partition of decision class D'. Given a fuzzy similarity relation S_r, the definitions of lower and upper approximations for each decision class $D'_i \in U' \setminus D'$ can be given as follows:

$$(S_r \downarrow_{A'} D'_i)(x) = \inf_{y \in U'} \max(S_{rA'}(x,y), D'_i(y)) \tag{9.2}$$

$$(S_r \uparrow_{A'} D'_i)(x) = \sup_{y \in U'} \min(S_{rA'}(x,y), D'_i(y)) \tag{9.3}$$

The pair $(S_r \downarrow_{A'} D'_i, S_r \uparrow_{A'} D'_i)$ is known as a fuzzy rough set, which further reduces to the following equations:

$$S_r \downarrow_{A'} D'_i(x) = \inf_{y \notin D'_i} 1 - S_{rA'}(x,y) \tag{9.4}$$

$$S_r \uparrow_{A'} D'_i(x) = \sup_{y \in D'_i} S_{rA'}(x,y) \tag{9.5}$$

Now, a positive region can be easily computed by the following formula:

$$\text{Pos}_{A'}(x) = \sup_{D'_i \in U' \setminus D'} \mu_{S_r \downarrow_{A'} D'_i}(x) \tag{9.6}$$

Now, the degree of dependency can be given by

$$\Upsilon'_{A'} = \frac{\sum_{x \in U'} |\text{Pos}_{A'}(x)|}{|U'|} \tag{9.7}$$

However, the fuzzy set only takes membership of an instance into consideration. In situations like polling, some person may vote in favor, some against, while some may remain abstained. Intuitionistic fuzzy set (IFS) is an effective way to handle problems involving membership and nonmembership. For an information system $(U', C' \cup D')$, an IFS $A' \in U'$ is represented as

$$A' = \{\langle \mu_{A'}(x), \nu_{A'}(x) \rangle \mid x \in U'\}$$

where $\mu_{A'} : U' \longrightarrow [0,1]$ and $\nu_{A'} : U' \longrightarrow [0,1]$ are the membership and nonmemberships of an instance x, respectively, satisfying

$$0 \le \mu_{A'}(x) + \nu_{A'}(x) \le 1, \quad \forall x \in U'$$
$$\pi_{A'}(x) = 1 - \mu_{A'}(x) - \nu_{A'}(x)$$

where $\pi_{A'}(x)$ represents the degree of hesitancy of x to A'. It is obvious from the above discussions that

$$0 \le \pi_{A'}(x) < 1, \quad \forall x \in U',$$

and $|A'| = \sum_{x \in U'} \frac{1 - \mu_{A'}(x) - \nu_{A'}(x)}{2}$.

Let $R_s = \{\langle \mu_{R_s}(x,y), \nu_{R_s}(x,y)\rangle \mid x,y \in U'\}$ be intuitionistic fuzzy relation induced on the system. R_s is the intuitionistic fuzzy similarity relation if it satisfies:

Reflexivity: $\mu_{R_s}(x,x) = 1$, $\nu_{R_s}(x,x) = 0$, $\forall x \in U'$
Symmetry: $\mu_{R_s}(x,y) = \mu_{R_s}(y,x)$, $\nu_{R_s}(x,y) = \nu_{R_s}(y,x)$, $\forall x,y \in U'$

Based on the IF relation and crisp partitioning of the decision class, $U' \setminus D' = \{D'_1, D'_2, \dots, D'_k\}$, IF lower and upper approximation are defined as follows:

$$R_s \downarrow_{A'} D'_i(x) = \left\langle \inf_{y \in U'} \max(\nu_{R_{sA'}}(x,y), \mu_{D'_i}(y)), \sup_{y \in U'} \min(\mu_{R_{sA'}}(x,y), \nu_{D'_i}(y)) \right\rangle \tag{9.8}$$

$$R_s \uparrow_{A'} D'_i(x) = \left\langle \sup_{y \in U'} \min(\mu_{R_{sA'}}(x,y), \mu_{D'_i}(y)), \inf_{y \in D'} \max(\nu_{R_{sA'}}(x,y), \nu_{D'_i}(y)) \right\rangle \tag{9.9}$$

which reduces to the following equation [36]:

$$R_s \downarrow_{A'} D'_i(x) = \begin{cases} \langle \inf_{y \notin D'_i} \nu_{R_{sA'}}(x,y), \sup_{y \notin D'_i} \mu_{R_{sA'}}(x,y)\rangle, & x \in D'_i \\ \langle 0, 1\rangle, & x \notin D'_i \end{cases} \tag{9.10}$$

$$R_s \uparrow_{A'} D'_i(x) = \begin{cases} \langle \sup_{y \in D'_i} \mu_{R_{sA'}}(x,y), \inf_{y \in D'_i} \nu_{R_{sA'}}(x,y)\rangle, & x \in D'_i \\ \langle 0, 1\rangle, & x \notin D'_i \end{cases} \tag{9.11}$$

The degree of certainty with which an instance $x \in U'$ belongs to a decision class is given by a positive region as

$$\mathrm{Pos}_{A'}(x) = \left\langle \max_{D'_i \in U' \setminus D'} \mu_{R_s \downarrow_{A'} D'_i}(x), \min_{D'_i \in U' \setminus D'} \nu_{R_s \downarrow_{A'} D'_i}(x) \right\rangle \tag{9.12}$$

Using the intuitionistic fuzzy positive region, dependency function can be computed as follows:

$$\tau'_{A'} = \frac{\sum_{x \in U'} |\mathrm{Pos}_{A'}(x)|}{|U'|} \tag{9.13}$$

Dependency function is expressed as the ratio of positive region size and the size of overall samples in feature space.

A forward greedy quick reduct algorithm using above defined fuzzy and IF rough sets are employed for feature selection. At each step, one attribute is included in the potential reduct set and the degree of dependency of decision feature over a new obtained subset of conditional features is computed. If there is no increment in the degree of dependency, this process terminates and the algorithm produces the required reduct set.

The reduct algorithm can be given by:

Reduct Algorithm (C', D')
C', the set of all conditional attributes;
D', the set of decision attributes.
$R \leftarrow \{\}; \tau'_{\text{best}} = 0; \tau'_{\text{prev}} = 0$
do
$T \leftarrow R$
$\tau'_{\text{prev}} = \tau'_{\text{best}}$
for each $x \in (C' - R)$
if $(\tau'_{R\cup\{x\}})(D') > (\tau'_T)(D')$
$T \leftarrow R \cup \{x\}$
end-if
end-for
$\tau'_{\text{best}} = (\tau'_T)(D')$
$R \leftarrow T$
until $\tau'_{\text{best}} == \tau'_{\text{prev}}$
return R

9.2.4 Classification protocol

Our experiments are conducted independently based on nine different machine learning algorithms, which can be extensively applied for classification and prediction tasks. From the conducted experiments, it can be easily observed that Kernel logistic regression (KLR) [112, 113] is the better performing algorithm. A brief description of KLR is given below.

Kernel logistic regression (KLR), a nonlinear form of logistic regression, is applied to transfer the original input space into a high-dimensional feature space based on the concept of kernel functions. The main objective of KLR is to classify the data which cannot be differentiated in current dimensional space by generating a linear logistic regression model in a high-dimensional space. This can be performed by using a nonlinear form of logistic regression as follows:

$$\text{logit}\{p\} = w \times \phi(x) + k \tag{9.14}$$

where x represents a vector of input variables, $\phi(.)$ performs a nonlinear transformation to each input variable, w is a weight vector and k is a vector of constant. A logit function can be given by

$$\text{logit}\{p\} = \log_e\left(\frac{p}{p-1}\right) \tag{9.15}$$

Now, transformation of the above equation can be written as follows:

$$p = \frac{1}{1 + \exp(-w \times \phi(x) - k)} \tag{9.16}$$

In KLR, the nonlinear transformation is defined as the kernel function, which is used to complete the entire classification task.

9.2.5 Performance evaluation metrics

The performance of different machine learning algorithms is assessed and compared by using threshold-dependent and threshold-independent parameters. These parameters are computed from the values of confusion matrix namely: true positives (T_P), false negatives (F_N), true negatives (T_N) and false positives (F_P). (T_P) represents the number of correctly predicted fraud application, (F_N) specifies the number of incorrectly predicted fraud application, (T_N) denotes the number of correctly predicted nonfraud applications and (F_P) indicates the number of incorrectly predicted nonfraud applications.

Sensitivity: It is the percentage of correctly predicted fraud applications and is given by

$$\text{Sensitivity} = \frac{T_P}{(T_P + F_N)} \times 100 \tag{9.17}$$

Specificity: This represents the percentage of correctly predicted nonfraud applications and is calculated by

$$\text{Specificity} = \frac{T_N}{(T_N + F_P)} \times 100 \tag{9.18}$$

Accuracy: This parameter indicates percentage of correctly predicted fraud and nonfraud applications and is computed as follows:

$$\text{Accuracy} = \frac{(T_P + T_N)}{(T_P + F_P + T_N + F_N)} \times 100 \tag{9.19}$$

AUC: It is applied to depict the area under the receiver operating characteristic curve (ROC). The more close is its value to 1, the better is the predictor of fraud applications. It is examined as one of the evaluation parameters which are robust to the imbalance characters of the data sets.

MCC (Mathew's correlation coefficient): Mathew's correlation coefficient is evaluated by using the following equation:

$$\mathrm{MCC} = \frac{(T_P \times T_N - F_P \times F_N)}{\sqrt{(T_P + F_P)(T_P + F_N)(T_N + F_P)(T_N + F_N)}} \tag{9.20}$$

It is widely used performance evaluation metric to perform binary classifications. An MCC value of 1 is taken into account as the best for a prediction of fraud applications.

9.3 Result and discussion

In this chapter, we have conducted the experiments by using nine classifiers namely: KLR, PART, J48, Multilayer Perceptron (MLP), sequential minimization optimization (SMO) with puk kernel, Nearest Neighbor method (IBk), boosted random forest (RARF), rotation forest (ROF) and random forest (RF) to evaluate the performance of our proposed methodology. We apply percentage split of 80:20 validation technique to evaluate our proposed methodology, where 80 % of the instances from the Australian data set are randomly selected for building a prediction model and the remaining 20 % of the instances are used for model evaluation. First, we apply the fuzzy rough and intuitionistic fuzzy rough sets based feature selection technique to eliminate redundant and irrelevant features. Consequently, we obtain 13 features and 12 features out of 14 features by using intuitionistic fuzzy rough and fuzzy rough sets assisted methods, respectively. Then we apply SMOTE to obtain balanced data sets for both the reduced data sets. Thereafter, we compare the learning performances of the different classifiers using different evaluation metrics namely: sensitivity, specificity, accuracy, AUC and MCC for both original as well as reduced balanced data sets. From the experimental results, it is observed that performances of the different learning algorithms have improved for reduced balanced data sets when compared with the original data sets. KLR is found to be the best performing algorithm with sensitivity of 93.5 %, specificity of 89.5 %, accuracy of 91.5 %, AUC of 0.971 and MCC of 0.831 for reduced (by IFRFS) balanced data set. Performances of different learning algorithms have been recorded in Tables 9.1–9.3 for original and reduced (by FRFS and IFRFS) balanced data sets, respectively. Intuitionistic fuzzy rough feature selection technique has been implemented in Matlab 2018a on hardware platform with Intel(R) Core(TM) i3-5005U CPU @ 2.00 GHz with 4.00 GB RAM. Fuzzy rough feature selection, balancing by SMOTE and performance evaluation of different learning algorithms have been performed by using Waikato Environment for Knowledge Analysis (WEKA), which is a workbench for machine learning and implements the majority of data mining, data visualization, data preprocessing and filtering techniques. ROC is an interesting ap-

Table 9.1: Performance evaluation metrics for original Australian data set.

Learning Algorithms	Imbalanced Data set				
	Sensitivity	Specificity	Accuracy	AUC	MCC
KLR	91.2	84.3	87.7	0.926	0.756
PART	88.2	87.1	87.7	0.868	0.754
J48	88.2	85.7	87.0	0.884	0.754
MLP	83.8	90.0	87.0	0.916	0.740
SMO	92.6	80.0	86.2	0.863	0.731
IBK	79.4	82.9	81.2	0.811	0.623
ROF	91.2	85.7	88.4	0.940	0.770
RF	89.7	85.7	87.7	0.940	0.754
RARF	89.7	84.3	87.0	0.935	0.741

Table 9.2: Performance evaluation metrics for reduced Australian data set based on FRFS.

Learning Algorithms	Reduced Data set				
	Sensitivity	Specificity	Accuracy	AUC	MCC
KLR	93.5	88.2	90.8	0.963	0.818
PART	85.7	93.4	89.5	0.908	0.793
J48	81.8	96.1	88.9	0.947	0.786
MLP	76.6	90.8	83.7	0.942	0.681
SMO	97.4	85.5	91.5	0.915	0.836
IBK	85.7	81.6	83.7	0.836	0.674
ROF	89.6	90.8	90.2	0.970	0.804
RF	88.3	90.8	89.5	0.791	0.949
RARF	87.0	88.2	87.6	0.948	0.752

Table 9.3: Performance evaluation metrics for reduced Australian data set based on IFRFS.

Learning Algorithms	Reduced Data set				
	Sensitivity	Specificity	Accuracy	AUC	MCC
KLR	93.5	88.2	90.8	0.963	0.818
PART	85.7	93.4	89.5	0.908	0.793
J48	81.8	96.1	88.9	0.947	0.786
MLP	76.6	90.8	83.7	0.942	0.681
SMO	97.4	85.5	91.5	0.915	0.836
IBK	85.7	81.6	83.7	0.836	0.674
ROF	89.6	90.8	90.2	0.970	0.804
RF	88.3	90.8	89.5	0.791	0.949
RARF	87.0	88.2	87.6	0.948	0.752

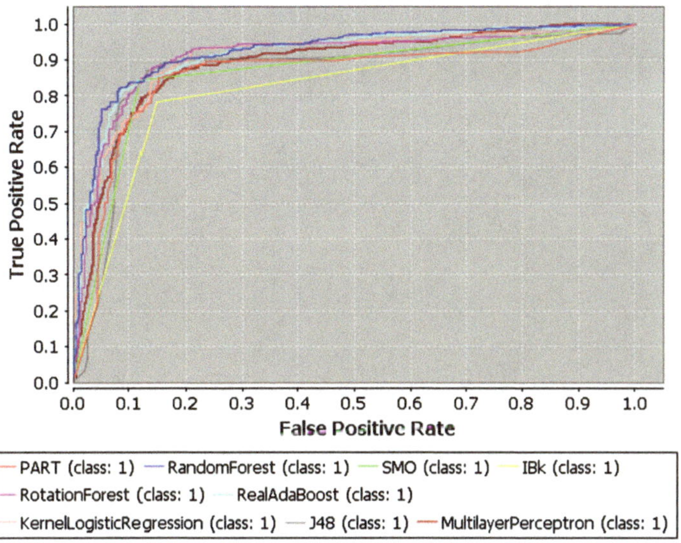

Figure 9.2: AUC for nine machine learning algorithms based on original data set.

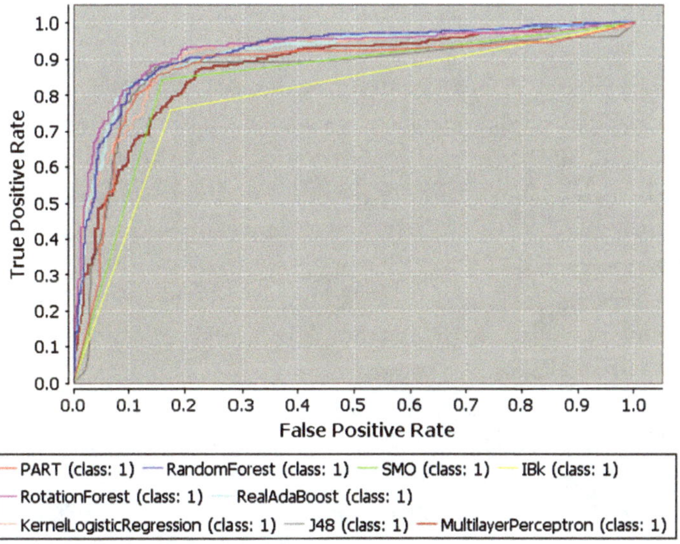

Figure 9.3: AUC for nine machine learning algorithms based on reduced data set by FRFS.

proach to visualize the performances of different classification algorithms. Figure 9.2 is a plot of ROC for original Australian data set, while ROC for reduced balanced data sets is presented in Figures 9.3–9.4. The entire visualization process is performed by using WEKA [111].

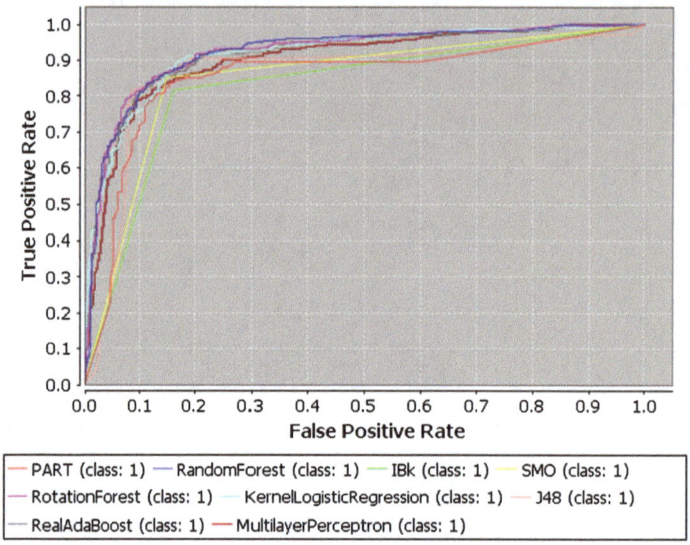

Figure 9.4: AUC for nine machine learning algorithms based on reduced data set by IFRFS.

9.4 Conclusion

In the last few years, a large volume of credit card transactions data has been created due to the fast circulation of credit cards and the advancement of e-services, including e-finance, e-commerce and mobile payments. Credit card fraud causes billion dollar losses due to massive use of credit cards and the various transaction schemes without strong verification and supervision inevitably. In the literature, different machine learning based approaches are successfully implemented for credit card fraud detection. In the recent years, various approaches are introduced to enhance the performances of machine learning algorithms to predict the credit fraud applications. There are various factors, which directly affect the real learning of the machine learning algorithms. The availability of irrelevant and redundant features in the data sets and class imbalance are the key factors among them. In the current study, we applied fuzzy and intuitionistic fuzzy rough set based techniques to remove the irrelevant and redundant features. Moreover, SMOTE was applied to produce the optimally balanced data set by suitably modifying the class distribution. Then the performances of various classifiers were explored on the optimally balanced reduced data set. From the experimental results, it can be observed that the performances of various machine learning algorithms to predict the credit fraud applications has improved after applying fuzzy and intuitionistic fuzzy rough feature selection techniques followed by SMOTE. The best result is produced by KLR with sensitivity of 93.5, specificity of 89.5, accuracy of 91.5, AUC of 0.971 and MCC of 0.831, which is better than the already reported results of the current date.

Bibliography

[1] Wang C, Han D. Credit card fraud forecasting model based on clustering analysis and integrated support vector machine. Clust Comput. 2018;22(S6):13861–6. 10.1007/s10586-018-2118-y.

[2] CyberSource. Online fraud report (Latin America edition). Tech rep. CyberSource Corporation, a Visa Company. 2016.

[3] Yee OS, Sagadevan S, Malim NHAH. Credit card fraud detection using machine learning as data mining technique. J Telecommun Electron Comput Eng. 2018;JTEC(1–4):23–7. 10.

[4] Ramentol E, Vluymans S, Verbiest N, Caballero Y, Bello R, Cornelis C et al.. IFROWANN: Imbalanced Fuzzy-Rough Ordered Weighted Average Nearest Neighbor classification. IEEE Trans Fuzzy Syst. 2015;23(5):1622–37. 10.1109/tfuzz.2014.2371472.

[5] Department of Justice. Office of Public Affairs. Bank of America to pay 16.65 Billion Dollar in Historic Justice Department Settlement for Financial Fraud Leading up to and During the Financial Crisis. 2014.

[6] Singh A, Jain A. Study of cyber attacks on cyberphysical system. SSRN Electron J. 2018. 10.2139/ssrn.3170288.

[7] Singh A, Jain A. Adaptive credit card fraud detection techniques based on feature selection method. In: Advances in intelligent systems and computing. Singapore: Springer; 2019. p. 167–78. 10.1007/978-981-13-6861-5_15.

[8] Randhawa K, Loo CK, Seera M, Lim CP, Nandi AK. Credit card fraud detection using AdaBoost and majority voting. IEEE Access. 2018;6:14277–84. 10.1109/access.2018.2806420.

[9] Somasundaram A, Reddy S. Parallel and incremental credit card fraud detection model to handle concept drift and data imbalance. Neural Comput Appl. 2018;31(S1):3–14. 10.1007/s00521-018-3633-8.

[10] de Sá AGC, Pereira ACM, Pappa GL. A customized classification algorithm for credit card fraud detection. Eng Appl Artif Intell. 2018;72:21–9. 10.1016/j.engappai.2018.03.011.

[11] Jurgovsky J, Granitzer M, Ziegler K, Calabretto S, Portier PE, He-Guelton L et al.. Sequence classification for credit-card fraud detection. Expert Syst Appl. 2018;100:234–45. 10.1016/j.eswa.2018.01.037.

[12] Liu Z, Pan S. Fuzzy-rough instance selection combined with effective classifiers in credit scoring. Neural Process Lett. 2017;47(1):193–202. 10.1007/s11063-017-9641-3.

[13] Ghobadi F, Rohani M. Cost sensitive modeling of credit card fraud using neural network strategy. In: 2016 2nd international conference of signal processing and intelligent systems (ICSPIS). IEEE; 2016. 10.1109/icspis.2016.7869880.

[14] Maes S, Tuyls K, Vanschoenwinkel B, Manderick B. Credit card fraud detection using Bayesian and neural networks. In: Proceedings of the 1st international NAISO congress on neuro fuzzy technologies. 2002. p. 261–70.

[15] Makki S, Assaghir Z, Taher Y, Haque R, Hacid MS, Zeineddine H. An experimental study with imbalanced classification approaches for credit card fraud detection. IEEE Access. 2019;7:93010–22. 10.1109/access.2019.2927266.

[16] Kumari P, Mishra SP. Analysis of credit card fraud detection using fusion classifiers. In: Advances in intelligent systems and computing. Singapore: Springer; 2018. p. 111–22. 10.1007/978-981-10-8055-5_11.

[17] Su P, Shen Q, Chen T, Shang C. Ordered weighted aggregation of fuzzy similarity relations and its application to detecting water treatment plant malfunction. Eng Appl Artif Intell. 2017;66:17–29. 10.1016/j.engappai.2017.08.009.

[18] Tran PH, Tran KP, Huong TT, Heuchenne C, HienTran P, Le MH. Real time data-driven approaches for credit card fraud detection. In: Proceedings of the 2018 international

conference on e-business and applications – ICEBA 2018. ACM Press; 2018. 10.1145/3194188.3194196.

[19] Langley P. Selection of relevant features in machine learning. 1994. 10.21236/ada292575.

[20] Pawlak Z. Rough sets. Int J Comput Inf Sci. 1982;11(5):341–56. 10.1007/bf01001956.

[21] Çoker D. Fuzzy rough sets are intuitionistic L-fuzzy sets. Fuzzy Sets Syst. 1998;96(3):381–3. 10.1016/s0165-0114.

[22] Cornelis C, Cock MD, Vaguely RAM. Quantified rough sets. In: Lecture notes in computer science. Berlin, Heidelberg: Springer; 2007. p. 87–94. 10.1007/978-3-540-72530-5_10.

[23] Intelligent decision support. Handbook of applications and advances of the rough sets theory. Fuzzy Sets Syst. 1993;57(3):396. 10.1016/0165-0114(93)90040-O.

[24] Amiri M, Jensen R. Missing data imputation using fuzzy-rough methods. Neurocomputing. 2016;205:152–64. 10.1016/j.neucom.2016.04.015.

[25] An S, Hu Q, Pedrycz W, Zhu P, Tsang ECC. Data-distribution-aware fuzzy rough set model and its application to robust classification. IEEE Trans Cybern. 2015;46(12):3073–85. 10.1109/tcyb.2015.2496425.

[26] Anaraki JR, Samet S, Eftekhari M, Ahn CW. A fuzzy-rough based binary shuffled frog leaping algorithm for feature selection. 2018. arXiv preprint. arXiv:1808.00068.

[27] Arunkumar C, Ramakrishnan S. Prediction of cancer using customised fuzzy rough machine learning approaches. Healthc Technol Lett. 2019;6(1):13–8. 10.1049/htl.2018.5055.

[28] Badria FA, Habib MMA, Shoaip N, Elmogy M. A framework for harmla alkaloid extraction process development using fuzzy-rough sets feature selection and J48 classification. Int J Adv Comput Res. 2017;7(33):213–22. 10.19101/ijacr.2017.733022.

[29] Bhatt RB, Gopal M. FRCT: fuzzy-rough classification trees. PAA Pattern Anal Appl. 2007;11(1):73–88. 10.1007/s10044-007-0080-z.

[30] Jensen R, Cornelis C. Fuzzy-rough nearest neighbour classification and prediction. Theor Comput Sci. 2011;412(42):5871–84. 10.1016/j.tcs.2011.05.040.

[31] Jensen R, Parthaláin NM. Towards scalable fuzzy–rough feature selection. Inf Sci. 2015;323:1–15. 10.1016/j.ins.2015.06.025.

[32] Jensen R, Shen Q. Fuzzy–rough attribute reduction with application to web categorization. Fuzzy Sets Syst. 2004;141(3):469–85. 10.1016/s0165-0114.

[33] Jensen R, Shen Q. Fuzzy-rough data reduction with ant colony optimization. Fuzzy Sets Syst. 2005;149(1):5–20. 10.1016/j.fss.2004.07.014.

[34] Jensen R, Fuzzy-Rough SQ. Sets assisted attribute selection. IEEE Trans Fuzzy Syst. 2007;15(1):73–89. 10.1109/tfuzz.2006.889761.

[35] Lu Y, Lei Y, Hua J. Attribute reduction based on intuitionistic fuzzy rough set. Control Decis. 2009;3:003.

[36] Tan A, Wu WZ, Qian Y, Liang J, Chen J, Li J. Intuitionistic fuzzy rough set-based granular structures and attribute subset selection. IEEE Trans Fuzzy Syst. 2019;27(3):527–39. 10.1109/tfuzz.2018.2862870.

[37] Tiwari AK, Shreevastava S, Shukla KK, Subbiah K. New approaches to intuitionistic fuzzy-rough attribute reduction. J Intell Fuzzy Syst. 2018;34(5):3385–94. 10.3233/jifs-169519.

[38] Tiwari AK, Shreevastava S, Som T, Shukla KK. Tolerance-based intuitionistic fuzzy-rough set approach for attribute reduction. Expert Syst Appl. 2018;101:205–12. 10.1016/j.eswa.2018.02.009.

[39] Chen D, He Q, Wang X. FRSVMs: fuzzy rough set based support vector machines. Fuzzy Sets Syst. 2010;161(4):596–607. 10.1016/j.fss.2009.04.007.

[40] Chen H, Yang H. One new algorithm for intuitiontistic fuzzy-rough attribute reduction. J Chin Comput Syst. 2011;32(3):506–10.

[41] Cheruku R, Edla DR, Kuppili V, Dharavath R. RST-BatMiner: a fuzzy rule miner integrating

rough set feature selection and Bat optimization for detection of diabetes disease. Appl Soft Comput. 2018;67:764–80. 10.1016/j.asoc.2017.06.032.
[42] Chinnaswamy A, Srinivasan R. Hybrid information gain based fuzzy roughset feature selection in cancer microarray data. In: 2017 innovations in power and advanced computing technologies (i-PACT). IEEE; 2017. 10.1109/ipact.2017.8244875.
[43] Dai J, Tian H. Fuzzy rough set model for set-valued data. Fuzzy Sets Syst. 2013;229:54–68. 10.1016/j.fss.2013.03.005.
[44] Dai J, Xu Q. Attribute selection based on information gain ratio in fuzzy rough set theory with application to tumor classification. Appl Soft Comput. 2013;13(1):211–21. 10.1016/j.asoc.2012.07.029.
[45] Dai J, Hu H, Wu WZ, Qian Y, Huang D. Maximal-discernibility-pair-based approach to attribute reduction in fuzzy rough sets. IEEE Trans Fuzzy Syst. 2018;26(4):2174–87. 10.1109/tfuzz.2017.2768044.
[46] He Q, Wu C, Chen D, Zhao S. Fuzzy rough set based attribute reduction for information systems with fuzzy decisions. Knowl-Based Syst. 2011;24(5):689–96. 10.1016/j.knosys.2011.02.009.
[47] Hu Q, Zhang L, An S, Zhang D, Yu D. On robust fuzzy rough set models. IEEE Trans Fuzzy Syst. 2012;20(4):636–51. 10.1109/tfuzz.2011.2181180.
[48] Zadeh LA. The concept of a linguistic variable and its application to approximate reasoning – I. Inf Sci. 1975;8(3):199–249. 10.1016/0020-0255.
[49] Zadeh LA, Klir GJ, Fuzzy Sets YB. Fuzzy logic, and fuzzy systems. World Scientific; 1996. 10.1142/2895.
[50] Shen Q, Jensen R. Selecting informative features with fuzzy-rough sets and its application for complex systems monitoring. Pattern Recognit. 2004;37(7):1351–63. 10.1016/j.patcog.2003.10.016.
[51] Bhatt RB, Gopal M. On fuzzy-rough sets approach to feature selection. Pattern Recognit Lett. 2005;26(7):965–75. 10.1016/j.patrec.2004.09.044.
[52] Hu Q, Yu D, Xie Z. Information-preserving hybrid data reduction based on fuzzy-rough techniques. Pattern Recognit Lett. 2006;27(5):414–23. 10.1016/j.patrec.2005.09.004.
[53] Jensen R, Shen Q. Tolerance-based and fuzzy-rough feature selection. In: 2007 IEEE international fuzzy systems conference. IEEE; 2007. 10.1109/fuzzy.2007.4295481.
[54] Jensen R, Cornelis C, Shen Q. Hybrid fuzzy-rough rule induction and feature selection. In: 2009 IEEE international conference on fuzzy systems. IEEE; 2009. 10.1109/fuzzy.2009.5277058.
[55] Jensen R, Shen Q. Computational intelligence and feature selection: rough and fuzzy approaches. vol. 8. John Wiley & Sons; 2008.
[56] Hu Q, An S, Yu D. Soft fuzzy rough sets for robust feature evaluation and selection. Inf Sci. 2010;180(22):4384–400. 10.1016/j.ins.2010.07.010.
[57] Chen D, Yang Y. Attribute reduction for heterogeneous data based on the combination of classical and fuzzy rough set models. IEEE Trans Fuzzy Syst. 2014;22(5):1325–34. 10.1109/tfuzz.2013.2291570.
[58] Parthaláin NM, Jensen R. Measures for unsupervised fuzzy-rough feature selection. Int J Hybrid Intell Syst. 2010;7(4):249–59. 10.3233/his-2010-0118.
[59] Parthalain NM, Jensen R. Simultaneous feature and instance selection using fuzzy-rough bireducts. In: 2013 IEEE international conference on fuzzy systems (FUZZ-IEEE). IEEE; 2013. 10.1109/fuzz-ieee.2013.6622500.
[60] Parthaláin NM, Jensen R. Unsupervised fuzzy-rough set-based dimensionality reduction. Inf Sci. 2013;229:106–21. 10.1016/j.ins.2012.12.001.
[61] Derrac J, Verbiest N, García S, Cornelis C, Herrera F. On the use of evolutionary feature

selection for improving fuzzy rough set based prototype selection. Soft Comput. 2012;17(2):223–38. 10.1007/s00500-012-0888-3.
[62] Cornelis C, Medina J, Verbiest N. Multi-adjoint fuzzy rough sets: definition, properties and attribute selection. Int J Approx Reason. 2014;55(1):412–26. 10.1016/j.ijar.2013.09.007.
[63] Inuiguchi M, Wu WZ, Cornelis C, Verbiest N. Fuzzy-rough hybridization. In: Springer handbook of computational intelligence. Berlin Heidelberg: Springer; 2015. p. 425–51. 10.1007/978-3-662-43505-2_26.
[64] Jensen R, Vluymans S, Parthaláin NM, Cornelis C, Saeys Y. Semi-supervised fuzzy-rough feature selection. In: Lecture notes in computer science. Springer; 2015. p. 185–95. 10.1007/978-3-319-25783-9_17.
[65] Onan A. A fuzzy-rough nearest neighbor classifier combined with consistency-based subset evaluation and instance selection for automated diagnosis of breast cancer. Expert Syst Appl. 2015;42(20):6844–52. 10.1016/j.eswa.2015.05.006.
[66] Parthalain NM, Jensen R. Fuzzy-rough feature selection using flock of starlings optimisation. In: 2015 IEEE international conference on fuzzy systems (FUZZ-IEEE). IEEE; 2015. 10.1109/fuzz-ieee.2015.7338023.
[67] Qian Y, Wang Q, Cheng H, Liang J, Dang C. Fuzzy-rough feature selection accelerator. Fuzzy Sets Syst. 2015;258:61–78. 10.1016/j.fss.2014.04.029.
[68] Vluymans S, DS T, Saeys Y, Cornelis C, Herrera F. Fuzzy rough classifiers for class imbalanced multi-instance data. Pattern Recognit. 2016;53:36–45. 10.1016/j.patcog.2015.12.002.
[69] Zeng A, Li T, Liu D, Zhang J, Chen H. A fuzzy rough set approach for incremental feature selection on hybrid information systems. Fuzzy Sets Syst. 2015;258:39–60. 10.1016/j.fss.2014.08.014.
[70] Arunkumar C, Ramakrishnan S. A hybrid approach to feature selection using correlation coefficient and fuzzy rough quick reduct algorithm applied to cancer microarray data. In: 2016 10th international conference on intelligent systems and control (ISCO). IEEE; 2016. 10.1109/isco.2016.7726921.
[71] Diao R, Chao F, Peng T, Snooke N, Feature SQ. Selection inspired classifier ensemble reduction. IEEE Trans Cybern. 2014;44(8):1259–68. 10.1109/tcyb.2013.2281820.
[72] Guo Q, Qu Y, Deng A, Yang L. A new fuzzy-rough feature selection algorithm for mammographic risk analysis. In: 2016 12th international conference on natural computation, fuzzy systems and knowledge discovery (ICNC-FSKD). IEEE; 2016. 10.1109/fskd.2016.7603303.
[73] Lasisi A, Ghazali R, Deris MM, Herawan T, Extracting LF. Information in agricultural data using fuzzy-rough sets hybridization and clonal selection theory inspired algorithms. Int J Pattern Recognit Artif Intell. 2016;30(09):1660008. 10.1142/s0218001416600089.
[74] Zhang X, Mei C, Chen D, Li J. Feature selection in mixed data: a method using a novel fuzzy rough set-based information entropy. Pattern Recognit. 2016;56:1–15. 10.1016/j.patcog.2016.02.013.
[75] Kumar CA, Sooraj MP, Ramakrishnan S. A comparative performance evaluation of supervised feature selection algorithms on microarray datasets. Proc Comput Sci. 2017;115:209–17. 10.1016/j.procs.2017.09.127.
[76] Qu Y, Rong Y, Deng A, Yang L. Associated multi-label fuzzy-rough feature selection. In: 2017 joint 17th world congress of international fuzzy systems association and 9th international conference on soft computing and intelligent systems (IFSA-SCIS). IEEE; 2017. 10.1109/ifsa-scis.2017.8023335.
[77] Wang C, Qi Y, Shao M, Hu Q, Chen D, Qian Y et al.. A fitting model for feature selection with fuzzy rough sets. IEEE Trans Fuzzy Syst. 2017;25(4):741–53. 10.1109/tfuzz.2016.2574918.
[78] Lin Y, Li Y, Wang C, Chen J. Attribute reduction for multi-label learning with fuzzy rough set.

Knowl-Based Syst. 2018;152:51–61. 10.1016/j.knosys.2018.04.004.
[79] Arunkumar C, Ramakrishnan S. Attribute selection using fuzzy roughset based customized similarity measure for lung cancer microarray gene expression data. Future Comput Inform J. 2018;3(1):131–42. 10.1016/j.fcij.2018.02.002.
[80] Dai J, Yan Y, Li Z, Liao B. Dominance-based fuzzy rough set approach for incomplete interval-valued data. J Intell Fuzzy Syst. 2018;34(1):423–36. 10.3233/jifs-17178.
[81] Han X, Qu Y, Deng A. A Laplace distribution-based fuzzy-rough feature selection algorithm. In: 2018 tenth international conference on advanced computational intelligence (ICACI). IEEE; 2018. 10.1109/icaci.2018.8377559.
[82] Hu Q, Zhang L, Zhou Y, Pedrycz W. Large-scale multimodality attribute reduction with multi-kernel fuzzy rough sets. IEEE Trans Fuzzy Syst. 2018;26(1):226–38. 10.1109/tfuzz.2017.2647966.
[83] Javidi MM. Mansoury S. Diagnosis of the disease using an ant colony gene selection method based on information gain ratio using fuzzy rough sets. J Part Sci Technol. 2017;3(4):175–86.
[84] Li Y, Lin Y, Liu J, Weng W, Shi Z, Wu S. Feature selection for multi-label learning based on kernelized fuzzy rough sets. Neurocomputing. 2018;318:271–86. 10.1016/j.neucom.2018.08.065.
[85] Sheeja TK, Kuriakose AS. A novel feature selection method using fuzzy rough sets. Comput Ind. 2018;97:111–6. 10.1016/j.compind.2018.01.014.
[86] Zhang X, Liu X, Yang Y. A fast feature selection algorithm by accelerating computation of fuzzy rough set-based information entropy. Entropy. 2018;20(10):788. 10.3390/e20100788.
[87] Zhang X, Liu X, Yang Y. A fast feature selection algorithm by accelerating computation of fuzzy rough set-based information entropy. Entropy. 2018;20(10):788. 10.3390/e20100788.
[88] Zhang X, Mei C, Chen D, Yang Y. A fuzzy rough set-based feature selection method using representative instances. Knowl-Based Syst. 2018;151:216–29. 10.1016/j.knosys.2018.03.031.
[89] Wang C, Huang Y, Shao M, Fan X. Fuzzy rough set-based attribute reduction using distance measures. Knowl-Based Syst. 2019;164:205–12. 10.1016/j.knosys.2018.10.038.
[90] Zhao R, Gu L, Zhu X. Combining fuzzy C-means clustering with fuzzy rough feature selection. Appl Sci. 2019;9(4):679. 10.3390/app9040679.
[91] Atanassov KT. Intuitionistic fuzzy sets. In: Intuitionistic fuzzy sets. Heidelberg: Physica-Verlag; 1999. p. 1–137. 10.1007/978-3-7908-1870-3_1.
[92] Atanassov KT. More on intuitionistic fuzzy sets. Fuzzy Sets Syst. 1989;33(1):37–45. 10.1016/0165-0114.
[93] Atanassov KT. Intuitionistic fuzzy sets. In: Intuitionistic fuzzy sets. Heidelberg: Physica-Verlag; 1999. p. 1–137. 10.1007/978-3-7908-1870-3_1.
[94] Jena S, Ghosh S, Tripathy B. Intuitionistic fuzzy rough sets. Notes IFS. 2002;8(1):1–18.
[95] Chakrabarty K, Gedeon T, Koczy L. Intuitionistic fuzzy rough set. In: Proceedings of 4th joint conference on information sciences (JCIS). Durham, NC. 1998. p. 211–4.
[96] Nanda S, Majumdar S. Fuzzy rough sets. Fuzzy Sets Syst. 1992;45(2):157–60. 10.1016/0165-0114.
[97] Cornelis C, Cock MD, Kerre EE. Intuitionistic fuzzy rough sets: at the crossroads of imperfect knowledge. Expert Syst. 2003;20(5):260–70. 10.1111/1468-0394.00250.
[98] Esmail H, Maryam J, Habibolla HL. Rough set theory for the intuitionistic fuzzy information systems. Int J Mod Math Sci. 2013;6(3):132–43.
[99] Samanta S, Mondal T. Intuitionistic fuzzy rough sets and rough intuitionistic fuzzy sets. J Fuzzy Math. 2001;9(3):561–82.
[100] Huang B, Zhuang Y-l, Li H-x, Wei D-k. A dominance intuitionistic fuzzy-rough set approach and its applications. Appl Math Model. 2013;37(12–13):7128–41. 10.1016/j.apm.2012.12.009.

[101] Shreevastava S, Tiwari AK, Intuitionistic ST. Fuzzy neighborhood rough set model for feature selection. Int J Fuzzy Syst Appl. 2018;7(2):75–84. 10.4018/ijfsa.2018040104.
[102] Singh S, Shreevastava S, Som T, Jain P. Intuitionistic fuzzy quantifier and its application in feature selection. Int J Fuzzy Syst. 2019;21(2):441–53. 10.1007/s40815-018-00603-9.
[103] Jain P, Tiwari AK, Som T. A fitting model based intuitionistic fuzzy rough feature selection. Eng Appl Artif Intell. 2020;89:103421. 10.1016/j.engappai.2019.103421.
[104] Blake CL, Merz CJ. UCI repository of machine learning databases. 1998.
[105] Gao M, Hong X, Chen S, Harris CJ. A combined SMOTE and PSO based RBF classifier for two-class imbalanced problems. Neurocomputing. 2011;74(17):3456–66. 10.1016/j.neucom.2011.06.010.
[106] Japkowicz N, Stephen S. The class imbalance problem: a systematic study. Intell Data Anal. 2002;6(5):429–49. 10.3233/ida-2002-6504.
[107] Han H, Wang WY, Mao BH. Borderline-SMOTE: a new over-sampling method in imbalanced data sets learning. In: Lecture notes in computer science. Berlin, Heidelberg: Springer; 2005. p. 878–87. 10.1007/11538059_91.
[108] Tiwari AK, Shreevastava S, Subbiah K, Som T. Enhanced prediction for piezophilic protein by incorporating reduced set of amino acids using fuzzy-rough feature selection technique followed by SMOTE. In: Mathematics and computing. Singapore: Springer; 2018. p. 185–96. 10.1007/978-981-13-2095-8_15.
[109] Chawla NV. Data mining for imbalanced datasets: an overview. In: Data mining and knowledge discovery handbook. Springer; 2009. p. 875–86. 10.1007/978-0-387-09823-4_45.
[110] Chawla NV, Bowyer KW, Hall LO, Kegelmeyer WP. SMOTE: synthetic minority over-sampling technique. J Artif Intell Res. 2002;16:321–57. 10.1613/jair.953.
[111] Hall M, Frank E, Holmes G, Pfahringer B, Reutemann P, Witten IH. The WEKA data mining software. ACM SIGKDD Explor Newsl. 2009;11(1):10. 10.1145/1656274.1656278.
[112] Roth V. Probabilistic discriminative kernel classifiers for multi-class problems. In: Lecture notes in computer science. Berlin, Heidelberg: Springer; 2001. p. 246–53. 10.1007/3-540-45404-7_33.
[113] Zhu J, Kernel HT. Logistic regression and the import vector machine. J Comput Graph Stat. 2005;14(1):185–205. 10.1198/106186005x25619.

Shruti Kaushik, Abhinav Choudhury, Nataraj Dasgupta,
Sayee Natarajan, Larry A. Pickett, and Varun Dutt

10 Evaluating single- and multi-headed neural architectures for time-series forecasting of healthcare expenditures

Abstract: Artificial neural networks (ANNs) are increasingly being used in the healthcare domain for time-series predictions. However, for multivariate time-series predictions in the healthcare domain, the use of multi-headed neural network architectures has been less explored in the literature. Multi-headed architectures work on the idea that each independent variable (input series) can be handled by a separate ANN model (head) and the output of each of the of these ANN models (heads) can be combined before a prediction is made about a dependent variable. In this paper, we present three multi-headed neural network architectures and compare them with the corresponding single-headed neural network architectures to predict patients' weekly average expenditures on certain pain medications. A multi-headed multilayer perceptron (MLP) model, a multi-headed long short-term memory (LSTM) model and a multi-headed convolutional neural network (CNN) model were calibrated along with their single-headed counterparts to predict patients' weekly average expenditures on medications. Results revealed that the multi-headed models outperformed the single-headed models and the multi-headed LSTM model outperformed the multi-headed MLP and CNN models across both pain medications. We highlight the utility of developing multi-headed neural architectures for prediction of patient-related expenditures in the healthcare domain.

Keywords: Time-series forecasting, multilayer perceptron (MLP), long short-term memory (LSTM), convolutional neural network (CNN), healthcare, multi-head neural networks

10.1 Introduction

The availability of electronic health records (EHRs) and advancement in data-driven machine learning (ML) architectures has led to several ML applications in the health-

Shruti Kaushik, Abhinav Choudhury, Varun Dutt, Applied Cognitive Science Laboratory, Indian Institute of Technology Mandi, 175005 Mandi, Himachal Pradesh, India, e-mails: shruti_kaushik@students.iitmandi.ac.in, abhinav_choudhury@students.iitmandi.ac.in, varun@iitmandi.ac.in
Nataraj Dasgupta, Sayee Natarajan, Larry A. Pickett, RxDataScience, Inc., Research Triangle Park (RTP) 27709, NC, USA, e-mails: nd@rxdatascience.com, sayee@rxdatascience.com, larry@rxdatascience.com

https://doi.org/10.1515/9783110671353-010

care domain [1, 4]. EHR data is often comprised of multivariate observations which are generally available over time [2]. ML offers a wide range of techniques to predict patients' expenditures and other healthcare outcomes over time using EHRs and different digital health records [4]. For example, literature has developed autoregressive integrated moving average (ARIMA), multilayer perceptron (MLP), long short-term memory (LSTM) and convolutional neural network (CNN) models to predict the patient related outcomes [3, 4, 5, 6, 7]. Researchers have also utilized traditional approaches like k-nearest neighbor (knn) and support vector machines frameworks for long-term time-series predictions in transportation domain [8]. However, the time-series data may be non-stationary (i. e., having seasonality or trend) or non-linear [10]. Thus, the non-stationary dynamics of the time-series may pose major challenges in predicting EHR data accurately [9]. Additionally, traditional ML algorithms (e. g., knn and linear regression) may ignore the temporal and sequential relationships in time-series datasets [11].

Literature has shown the advantages of using neural network architectures for performing time-series predictions considering their capabilities to handle the nonlinear relationships in time-series data [9]. There are several neural network architectures which can also handle the temporal sequence of clinical variables [3, 4]. For example, researchers have used MLP to predict morbidity of tuberculosis [10], LSTMs to predict patients' expenditures and patient-related diagnoses [6, 12] and CNN [7] for predicting patients' length of stay and hospital costs using time-series data. However, prior research has not developed or investigated multi-headed counterparts of the MLP, LSTM and CNN architectures for making predictions in healthcare data. In a multi-headed architecture, each independent variable (input series) could be handled by a separate neural network model (head) and the output of each of these models (heads) could be combined before a prediction is made about a dependent variable [13]. Thus, in a multi-headed architecture, a one-dimensional time-series is given to a separate machine-learning model (head), which may be able to learn features from the inputted time-series. This kind of architecture may be helpful in multivariate data where the predicted value at a time-step is a function of the inputs at prior time-steps across multiple features, and not just the feature being predicted.

Prior research has proposed some multi-headed neural network architectures in domains beyond healthcare [20]. For example, [20] has used multi-headed CNNs for waveform synthesis from spectrograms in speech data. These researchers demonstrated promising results from multi-headed CNNs for high quality speech syntheses. However, a comprehensive evaluation of different multi-headed architectures across MLP, LSTM and CNN networks against their single-headed counterparts has yet to be undertaken. Also, such an evaluation has yet to be undertaken for non-stationary and non-linear EHR data in the healthcare domain. This evaluation will be helpful to several stakeholders (patients, pharmacies and hospitals) and it will allow the research community to consider multi-headed architectures for predicting healthcare variables in future.

The primary objective of this research is to address the gaps in literature highlighted above. Specifically, in this research, we comprehensively evaluate single- and multi-headed architectures involving MLP, LSTM and CNN models in EHR data. For performing our evaluation, we predicted patients' average daily expenditures on two prescription-based pain medications.[1] Beyond the average daily expenditures, the EHR data consists of patients' demographic and other features that are inputted into separate heads (models) in the multi-headed architectures.

In what follows, we first provide a brief review of related literature involving single- and multi-headed architectures. In Section 10.3, we explain the methodology of applying different single- and multi-headed neural network architectures for multivariate time-series prediction of healthcare expenditures using two medicines' time-series datasets. In Section 10.4, we present the experimental results, where we compare results of different single- and multi-headed models on time-series data of two pain medicines. Finally, we discuss the conclusions from our research, its implication and the future scope.

10.2 Background

In recent years, neural architectures have gained lot of attention in almost every domain [6, 9, 14, 15, 16]. The neural networks can automatically learn the complex and arbitrary mappings from inputs to outputs [9]. In the healthcare domain, prior research has used single-headed LSTMs to find patterns in multivariate time-series data. Specifically, [12] performed multilabel classification given 128 diagnoses in a pediatric intensive care unit (PICU) dataset. These authors also compared single-headed LSTM models against single-headed MLP models and found the LSTMs to surpass the performance of the MLPs for classifying diagnoses related to PICU patients. Similarly, single-headed CNN architectures have also been used in the healthcare domain [17, 18, 19]. For example, medical imaging has greatly benefited from the advancement in classification using CNNs [17, 18, 19]. Several studies have demonstrated promising results in radiology [17], pathology [18] and in genomics where CNN was used to find relevant patterns on DNA sequences [19].

Recently, certain multi-headed neural network models have been proposed in literature [13, 20]. In these models, a head (a neural network model) is used for each independent variable and outputs of each head are combined to give the final prediction for the dependent variable [13, 20]. Prior researchers have used the multi-headed neu-

1 Pain medications were chosen as they cut across several patient-related ailments.

ral network architectures in the signal-processing [20] and natural language processing [13] domains. For example, [20] evaluated multi-headed CNN for waveform synthesis from spectrograms. Researchers demonstrated promising results from multi-headed CNNs for high quality speech syntheses. Similarly, [13] used multi-headed recurrent neural network models to predict the language of several documents with unknown authors by clustering documents.

To the best of author's knowledge, multi-headed neural network architectures of MLPs, LSTMs and CNNs have not yet been evaluated in the healthcare domain. Also, a comprehensive evaluation of these architectures across single-headed and multi-headed configurations has not been undertaken yet. In this paper, we attend to these gaps in literature by developing multi-headed MLP, LSTM and CNN models to perform time-series predictions for predicting patients' expenditures for two different pain medications. To evaluate the ability of multi-headed architectures, we also develop corresponding single-headed counterparts of these MLP, LSTM and CNN multi-headed architectures. Based upon the literature above, we expect the multi-headed models to perform better compared to the single-headed models because each head (model) will likely learn from individual features and future expenditure is some function of these individual features at prior time-steps.

10.3 Method

10.3.1 Data

In this paper, we selected two pain medications (named "A" and "B") from the Truven MarketScan dataset for our analyses [10].[2] These two pain medications were among the top-ten most prescribed pain medications in the US [21]. Data for both medications range between 2 January 2011 and 15 April 2015 (1565 days). For our analyses, across both pain medications, we used the dataset between 2 January 2011 and 30 July 2014 (1306 days) for model training and the dataset between 31 July 2014 and 15 April 2015 (259 days) for model testing. Every day, on average, about 1,428 patients refilled medicine A and about 550 patients refilled medicine B. For both medicines, we prepared a multivariate timeseries containing the daily average expenditures by patients on these medications, respectively. We used 20 attributes for performing multivariate time-series analyses. These attributes provide information regarding the number of patients of a particular gender (male, female), age group (0–17, 18–34, 35–44, 45–54 and 55–65), region (south, northeast, north central, west and unknown), health-plan (two type of health plans), and different diagnoses and procedure codes (six ICD-9 codes) who consumed medicine on a particular day. These 6 ICD-9 codes were selected

2 To maintain privacy, the actual names of the two pain medications have not been disclosed.

from the frequent pattern mining using Apriori algorithm [22]. The 21st attribute was the average expenditure per patient for a medicine on the tth day and was defined as per the following equation:

$$\text{Daily Average Expenditure}_t = i_t / j_t \tag{10.1}$$

where i_t was the total amount spent in a day t on the medicine across all patients and j_t was the total number of patients who refilled the medicine in day t. This daily average expenditure on a medicine along with the 20 other attributes were used to compute the weekly average expenditure, where the weekly average expenditure was used to evaluate model performance.

10.3.2 Evaluation metrics

All the models were fit to data at a weekly level using the Root Mean Square Error (RMSE; error) and R-square (R^2; trend) [23]. As weekly average expenditure predictions were of interest, the RMSE and R^2 scores and visualizations for weekly average expenditures were computed in weekly blocks of 7 days. Thus, the daily average expenditures per patient were summed across 7 days in a block for both training and test datasets. This resulted in the weekly average expenditure across 186 blocks of training data (1306 days were used in training) and 37 blocks of test data (259 days were used in testing). We calibrated all models to reduce error and capture trend in data. Thus, all models were calibrated using an objective function that was defined as the following: $(\text{RMSE}/10 + (1 - R^2))$.[3] This objective function ensured that the obtained parameters minimized the error (RMSE) and maximized the trend (R^2) on the weekly average expenditure per patient between model and actual data. The R^2 (between 0 and 1) accounts for whether the model's predictions follow the same trend as that present in the actual data; the larger the R^2 (closer to 1), the larger the ability of the model to predict the trend in actual data.

We performed the augmented Dickey–Fuller (ADF) test [24] to determine the stationarity of a time-series. As shown in Figure 10.1(A), the time-series for medicine A was stationary (ADF statistics = −10.10, $p < 0.05$). Figure 10.1(A) shows the weekly expenditure data for medicine A. In Figure 10.1, the first 186 blocks correspond to training data and the last 37 blocks correspond to the test data. The x-axis shows the weekly blocks, and the y-axis shows the weekly average expenditure (in USD per patient). As shown in Figure 10.1(B), medicine B was non-stationary (ADF statistics = −2.20, *ns*). Thus, while training models for medicine B, we first made the time-series stationary using first-order differencing (ADF statistics after one time differencing = −13.07,

3 RMSE was divided by 10 in order to bring both RMSE and $1 - R^2$ on the same scale. RMSE captures the error and R^2 captures the trend.

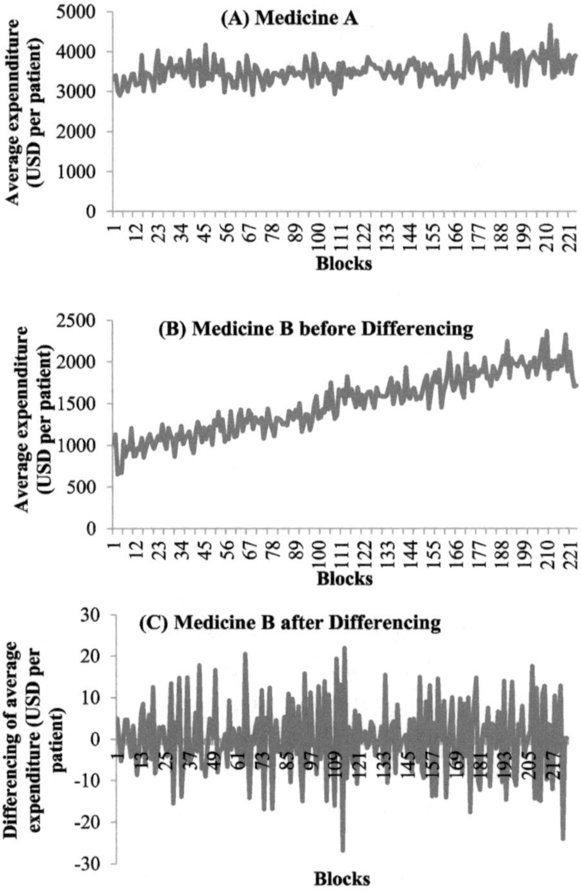

Figure 10.1: The weekly average expenditure (in USD per patient) for medicine A without differencing (A), for medicine B before differencing (B), and for medicine B after differencing (C).

$p < 0.05$) (see Figure 10.1(C)). We used stationary data across both medicines to train the models.

Figures 10.1(B) and 10.1(C) show the weekly expenditure data for medicine B before and after differencing, respectively. The predictions obtained from models for medicine B were first transformed to the non-stationary data before calculating the value of the objective function.

10.3.3 Experiment design for single-headed architectures

To implement the single-headed architectures across MLP, LSTM and CNN, we inserted all the features on prior time-steps (e. g., $t-2$ and $t-1$) together in one head as an input, to predict the dependent variable at time-step t. Figure 10.2 shows the

Figure 10.2: Single-headed architecture.

single-headed architecture used across all the three models in this paper where all the 21 features were inserted together in the model to predict the 21st feature. For both medicines, we used the following set of hyperparameters for performing one-step-ahead multivariate time-series forecasting in order to train all the three single-headed neural network architectures (i. e., MLP, LSTM and CNN): hidden layers (1, 2, 3 and 4), number of neurons in a layer (4, 8, 16, 32, 64 and 128), batch size (4 to 20), number of epochs (8, 16, 32, 64, 128, 256 and 512), lag/look-back period (2 to 8), activation function (tanh, relu and sigmoid), and dropout rate (20 % to 60 %).[4] All the models were trained to predict the daily average expenditures (21st feature). For training the CNN model, in order to apply the convolution operations, we also varied the filters and kernel size. Convolution is a mathematical operation which is performed on the input data with the use of a filter (a matrix) to produce a feature map [15]. We passed (32, 64 or 128) filters with different kernel size (1, 3, 5 and 7) to perform the convolution operation. The output of the convolution operations was then passed through different fully connected or dropout layers. These layers were decided by varying the above mentioned hyperparameters. We used grid search procedure for hyperparameter optimization of all the three models to perform time-series forecasting using single-headed neural architectures.

10.3.4 Experiment design for multi-headed architectures

Figures 10.3(A) and 10.3(B) shows the multi-headed MLP and LSTM architecture, respectively, which are used in this paper. In Figure 10.3, the first layer across all heads is the input layer where mini-batches of each feature in data are put into a separate head. As shown in Figure 10.3, for training the multi-headed MLP and LSTM on a medicine, each variable (20 independent variables and 1 dependent variable) for the medicine was put into a separate MLP/LSTM model (head) to produce a single combined concatenated output. The dense (output) layer contained 1 neuron which gave the expenditure prediction about the medicine for a time period. We used a grid search procedure to find optimum parameters of all three multi headed architectures. The hyperparameters used and their range of variation in the grid search were the following:

4 A 20 % dropout rate means that 20 % nodes will be dropped randomly from this layer.

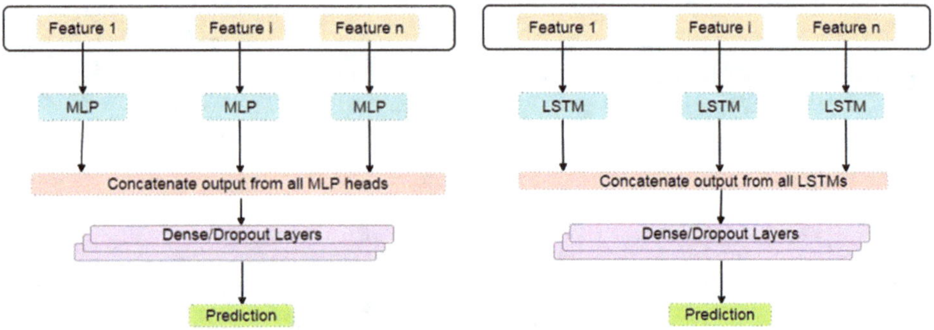

Figure 10.3: (A) Multi-headed MLP and (B) Multi-headed LSTM.

hidden layers (1, 2, 3 and 4), number of neurons in a layer (4, 8, 16, 32, 64 and 128), batch size (4 to 20), number of epochs (8, 16, 32, 64, 128, 256 and 512), lag/look-back period (2 to 8), activation function (tanh, relu, sigmoid), and dropout rate (20 % to 60 %).

The multi-headed CNN architecture was also trained exactly in a same manner as multi-headed MLP and LSTM. However, CNN model also includes convolution operations for which we passed (32, 64 or 128) filters with different kernel size (1, 3, 5 and 7). Figure 10.4 shows the example of a multi-headed CNN architecture in which the first layer across all heads is the input layer where mini-batches of each feature in data are put into a separate head. The input was then processed through convolution operation in Conv1D layer. The output of this Conv1D layer was passed from a maxpool layer. The output of maxpool layer was flattened, which will pass through different fully connected or dropout layers (these were decided by varying the hyperparameters as we did for MLP and LSTM architectures). At last, the output from each CNN head was then concatenated to predict the expenditure on a medicine on a day. The dense (output) layer at the end contained 1 neuron which gave the expenditure prediction about the medicine for a time period.

10.4 Results

10.4.1 Single-headed MLP model

Table 10.1 shows the RMSE and R^2 on training and test data on medicines A and B from all the single-headed architectures. As shown in Table 10.1, we obtained RMSE (= USD 380.38 per patient) on test data for medicine A using MLP and this model was trained with 2 lag period, 64 epochs, 5 batch size and tanh activation function. The architecture description is as follows: first hidden layer with 128 neurons, dropout layer with 20 % dropout rate, second dense layer with 128 neurons, second dropout

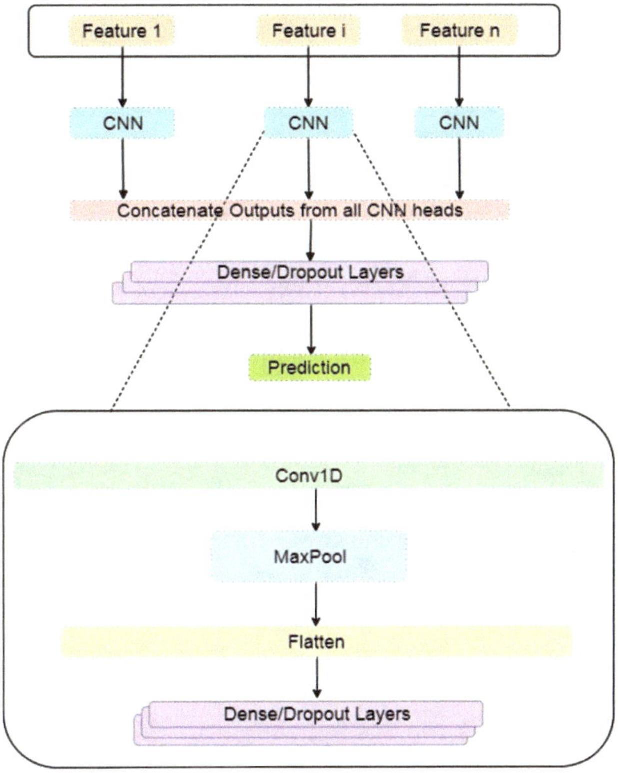

Figure 10.4: Multi-headed CNN.

Table 10.1: Single-headed model results during training and test.

Medicine Name	Model Name	Train RMSE	Train R^2	Test RMSE	Test R^2
A	MLP	180.31	0.20	380.38	0.01
	LSTM	181.99	0.61	338.04	0.02
	CNN	262.32	0.23	411.36	0.02
B	MLP	44.28	0.98	49.68	0.86
	LSTM	44.13	0.98	42.92	0.89
	CNN	66.65	0.91	89.95	0.79

layer with 20 % dropout rate and finally the dense (output) layer with 1 neuron. On medicine B, we obtained RMSE (= USD 49.68 per patient) on test data. The corresponding MLP architecture contained 2 fully connected hidden layers, 1 dropout layer, and an output layer at the end. The detailed description of architecture in sequence: first hidden layer with 8 neurons, dropout layer with 20 % dropout rate, second hidden layer with 8 neurons, batch normalization layer, and finally the output layer with

1 neuron. This architecture was trained with 2 look-back period on differenced series, 16 epochs, 8 batch size, relu activation function, and adam optimizer.

10.4.2 Single-headed LSTM model

As shown in Table 10.1, we obtained RMSE (= USD 338.04 per patient) on test data using LSTM model for medicine A and this model was trained with 2 lag period, 128 epochs, 8 batch size, relu activation function, and adam optimizer. The architecture contained first hidden layer with 8 neurons and then the output layer with 1 neuron. On medicine B, we obtained RMSE (= USD 42.92 per patient) on test data using LSTM model. The corresponding LSTM architecture contained 2 hidden layers, 1 dropout layer, and an output layer at the end. The detailed description of architecture in sequence: LSTM layer with 8 neurons, dropout layer with 20 % dropout rate, second LSTM layer with 8 neurons, the dense (output) layer with 1 neuron. This architecture was trained with 2 look-back period on differenced series, 5 epochs, 5 batch size, relu activation function, and adam optimizer.

10.4.3 Single-headed CNN model

As shown in Table 10.1, from CNN model, we obtained RMSE (= USD 411.36 per patient) on test data for medicine A and this architecture was trained with 2 lag period, 128 epochs, 5 batch size, tanh activation function, and adam optimizer. The model comprised of 1D convolution layer with 128 filters having 3 kernel size, followed by one dropout layer with 30 % dropout rate, a maxpool layer (pool size = 3), and flattened layer. The output of the flattened layer was passed to the dense layer with 16 neurons, and finally the output layer having 1 neuron. On medicine B, we obtained RMSE (= USD 89.95 per patient) on test data. The corresponding CNN model possessed 1D convolution layer having 32 filters with 3 kernel size, and relu activation, followed by a maxpool layer (pool size = 2), and flattened layer. The output of the flattened layer was followed by a fully connected layer with 64 neurons having tanh activation, a dropout layer with 50 % dropout rate, another dense layer with 64 neurons, and tanh activation, second dropout layer with 50 % dropout rate, finally the dense (output) layer at last having 1 neuron. This architecture was trained with 2 lag periods on differenced series, 64 epochs, 10 batch size, and adam optimizer.

As can be seen from Table 10.1, single-headed LSTM performed best for both medicines. Figure 10.5 shows the model fits for best performing single-headed LSTM model for medicine A (Figure 10.5(A)) and medicine B (Figure 10.5(B)) in test data, respectively.

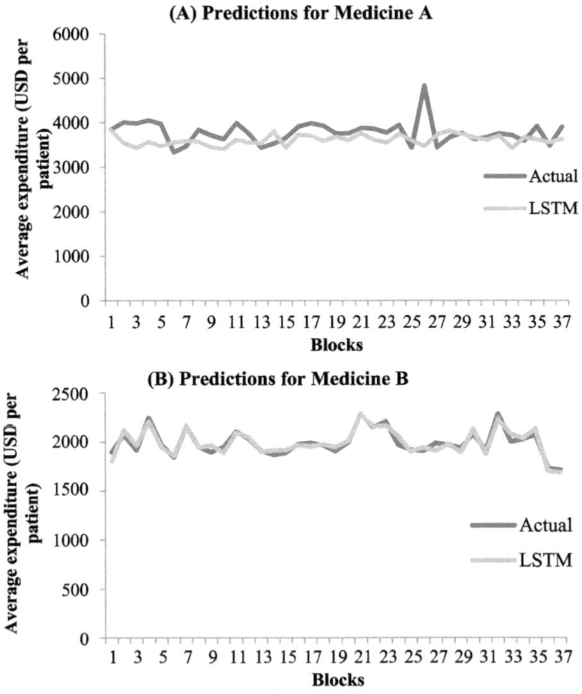

Figure 10.5: Average expenditure (in USD per patient) from the best single-headed model for medicine A (A), and for medicine B (B) in test data.

Table 10.2: Multi-headed model results during training and test.

Medicine Name	Model Name	Train RMSE	Train R^2	Test RMSE	Test R^2
A	MLP	207.16	0.54	325.85	0.03
	LSTM	237.81	0.45	318.82	0.05
	CNN	222.51	0.39	320.85	0.02
B	MLP	44.04	0.98	41.82	0.91
	LSTM	42.92	0.98	40.31	0.93
	CNN	58.51	0.96	66.71	0.76

10.4.4 Multi-headed MLP model

Table 10.2 shows the RMSE and R^2 on training and test data on medicines A and B from all the multi-headed architectures. As shown in Table 10.2, we obtained RMSE (= USD 325.85 per patient) on test data for medicine A using multi-headed MLP. The corresponding MLP architecture contained 2 fully connected layers in each head having 128 neurons, two dropout layers in between having 20 % dropout rate, and output layer with one neuron at the end. The output of each head was merged which was followed by one dense layer with 128 neurons, dropout layer with 20 % dropout, second dense layer with 128 neurons, second dropout layer with 20 % dropout, and finally the dense layer with one neuron. This architecture was trained with 2 lag values using actual time-series, 64 epochs, 5 batch size, adam optimizer, and tanh activation. On medicine B, we obtained RMSE (= USD 41.82 per patient) on test data. The corresponding MLP architecture contained 1 fully connected layer in each head. The output from each head (21 MLP models) were concatenated which was followed by 6 dense layers, 6 dropout layers, and finally the output layer at the end. The first dense layer contained 128 neurons, first dropout layer contained 60 % dropout, second dense layer contained 64 neurons, second dropout layer contained 60 % dropout, third dense layer contained 32 neurons, third dropout layer contained 60 % dropout, fourth dense layer contained 16 neurons, fourth dropout layer contained 60 % dropout, fifth dense layer contained 8 neurons, fifth dropout layer contained 60 % dropout, sixth dense layer contained 4 neurons, sixth dropout layer contained 60 % dropout, and finally the output layer with one neuron. This architecture was trained with 2 lag value using differenced series, 64 epochs, 15 batch size, adam optimizer, and relu activation.

10.4.5 Multi-headed LSTM model

As shown in Table 10.2, we obtained RMSE (= USD 318.82 per patient) on test data using multi-headed LSTM model for medicine A. The corresponding LSTM architecture was trained with 2 lag periods, 64 epochs, 20 batch size, and tanh as the activation function. All the heads contained 2 LSTM layers, 2 dropout layers, and one output layer. The architecture description is as follows: first LSTM layer with 64 neurons, dropout layer with 30 % dropout rate, second LSTM layer with 64 neurons, another dropout layer with 30 % dropout rate, and finally the output layer with 1 neuron. The output from each head was then merged using a dense layer with 64 neurons, a dropout layer with 30 % dropout rate, another dense layer with 64 neurons, dropout layer with 30 % dropout rate, and finally one output layer with 1 neuron (concatenated output). On medicine B, we obtained RMSE (= USD 40.31 per patient) on test data using LSTM model. The corresponding LSTM architecture was trained with 2 lag periods using differenced series, 64 epochs, 15 batch size, adam optimizer, and relu activation function. First, we trained all the 21 heads and then the output from each head was concatenated using 5 dense layers, 5 dropout layers, and finally one output layer at the end. The 20 heads contained one LSTM layer with 64 neurons, one dropout layer with 30 % dropout rate, and a second LSTM layer with 64 neurons. The 21st head contained four LSTM layers with 64 neurons and 3 dropout layers having 50 % dropout rate between each LSTM layer. The output from each head was concatenated using first dense layer containing 128 neurons, followed by one dropout layer with 60 % dropout rate, second dense layer with 64 neurons, second dropout layer with 40 % dropout rate, third dense layer with 32 neurons, third dropout layer with 60 % dropout rate, fourth dense layer with 16 neurons, fourth dropout layer with 60 % dropout rate, fifth dense layer with 8 neurons, fifth dropout layer with 60 % dropout rate, and finally the output layer with 1 neuron.

10.4.6 Multi-headed CNN model

As shown in Table 10.2, from multi-headed CNN model, we obtained RMSE (= USD 320.85 per patient) on test data for medicine A. All the 21 heads of CNN were trained with one Conv1D layer containing 128 filters with 3 kernel size. The conv1D layer in each head was followed by one dropout layer having 30 % dropout rate, one maxpool layer with pool size 2, flatten layer, one dense layer with 16 neurons, and finally the output layer with one neuron. The output from each head was then merged to predict the 21st feature. The concatenated output was followed by a dense layer having 64 neurons, followed by a dropout layer with 20 % dropout rate, and finally the dense (output) layer containing 1 neuron. This architecture was trained with 2 lag periods, 64 epochs, 5 batch size, adam optimizer, and tanh activation function. On medicine B, we obtained RMSE (= USD 66.71 per patient) on test data. All the 21 CNN heads were trained with one Conv1D layer containing 32 filters with 3 kernel size. The conv1D layer in each head was followed by maxpool layer with pool size 2, flatten layer, one dense layer with 64 neurons, one dropout layer with 20 % dropout rate, and finally the output layer with one neuron. The output from each head was then merged to predict the 21st feature. The concatenated output was followed by a dense layer having 64 neurons, followed by a dropout layer with 20 % dropout rate, and finally the dense (output) layer containing 1 neuron. This architecture was trained with 2 lag period using differenced series, 8 epochs, 5 batch size, adam optimizer, and tanh activation function.

As can be seen from Table 10.2, multi-headed LSTM performed best for both medicines. Figure 10.6 shows the model fits for best performing multi-headed LSTM model for medicine A (Figure 10.6(A)) and medicine B (Figure 10.6(B)) in test data, respectively.

10.5 Discussion and conclusions

Time-series architectures have gained popularity among researchers across various disciplines [12, 13, 14, 15]. Researchers have utilized the single-headed neural network architectures to predict the future time-series [7]. However, the potential of multi-headed neural network architectures need to be utilized for multivariate time-series predictions. In the multi-headed architectures, each head takes one variable as input and the finally output from each head (model) is merged to provide a single output for the variable of interest. Therefore, the primary objective of this research was to

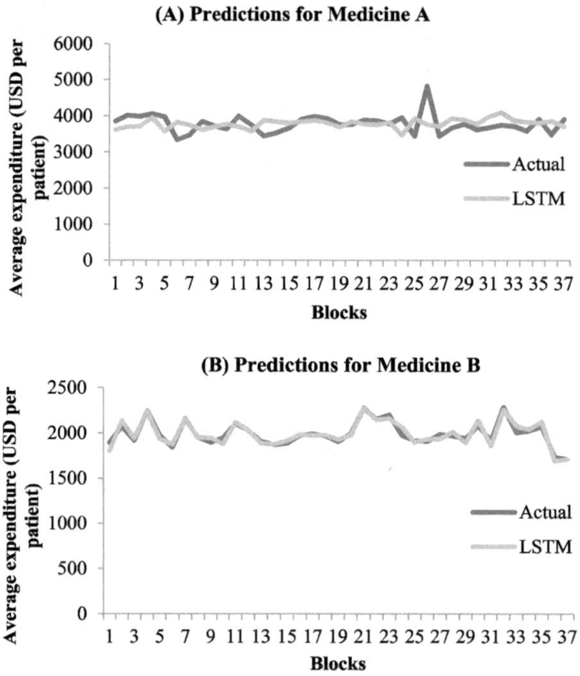

Figure 10.6: Average expenditure (in USD per patient) from the best multi-headed model for medicine A (A), and for medicine B (B) in test data.

evaluate the performance of multi-headed architectures of popular neural networks, i. e., MLP, LSTM and CNN to predict the weekly average expenditure by patients on two pain medications in an EHR dataset. The second objective was to compare the performance of multi-headed architectures with their single-headed counterparts.

First, as per our expectation, we found that the all the three multi-headed neural networks performed better than their single-headed counterparts. In the prior literature also, authors obtained promising results by using multi-headed CNN for speech synthesis [13]. The best value of test RMSE and test R^2 was obtained from the multi-headed architectures for both the medications. The likely reason behind this finding could be that all the single-headed architectures deal with the past time-steps of all the features simultaneously. This may be confusing for them to learn features' dependencies accurately. Whereas, in the multi-headed architectures, all the features are dealt separately, therefore, better feature representations are learnt.

Second, we found that the multi-headed LSTM performed better than other two architectures. A likely reason why multi-headed LSTM performed better could be because the convolution architectures are known for learning the spatial features representations in datasets (specially in image datasets where spatial characteristics are important) [17], whereas, in this paper, we dealt with only temporal features. Moreover, LSTMs are known for handling the temporal sequences in time-series datasets.

Also, in absence of recurrence relationships in MLP architecture, we obtained less accurate prediction accuracies from MLP than LSTMs for both medicines.

Third, we found that the results of medicine A were overfitted from all the models. In this paper, we tried to reduce overfitting using regularization technique, i. e., dropout [25]. However, adding dropout layers did not help much in case of medicine A. In the future, we plan to apply other regularization techniques such as l1 and l2 regularization [25]. These techniques add a regularization term to the cost function to penalize the model for having several parameters. The parameter reduction would lead to simpler models that likely reduce overfitting.

Overall, we believe that the multi-headed approaches could be helpful to caregivers, patients and pharmaceutical companies to predict per-patient expenditures where we can utilize the demographic details and other variables of patients in predicting their future expenditures. Predicting future expenditures is helpful for patients to manage their spending on healthcare and for pharmaceutical companies to optimize their manufacturing process in advance. In this paper, we performed the one time-step ahead forecasting. Prior literature has shown that it is difficult to perform long-term time-series predictions [8]. Therefore, in the future, we plan to perform long-term predictions using the proposed multi-headed neural network architectures. Also, we plan to evaluate other networked architectures (e. g., generative adversarial networks) and their ensembles for time-series forecasting of healthcare expenditure data.

Acknowledgement

The project was supported by grant (award: # IITM/CONS/RxDSI/VD/33 to Varun Dutt).

Bibliography

[1] Song H, Rajan D, Thiagarajan JJ, Spanias A. Attend and diagnose: clinical time series analysis using attention models. In: Thirty-second AAAI conference on artificial intelligence. 2018 Apr 29.

[2] Danielson E. Health research data for the real world: the MarketScan® Databases. Ann Arbor, MI: Truven Health Analytics. 2014 Jul 7.

[3] Pham T, Tran T, Phung D, Venkatesh S. Deepcare: a deep dynamic memory model for predictive medicine. In: Pacific-Asia conference on knowledge discovery and data mining, 2016 Apr 19. Cham: Springer; 2016. p. 30–41.

[4] Hunter J. Adopting AI is essential for a sustainable pharma industry. Drug Discov. World. 2016:69–71.

[5] Xing Y, Wang J, Zhao Z. Combination data mining methods with new medical data to predicting outcome of coronary heart disease. In: 2007 international conference on convergence information technology (ICCIT 2007), 2007 Nov 21. IEEE; 2007. p. 868–72.

[6] Kaushik S, Choudhury A, Dasgupta N, Natarajan S, Pickett LA, Dutt V. Using LSTMs for predicting patient's expenditure on medications. In: 2017 international conference on machine learning and data science (MLDS), 2017 Dec 14. IEEE; 2017. p. 120–7.
[7] Feng Y, Min X, Chen N, Chen H, Xie X, Wang H, Chen T. Patient outcome prediction via convolutional neural networks based on multi-granularity medical concept embedding. In: 2017 IEEE international conference on bioinformatics and biomedicine (BIBM), 2017 Nov 13. IEEE. p. 770–7.
[8] Huang Z, Shyu ML. Long-term time series prediction using k-NN based LS-SVM framework with multi-value integration. In: Recent trends in information reuse and integration. Vienna: Springer; 2012. p. 191–209.
[9] Gamboa JC. Deep learning for time-series analysis. arXiv preprint. arXiv:1701.01887 (2017 Jan 7).
[10] Eswaran C, Logeswaran R. An adaptive hybrid algorithm for time series prediction in healthcare. In: 2010 second international conference on computational intelligence, modelling and simulation. 2010, Sep 28. IEEE; 2010. p. 21–6.
[11] Choi E, Bahadori MT, Sun J, Kulas J, Schuetz A, Stewart W. Retain: an interpretable predictive model for healthcare using reverse time attention mechanism. In: Advances in neural information processing systems. 2016. p. 3504–12.
[12] Lipton ZC, Kale DC, Elkan C, Wetzel R. Learning to diagnose with LSTM recurrent neural networks. arXiv preprint. arXiv:1511.03677 (2015 Nov 11).
[13] Bagnall D. Authorship clustering using multi-headed recurrent neural networks. arXiv preprint. arXiv:1608.04485 (2016 Aug 16).
[14] Zhao Z, Chen W, Wu X, Chen PC, Liu J. LSTM network: a deep learning approach for short-term traffic forecast. IET Intell Transp Syst. 2017 Mar 9;11(2):68–75.
[15] Xingjian SH, Chen Z, Wang H, Yeung DY, Wong WK, Woo WC. Convolutional LSTM network: a machine learning approach for precipitation nowcasting. In: Advances in neural information processing systems. 2015. p. 802–10.
[16] Lin T, Guo T, Aberer K. Hybrid neural networks for learning the trend in time series. In: Proceedings of the twenty-sixth international joint conference on artificial intelligence. 2017. p. 2273–9.
[17] Cicero M, Bilbily A, Colak E, Dowdell T, Gray B, Perampaladas K, Barfett J. Training and validating a deep convolutional neural network for computer-aided detection and classification of abnormalities on frontal chest radiographs. Invest Radiol. 2017 May 1;52(5):281–7.
[18] Liu Y, Gadepalli K, Norouzi M, Dahl GE, Kohlberger T, Boyko A, Venugopalan S, Timofeev A, Nelson PQ, Corrado GS, Hipp JD. Detecting cancer metastases on gigapixel pathology images. arXiv preprint. arXiv:1703.02442 (2017 Mar 3).
[19] Alipanahi B, Delong A, Weirauch MT, Frey BJ. Predicting the sequence specificities of DNA-and RNA-binding proteins by deep learning. Nat Biotechnol. 2015 Aug;33(8):831.
[20] Arık SÖ, Jun H, Diamos G. Fast spectrogram inversion using multi-head convolutional neural networks. IEEE Signal Process Lett. 2018 Nov 9;26(1):94–8.
[21] Scott G. Top 10 painkillers in the US. MD Magazine [Internet]. 2014, October 6. Available from: https://www.mdmag.com/medical-news/top-10-painkillers-in-us.
[22] Kaushik S, Choudhury A, Dasgupta N, Natarajan S, Pickett LA, Dutt V. Evaluating frequent-set mining approaches in machine-learning problems with several attributes: a case study in healthcare. In: International conference on machine learning and data mining in pattern recognition, 2018 Jul 15. Cham: Springer; 2018. p. 244–58.
[23] Yilmaz I, Erik NY, Kaynar O. Different types of learning algorithms of artificial neural network (ANN) models for prediction of gross calorific value (GCV) of coals. Sci Res Essays. 2010 Aug 18;5(16):2242–9.

[24] Dickey DA, Fuller WA. Likelihood ratio statistics for autoregressive time series with a unit root. Econometrica. 1981 Jul;49(4):1057–72.
[25] Srivastava N, Hinton G, Krizhevsky A, Sutskever I, Salakhutdinov R. Dropout: a simple way to prevent neural networks from overfitting. J Mach Learn Res. 2014 Jan 1;15(1):1929–58.

Ankita Mazumder, Dwaipayan Sen, and Chiranjib Bhattacharjee

11 Optimization of oily wastewater treatment process using a neuro-fuzzy adaptive model

Abstract: In the recent era, accelerated expansion of industrialization and urbanization leads to the release of a large volume of effluent comprising of wide varieties of toxic pollutants. Amidst these components, oily hydrocarbon is one of the alarming toxic pollutants causing severe health hazards to ecological species. For the treatment of oily wastewater discharged from various sources, membrane technology is an established separation process to deliver clean water before disposal in accordance with government environmental regulations. However, one of the main limitations of membrane technology is its economic feasibility for the long-run of treating heavy pollution load because of membrane fouling, which gradually declines the permeate flux rate. Moreover, real difficulties lie in evaluating the foulants' interaction with the membrane material, and henceforth, parameter manipulation, which is highly dependent on the in-depth chemistry associated with different unforeseen real-life consequences. Therefore, conventional mathematical models are often considered unsuitable for membrane separation processes as it does not incorporate any randomness of the system, but only dealing with known mass, energy and momentum balance. But stochastic models can proficiently handle real, complex system with inherent uncertainties, where one fails to predict any relation between the outcome and the input. Among these stochastic models, pattern mapping models such as Artificial Neural Network (ANN) is one of the most significant approach, which mainly relies on the experimental data and their accuracy level rather than developing any correlation in between the process parameters and outcome. However, one of the main intricacies involved with ANN is the unexpected occurrences of errors that predict unreliable output of the process. To compensate this limitation, research efforts has been directed toward the development of neuro-fuzzy hybrid models such as Adaptive neuro-fuzzy inference system (ANFIS) for modeling membrane separation process efficiently.

Keywords: Oily wastewater, fuzzy logic, pattern mapping, artificial neural network, adaptive neuro-fuzzy inference system, membrane separation, fouling

Ankita Mazumder, Chiranjib Bhattacharjee, Department of Chemical Engineering, Jadavpur University, 700032 Kolkata, India, e-mails: chemical.ankita91@gmail.com, c.bhatta@gmail.com
Dwaipayan Sen, Department of Chemical Engineering, Heritage Institute of Technology, 700107 Kolkata, India, e-mail: dwaipayanju@gmail.com

https://doi.org/10.1515/9783110671353-011

11.1 Introduction

Bulk volume of oily wastewater discharge from industrial sectors such as oil refinery, metal working plants, chemical plants, petrochemical plants, paint industry, food production facilities, etc. are the major problems nowadays that is adversely affecting the ecological balance of the environment. Oily effluent is primarily comprised of hydrocarbons of different chain length along with different hazardous components such as polyaromatic hydrocarbons (PAHs), polycyclic hydrocarbons (PCBs), etc., that are highly carcinogenic and mutagenic to human being. Therefore, proper treatment of oily effluent prior to its disposal to aquatic bodies is highly recommended to meet the safe permissible limit fixed by governmental authorities [1, 2, 3, 4]. There are several established conventional treatment methods such as gravity separation, dissolved air flotation, adsorption and biological treatment for treating oily wastewater. However, in addition to these technologies membrane technology nowadays has drawn a considerable attention in treating such oily wastewater coming out of different sources [5]. One of the primary intricacies in membrane technology is the sustenance of the process for a long run with heavy pollutants' load. Membrane fouling is the prime reason that limits the economic usage of the membrane technology for a long run [6, 7]. Manipulating the process parameters through predictive model formulation will help to restrict the buildup of foulants over the membrane, and thus aiding the membrane separation process with better performance. However, one of the real complexities in understanding the foulants' interaction with the membrane material, and henceforth, parameter manipulation, is the in-depth chemistry involved along with different unforeseen consequences. Hence, modeling of a membrane process becomes very crucial to clearly understand, analyze, simulate and even predict the behavior of that membrane system. During the conventional mathematical model, the response is precisely determined through a known relationship based on mass, momentum and energy balance. However, such approach does not include any randomness of the system, a characteristics in case with the stochastic model. Stochastic models generate an ensemble time average of varied outcomes as a function of time. Here, an uncertainty is included as the probability of possible alternate outcomes to the varying changes within the parameters. In the stochastic approach, the estimation is based on the historical data and the probability of occurrence of an event. Therefore, stochastic models are often considered more suitable for handling a real, complex system with inherent uncertainties, where one fails to predict any relation between the outcome and the input. In this context, the most suitable approach is the pattern mapping modeling such as Artificial Neural Network (ANN), which mainly relies on the experimental data and their accuracy level instead of drawing any correlation between the process parameters and outcome [8, 9]. However, one of the main limitations with ANN is that any inaccuracy in the experiment predicts erratic outcomes of the process. Such unexpected occurrences of errors associated with ANN modeling can be mitigated after appending the qualitative statement to the neural network to extrapolate the predictions. Such extension of network converts network architecture from binary to fuzzy system that helps to identify the partial impact of the process parameters on the process outcome. In this chapter, neuro-fuzzy model formulation for the membrane separation process will be subsequently elaborated after a brief introduction on the fuzzy logic and neural network.

11.2 Classical and fuzzy logic

Classical logic is a qualitative approach based on binary system (0 or 1) where 0 is considered equivalent to "completely false value" and 1 to "completely true value." Though classical strategy is quite significant in dealing the problems from computer science and mathematics, it is often not suitable for those applications, where decision making is mandatory as in the human mind. This qualitative classical logic saying "true or false" determination often lags behind to handle the complexities of our real life problems because of the inherent element of uncertainty that requires extrapolation. For example, if we have to define a set of items having value less than 15, a classical statement will be given as shown below:

$$S = \{x \mid x < 15\}, \tag{11.1}$$

where S represents a set which are less than 15. Unlike previous instance, where one deals with the discrete values, if one has to indicate a qualitative statement on the height of a person, classical set logic may not be an appropriate tool. Here for identification, the logic should be based on the relative mind processing activity such as if the height of the person is greater than the average height, he or she will be considered as tall. According to classical logic, if the set of tall persons includes those having the height more than 180 cm, then any person having height greater than 180 cm will be considered as tall. However, with the simple classic statement, any person of height 178 cm will not be included in this classical set of tall persons. On the contrary, if one has to specify the grade of tallness, it is difficult to extract this concept from this crisp set. For example, if suppose for two different persons those who are having heights of 185 cm and 215 cm, whether both will be considered tall persons or we need to specify the gradation in tallness also? A person with 185 cm is tall, but a person with 215 cm height is tallest. Classical set logic thus only depends on the fundamentals of either exclusion or inclusion.

Fuzzy logic approach has evolved as an effective and potential tool for addressing such noncrisp sets. The fuzzy logic operates as a connecting bridge in between the domain of qualitative and quantitative modeling. Though the fuzzy logic was first introduced by Professor L. A. Zadeh in 1965 [10], it gained recognition in 1974 after 10 years when Dr. E. H. Mamdani practically applied this logic in a real world operation to control an automatic steam engine in 1974 [11]. Numerous implementations of fuzzy logic in various fields such as industrial manufacturing, education system, libraries, hospitals, banks, automobile production, automatic control system, etc. have been observed since 1980 [12]. Fuzzy logic sets does not restrict within definite boundaries, but deals with any values in between 0 and 1 in the real interval {0, 1} for a decision-making system. Fuzzy logic being an extended version of Boolean logic introduces the concept of partial truth, and in turn, substitutes the Boolean truth-values concept of classical logic with some degree of truth. This degree of truth helps to reflect the imprecise modes of reasoning that plays a significant role in decision making capability for a process. Henceforth, this logic gained significant prominence in certain domain such as human thinking, especially in the field of pattern recognition. It defines uncertain characteristics of things without dealing with randomness. The fuzzy logic includes some fundamental elements such as fuzzy sets, linguistic variables and fuzzy rules. While in mathematical operations, variables are generally represented by numerical values, the uses of nonnumeric linguistic variables such as adjectives (small, large, little, medium, high, etc.) are quite common in fuzzy logic applications to facilitate the expression of rules and facts [13].

11.3 Basics of fuzzy set

Unlike the classical set, fuzzy set theory deals with a set, where the boundaries are not binomially specified. Henceforth, it appends the flexibility of utilizing nonbinomial variables with the crisp set. An element belonging to a fuzzy set can be expressed by the degree of truth or degree of membership through a membership function. For a given fuzzy set, the characteristic function enables the elements of the set to have different degrees of membership of values, i. e., a fractional value in between 0 and 1. Let us explain the difference between classical (crisp) set theory and fuzzy (noncrisp) set theory with an appropriate example highlighting the concept of "degree of membership" in fuzzy logic. For instance, if somebody asked, "Whether a person called Jack is honest"? According to classical or traditional binary logic concept, the answer will be either "HONEST" ($\approx$1) or "DISHONEST" ($\approx$0) (as shown in Figure 11.1(a)). Now in this case, if fuzzy set theory is applied, multiple answers depending on the degrees of membership are possible. In the present example, for the degree of honesty can be explained by the fuzzy set, as no person can be either fully honest or fully dishonest. Hence, different level of honesty is possible based on various degrees of membership in the range of 0 to 1. The degree of membership may be classified as "EXCESSIVE HONEST," "VERY HONEST," "MODERATE HONEST," "DISHONEST" and "EXCESSIVE DISHONEST" are considered as 1, 0.8, 0.6, 0.4, 0.2 and 0, respectively. Henceforth, by implementing fuzzy set theory it is possible to ascertain the level of honesty of Jack,

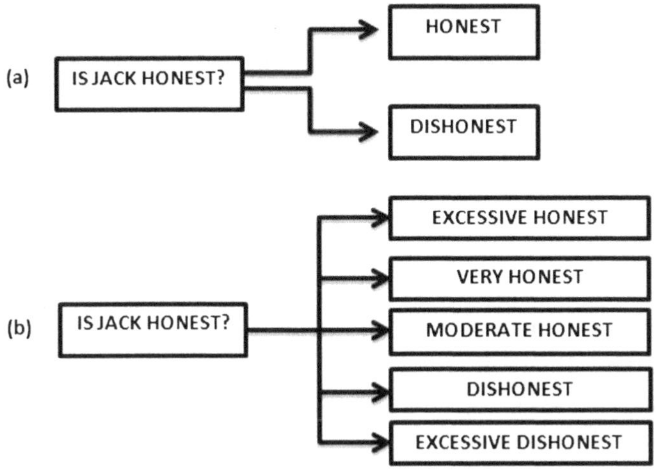

Figure 11.1: Boolean logic based on true or false concept (a); Level of honesty based on the degrees of truth or membership (1, 0.8, 0.6, 0.4, 0.2 or 0) concept of fuzzy logic (b).

whereas classical set theory enables us to only answer whether Jack is honest or dishonest.

In the fuzzy set, the membership function is represented as $\mu_X \in (0, 1)$, which signifies the grades of membership of the element "x" within set "X" will be any value in between 0 and 1. Some of the properties of membership function have been discussed below [14]:

(i) $\mu_{U-X} = 1 - \mu_X, x \in U$
(ii) $\mu_{X\cup Y} = \max\{\mu_X, \mu_Y\}$ or $\mu_X \vee \mu_Y, x \in U$
(iii) $\mu_{X\cap Y} = \min\{\mu_X, \mu_Y\}$ or $\mu_X \wedge \mu_Y, x \in U$,

where "U" is universe.

Let us explain the above mentioned three properties of membership function by considering an appropriate example. Suppose the grade system of our educational system qualitatively categorize the students on the basis that those who obtain marks above 90 are exceptional, in the range 80 to 89 students are outstanding, in the range 70 to 79 students are very good, in the range 60 to 69 students are good, in the range 50 to 59 students are average, in the range 40 to 49 students are marginal and below 40 students are failed (refer Table 11.1). So in the present case, seven degrees namely "*exceptional*," "*outstanding*," "*very good*," "*good*," "*average*," "*marginal*" and "*failed*" are selected to assess the academic quality of the students from broad scale of merits.

According to the property (ii), the percentage of students obtained marks less than 60 will be determined as max $\{0.07, 0.1, 0.03, 0.15, 0.08, 0.02\} = 0.15$. Therefore, the percentage of students obtained marks above 60, will be $(1 - 0.15) = 0.85$ (refer property (i)). Moreover percentage of students who obtained marks *at least* in the range

Table 11.1: Marks distribution of students.

Marks	Above 90	80–89	70–79	60–69	50–59	40–49	Below 40
Membership	exceptional	outstanding	very good	good	average	marginal	failed
Boys (%)	5	30	25	20	7	10	3
Girls (%)	10	20	15	30	15	8	2

of 60–69 will be $\min\{0.2, 0.3\} = 0.2$ (refer property (iii)) and percentage of students obtained marks in the range of 60–69 = $\max\{0.2, 0.3\} = 0.3$ (refer property (ii)).

Now in this current example, the students are assumed as universal set (*universe of discourse*) where boys and girls will be the subset of universal one. Let us elaborate the idea to a more extent, when we are saying about set (U) of students of "Don Bosco School" (x_1) and students of "Birla High School" (x_2). Let us consider G is the subset of universal set U, which denotes the girls among the students in both of the schools. So in case if both of the schools are specific to a particular gender, then either the girls belong to the schools or not, representing the set as crisp set possessing either "1" or "0" value. Now if both the schools are coeducational one, then say the girls' percentage of Don Bosco School and Birla High School are 50 % and 40 % of the student population, respectively. Then the sets can be represented in the given way:

$$U = \{x_1, x_2\}; \quad G = \{0.5|x_1, 0.4|x_2\} \quad \text{or} \quad G = \{x_1|0.5, x_2|0.4\},$$

where G signifies the membership function of set G. $G(x_1)$ and $G(x_2)$ are the grade of membership of x_1 and x_2 in the subset G.

If at least one grade of membership of an element belonging to the universal set is equal to 1, then this subset is termed as *normal* to the universal set. Otherwise, the set is a *subnormal* fuzzy set. In the above mentioned example, G is subnormal as any of either $G(x_1)$ or $G(x_2)$ is not equal to 1. However, say if Don Bosco School is only admitting girls, then $G(x_1)$ will be 1 and G will be then a normal fuzzy set.

The *height* of fuzzy set is defined as the maximum grade of membership. Here, it is calculated as $\text{height}(G) = 0.5$. The crisp or classical set of elements for which the value of membership grade will be nonzero is called *Support* of subset. For this example, *Support* of G, denoted as $\text{Supp}(G)$, is $\text{Supp}(G) = \{x_1, x_2\}$. Now if Birla High School will be a boys' school, then B can be expressed as either $G = \{0.5|x_1, 0|x_2\}$ or $G = \{x_1|0.5, x_2|0\}$. Then $\text{Supp}(G) = \{x_1\}$. Now to be more precise, let us elaborate the previous example by including another parameter called "the quality of the girls among the total population of students." These are assessed by another fuzzy set S, which is expressed as 15 % of the girls are good in sports from Don Bosco School and 7 % girls from Birla High School. Henceforth, we can write 10 % of $0.5x_1$ and 5 % of $0.4x_2$ are good in sports. Therefore, $S = \{0.05|x_1, 0.02|x_2\}$.

As $G \geq S$, then S is a subset of G and is expressed as $S \subset G$.

Theorem. *If M and N are two fuzzy subsets of X and if* $A = M \cup N$ *and* $B = M \cap N$*, then*
a) $B \subset A$
b) $M \subset A$ *and* $N \subset A$
c) $B \subset M$ *and* $B \subset N$

If M is a fuzzy subset of X, then negation or complement of M, denoted by $\overline{M}$ is expressed as the fuzzy subset $\overline{M} = X - M$. Hence for each $x \in M$, $M = 1 - M$. In the previous example, $\overline{G} = \{0.5|x_1, 0.6|x_2\}$ represents the boys in both of the schools.

Now if we consider M and N as two fuzzy subsets of two given sets X and Y, their cross product $(M \times N)$ is the relationship R on $X \times Y$, written as $R = M \times N$ and $R(x, y) = \min\{M, N(y)\}$.

Suppose from both Don Bosco and Birla High School, the girls and boys have scored a similar range of grades such as "O" grade, "A" grade and "B" grade. $Q = \{0.5|g_D, 0.5|b_D, 0.4|g_B, 0.6|b_B\}$ is the subset of students from Don Bosco and Birla High School.

Now, $QO = \{0.3|g_{DA}, 0.6|g_{DB}, 0.1|g_{DC}, 0.5|b_{DA}, 0.4|b_{DB}, 0.1|b_{DC}, 0.4|g_{BA}, 0.4|g_{BB}, 0.2|g_{BC}, 0.6|b_{BA}, 0.3|b_{BB}, 0.1|b_{BC}\}$ signifies the subset of the students' performance from both the mentioned schools. Subscript "B" and "D" stands for Birla High School and Don Bosco School, respectively. Therefore, the relationship set for the sets Q and QO will be

$$\begin{aligned} Q \times QO = \{&0.3|(g_D, g_{DA}), 0.5|(g_D, g_{DB}), 0.1|(g_D, g_{DC}), 0.5|(g_D, b_{DA}), 0.4|(g_D, b_{DB}),\\ &0.1|(g_D, b_{DC}), 0.4|(g_D, g_{BA}), 0.4|(g_D, g_{BB}), 0.2|(g_D, g_{BC}), 0.5|(g_D, b_{BA}),\\ &0.3|(g_D, b_{BB}), 0.1|(g_D, b_{BC}), 0.3|(b_D, g_{DA}), 0.5|(b_D, g_{DB}), 0.1|(b_D, g_{DC}),\\ &0.5|(b_D, b_{DA}), 0.4|(b_D, b_{DB}), 0.1|(b_D, b_{DC}), \ldots, 0.1|(b_B, b_{BC})\} \end{aligned}$$

11.4 Pattern mapping network

One of the basic utilities with the pattern mapping network is identification followed by memorizing information in order to understand the trends of a particular process for future retrieval. Process recalling pattern recognition is pattern mapping, where the card of similar colors is matched together. With a set of input, the corresponding outputs are mapped in order to derive an implicit mathematical correlation between them so that on introduction of a new set of inputs, the analogous output pattern can be recovered. Let us explain by one common example saying about a garments' stockroom, where distinct grades of jeans are aggregated in packets with stamped varying price. Say, the shopkeeper had received three grades – A, B and C, where the price of grade A, grade B and grade C are \$20, \$30 and \$40, respectively. After gradation based on prices, he has to keep the jeans in their respective warehouses. This way of storing the jeans of different grades in separate warehouses is a pattern mapping task. If this task could be done properly by the shopkeeper, a well-defined and outstanding trend will be defined, which will be later imitated by each worker of that shop. However, if he commits some mistake in the selection of proper warehouse, the subsequent arrangers will also follow the wrong trend as they were inaccurately trained about the pattern. Hence, one of the intricacies associated with the pattern mapping design is if improper recognition of pattern is carried out, then the whole system becomes chaotic and fails miserably. In case for separation of oil from oily wastewater using adsorption, once the experimental results are being accurate, it is expected that the pattern mapping algorithm provides a good relation between separation extent and the experimental parameters affecting the adsorption.

11.5 Artificial neural network (ANN)

The concept of ANN was first introduced in 1943 by Warren McCulloch, a neurophysiologist, and Walter Pitts, a young mathematician, when they had discussed the probable working principle of neurons and also modeled a simple neural network with the help of electrical circuits [15]. Following such a concept on signal transmission through neurotransmitter, a book entitled "*The Organization of Behavior*" was written by Donald Hebb in 1949 on the pattern mapping [16]. In 1956, the Dartmouth Summer Research Project on Artificial Intelligence organized by Professor John McCarthy had motivated research on artificial intelligence (AI), which is itself a milestone in the field of AI and a path creator toward initiating research on AI. Frank Rosenblatt, a neuro-biologist of Cornell Aeronautical Laboratory in Buffalo, introduced the concept of "Perceptron," which is still the oldest neural network in use [17]. Perceptron computes a weighted sum of the inputs, subtracts a threshold, and passes net result to the next layer of neurons. Subsequently, Professor Bernard Widrow from Stanford University and Dr. Marcian Hoff developed models, named ADAptive LINear Elements (ADALINE) and Multiple ADAptive LINear Elements (MADALINE), two different neural network algorithms in 1959 [18]. MADALINE was the first commercially applied neural network tool to remove echoes from telephone lines. After a long period of stagnation, in 1982, Professor John Hopfield presented a paper to the National Academy of Sciences saying that the neural network might also be capable of remembering history apart from mapping [19]. There he explained along with some supportive mathematical analysis that how neural networks could operate and what they could perform. In the period 1985 to 1989, several works were carried out on the artificial neural network and its importance in money flow and industrialization. Below, a detailed understanding will be elaborated on the topology along with the algorithm for ANN. However, in this present chapter we have restricted our scope to perceptron instead of understanding the network that can have the capacity of storing history.

11.5.1 Topology of artificial neural network

The topology of neural network was simulated with the working principle of human neural system, where the mapping of action–reaction is done through the brain. For example, *action* is "a man dips his fingers in a hot cup of tea." The person will immediately withdraw his fingers from the cup – a *reaction*. Now, one can ask – why the person had withdrawn his fingers. The answer must be "because the tea is hot" and the sense to withdraw hand once it gets contacted with the hot surface is learned by his brain or more specifically synaptic neural network. This is the way that the neural network actually operates. We need to make the network to be learned first, called *training*, and need the network to respond according to the training, called *validation*. The architecture of a neural network consists of three main parts namely input layer, hidden layer and output layer. Figure 11.2 displays a typical architecture for the neural network [20]. The input layer is a receptor nodes' layer receiving the inputs and subsequently transfer it to another layer called hidden layer using the weighted connectors. Then the signal gets processed there to generate an outcome that will be subsequently conveyed to output layer with the aid of connector again for response. One of the biggest limitations with the ANN is its response against known pattern, i. e., the network cannot extrapolate the response if the network was not trained earlier for the same. As the network fails to respond against the unknown signals, therefore, in the *validation* process one will get big deviation in the predicted and the actual values. Such deviation indicates the network's performance. In the final step of ANN, i. e., in "*testing*," some inputs will be introduced for which it is required to predict the responses, when the real values are unknown for those.

The layers are consisted of nodes, which receives the signal from the preceding layer (as shown in Figure 11.2). Each node in the input layer is dedicated to an individual process parameter, and hence total nodes in this layer are equal to total number of process parameters for which the response needs to be assessed. However, inclusion of nodes for hidden layer(s) is based on the overall complexity of the given system and the characteristics of the used neural network architecture. Again nodes of output layer are dependent on the number of responses from the process. The individual nodes for two subsequent layers are interconnected completely with a proper weightage assigned to these connections. The nodes of both hidden layer(s) and output layer possess biases, which ensure that no "*zero*" value will be processed from the nodes. Now, let us consider a process, where oily wastewater treatment is done using membrane separation technology. In a treatment plant, oily effluents are treated by separating oil or hydrocarbons from effluent using membrane separation process to get water with low organic load. Considering the hydrophobicity of hydrocarbons, a specific hydrophilic membrane of suitable molecular weight cut off (MWCO) is chosen. The operating parameters that are optimized for the process to purify water by separating hydrocarbons are transmembrane pressure (TMP) (ΔP), pH of the system

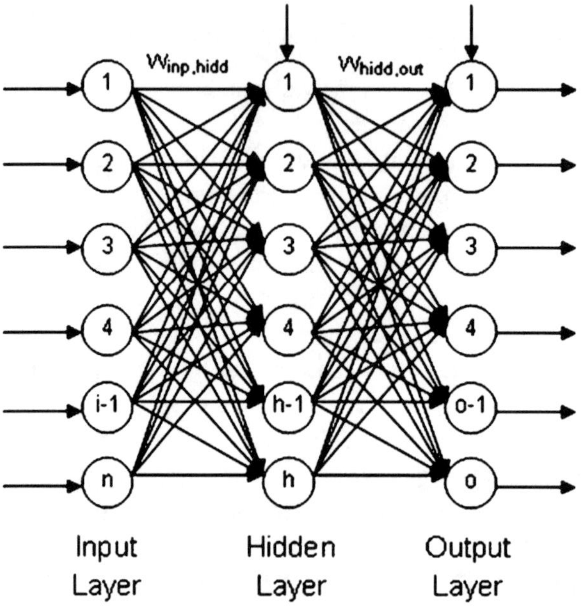

Figure 11.2: Artificial neural network architecture (Reprinted after the permission taken from Desalination, 273, 2011, 168–178 [20]).

and the temperature (T) as these three conditions have a significant effect on the separation. We want to assess the average flux (J_{avg}) from the membrane once the process will run for 4 h.

Condition 1: $\Delta P_i; i = 1, 2, 3, \ldots, N$
Condition 2: $\text{pH}_i; i = 1, 2, 3, \ldots, N$
Condition 3: $T_i; i = 1, 2, 3, \ldots, N$
Response 1: $J_{\text{avg},i}; i = 1, 2, 3, \ldots, N$

From total "N" number of observations, "n" numbers and "$N - n$" numbers are taken for the respective training and the validation of the network. Now, how much percentage of data will be considered for training and will be considered for validation cannot be explicitly elaborated. In general, the training dataset will be those sets, which is truly reflecting the trend of the process with every changing conditions. The general thumb rule is that the training set will be 60 % of the information, while 40 % will be for validation. In the below box, we have explicitly described the algorithm primarily used to train the perceptron network.

Training algorithm

(i) Note the observations on J_{avg} with the variation in ΔP, pH, and T of size N.
(ii) Isolate "n" number of observations from total "N" number of observations.

(iii) Assign initial values of weights for the input to hidden (W_{ih}), biases on hidden layers [20], weights between hidden and output (W_{ho}) and biases on output layers (b_o).
(iv) Assume $i = 3$, i. e., three number of nodes in input layer, which is equivalent to the number of experimental parameters (conditions) and $o = 1$, i. e., the number of nodes in output layer, which equals to response number.
(v) Assign appropriate transfer function for the hidden layer input $(= (b_h + \sum(W_{1h}\Delta P_i + W_{2h}\mathrm{pH}_i + W_{3h}T_i)))$.
(vi) Assign appropriate transfer function for the output layer input

$$\left(= b_o + \sum W_{1h} f\left(b_h + \sum(W_{1h}\Delta P_i + W_{2h}\mathrm{pH}_i + W_{3h}T_i)\right)\right).$$

(vii) Check the mean average error (MSE) between known observations and the response predicted by neural network:

$$\mathrm{MSE} = \frac{[J_{\mathrm{avg},i} - \{b_o + \sum W_{1h} f(b_h + \sum(W_{1h}\Delta P_i + W_{2h}\mathrm{pH}_i + W_{3h}T_i))\}]^2}{n}$$

(viii) If MSE is less than the tolerance, please call the validation model or else proceed to next step (ix).
(ix) Apply Levenberg–Marquardt algorithm to adjust weights and biases of the neural network. The adjustment will be done backwardly, i. e., at first the connections of output to hidden layer is adjusted followed by the adjustment of hidden to input layer connections. Go back to step (v) after adjustment.

Validation algorithm
(i) Assign trained values for weights and biases for all the connections and nodes, respectively.
(ii) Fit the input data for ΔP, pH, and T to evaluate the output with the assigned values for weights and biases.
(iii) Check MSE between known observations and network predicted response.
(iv) If MSE is less than the tolerance, the conclusion will be the training was done accurately otherwise retraining of the network is required with a new set of values indicating the trend.

It is evident from the algorithm analysis that the proper training of the network is one of the important steps to make the prediction correct. It is already discussed that ANN is a network allowable to interpolate the information. However, no qualitative statement is being used either in order to assess the response or to extrapolate the information. In order to do so, one needs the inclusion of fuzzy set with the pattern mapping tool. Henceforth, combined approach of both neural network and fuzzy logic will facilitate to examine any process more proficiently. In the next section, we will briefly discuss about the combined fuzzy and the neural network system, which is called as adaptive neuro-fuzzy inference system (ANFIS).

11.6 Overview of adaptive neuro-fuzzy inference system

Adaptive neuro-fuzzy inference system (ANFIS) is developed as a hybrid of two tools namely ANN and Fuzzy Inference System (FIS). It is a multilayer feed-forward network. As highlighted previously, though ANN is an efficient neural network based on generalization concept, but the intricacy arises because of its "black box" nature. A FIS executes a nonlinear mapping from its input to output domain, which offers the benefits of approximate reasoning, however, is not efficient in learning. Thus ANFIS is developed as an improved neuro-fuzzy system, which has the efficient computational capability of ANN combined with the outstanding reasoning ability of FIS. Unlike the black box system, ANFIS is an adaptive system and required modifications can be incorporated depending on user's preferences. Attributing to this property, ANFIS is the finest "function approximator" until today amidst various available neuro-fuzzy models [8, 21]. Among other models, the training speed and the learning algorithm of ANFIS is the most efficient for handling any nonlinear, complex real-world systems such as the membrane separation process [8]. Sargolzaei et al. [22] applied ANFIS to predict the performance of starch removal from the wastewater in a food processing industry using polyethersulfone membrane [1]. Better and accurate prediction on permeate flux and rejection was achieved by ANFIS in comparison to other techniques. In another study, Salahi et al. [23] used ANFIS model to predict the permeate flux in the treatment of oily wastewater using ultrafiltration process where TMP, filtration time and cross-flow velocity (CFV) were set as input parameters, while the average permeate flux as the output. The simulation results showed good compliance between the experimental model and data. Rahmanian et al. [24] utilized both ANN and ANFIS for the prediction of permeate and rejection of micellar-enhanced ultrafiltration process for removing lead ions from the water. According to their study, although regression coefficients with both the approaches showed good agreement in prediction, but the response of ANFIS was more rapid and accurate. In the subsequent section, a brief on the ANFIS architecture and working principle will be elaborated.

11.6.1 Architecture of ANFIS

ANFIS architecture is an adaptive neural network that involves supervised learning with a mathematical formulation of the qualitative statement either with Takagi–Sugeno approach or with the Mamdani approach. Takagi–Sugeno is more appropriate to mathematical quantification, while Mamdani is more limited to qualitative formulization. For example, inference on the numerical value out of the process will be dealt by Takagi–Sugeno, while opening/closing of an on-off valve may be dealt with the Mamdani model. In this chapter, we will limit our discussion based on the Takagi–Sugeno approach. Figure 11.3 displays the detailed architecture of ANFIS with the aid

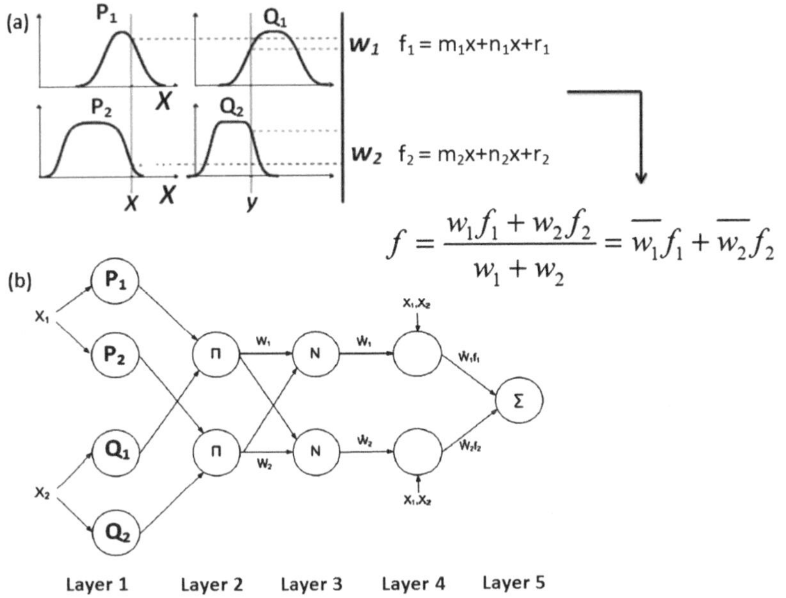

$$f = \frac{w_1 f_1 + w_2 f_2}{w_1 + w_2} = \overline{w_1} f_1 + \overline{w_2} f_2$$

Figure 11.3: A two input and one output Takagi–Sugeno fuzzy inference model with two rules (a); The detailed architecture of ANFIS showing five layers (b).

of fuzzy reasoning mechanism of the Takagi–Sugeno model [25]. Let us consider two inputs namely "x" and "y" from which one gets output "f." Each input possesses two fuzzy sets is related with "f" with a degree given by membership function. A set of two fuzzy rules were utilized in the logic "If-Then" for the Takagi–Sugeno model, as given below:

Rule 1: If x is P_1 AND y is Q_1 Then $f_1 = m_1x + n_1x + r_1$
Rule 2: If x is P_2 AND y is Q_2 Then $f_2 = m_2x + n_2x + r_2$

where P_1, P_2 and Q_1, Q_2 represents the membership functions of input x and y, respectively. f_1 and f_2 are the linear combinations of the inputs in the *consequent* part (m, n and r are called *consequent* parameters) of the Takagi–Sugeno fuzzy inference model. ANFIS architecture contains five layers, where first and fourth layers involve an adaptive node, while rest of the layers possesses a fixed node. Description of each layer is given briefly in the Table 11.2.

11.7 Application of ANFIS in modeling of membrane separation

One of the primary requisites for modeling any membrane separation process using ANFIS is the identification of the parameters on which the efficiency of the separa-

Table 11.2: Architecture of ANFIS model [26].

Layer No.	Description	Equations
1	**Fuzzification layer:** Each node of this layer indicates the input parameters, which is related to the linguistic phrase through a degree of membership called *membership function*. Output resulted from every node has a membership grade according to linguistic label set for the inputs. The membership function can be of different types such as Gaussian membership function, a generalized bell membership function, or other type of membership function	$\mu_{Ai}(x) = \exp[-(\frac{x-c}{2b})^2]$ $O_i^1 = \mu_{Ai}(x), i = 1, 2$ $O_i^1 = \mu_{Bi-2}(y), i = 3, 4$ where μ_{Ai} and μ_{Bi-2} are the degree of membership functions for the fuzzy sets A_i and B_i, respectively. $\{a_i, b_i, c_i\}$ are the parameters of a membership function that controls the shape of the membership function, also known as the *premise parameters*
2	**Rule layer:** This layer generates a combined effect of the decisions from the previous layer, which produces output based on the rule set	$O_i^2 = w_i = \mu_{Ai}(x) * \mu_{Bi}(y), i = 1, 2$ where w_i denotes the output that signifies the firing strength of each rule
3	**Normalization layer:** This layer helps to normalize the firing strength of each rule. Output from this layer is the ratio between the firing strength of ith rule and the sum of firing strengths of all rules. This final outcome is referred as normalized firing strength	$O_i^3 = \overline{w}_i = \frac{w_i}{\sum_i w_i}$
4	**Defuzzification layer:** This layer finds the product of a linear combination of the fuzzy rule output function and the normalized firing strength. Each node in this layer is adaptive to an output	$O_i^4 = \overline{w}_i f_i = \overline{w}_i(p_i x + q_i x + r_i)$ where $\overline{w}_i$ represent the normalized firing strength from the third layer and $(p_i x + q_i x + r_i)$ denotes a parameter in the node, also known as consequent parameter
5	**Output layer:** The single node in this layer generates the final output as the summation of all incoming signals from the previous node after defuzzification	$O_i^5 = \sum_i \overline{w}_i f_i$

tion process depends. Now in the case with the oily wastewater treatment through the UF membrane, one of the primary issues is the fouling of the membrane due to oil sticking over the membrane. Therefore, identification of the parameters such as TMP, fluid velocity over the membrane, pH of the system, etc. that is actually being considered in case of oil separation becomes intricate. This limits the predicted model formulation simply with a pattern mapping network. Perhaps, one has to take the help of some qualitative statement that can be deduced from the real runs carried out with the oily wastewater treatment using membrane separation technology.

To evaluate the performance of the separation process, it is required to know the permeate flux and degree of fouling on the membrane surface. As it is well known that fouling is an undeniable phenomenon associated with the membrane process, therefore, the fouling factor also plays a significant role in determining the performance of the process. As a result of fouling, deposition of feed constituents occurs on the membrane surface, which in turn results in the reduction of permeate flux constraining a membrane separation process. There are several types of membrane fouling that can be seen with the oily wastewater separation process. It may occur because of (1) blocking of the membrane pores, (2) adsorption inside the membrane pores, (3) concentration polarization phenomenon (increased concentration of foulants adjacent to the membrane surface), (4) deposition on the membrane surface forming a cake/gel layer and (5) compression of the cake/gel layer. During membrane separation, these fouling mechanisms may take place concurrently. Mainly, the degree of fouling during membrane separation of oily wastewater is dependent on three prime factors namely conditions of oily wastewater feed, operating conditions and nature of the membrane [8, 27]. During oily wastewater treatment, some of the low-chain hydrocarbons will pass through the UF membrane. One circumstance might arise that during the passage through the tortuous path of the membrane pores; the hydrocarbon molecules may either get attached at the surface or result in the pore plugging. Apart from that, another possibility is the adsorption of those molecules on the membrane surface depending on the nature of the membrane material. On increasing transmembrane pressure, formation of the cake or gel layer of higher density will be enhanced, ultimately leading to a complete pore blocking of the membrane. Thus membrane fouling can be the result of many such uncertain parameters. For better understanding of the permeation flux efficiency, several fouling mechanisms are proposed to assess the ability of simple cake filtration and to predict the permeation flux variation with respect to time. In standard pore blocking, oil droplets penetrate completely within the membrane pores. However, when pore blocking takes place outside the membrane pores, either the oil droplets block the pores partially (intermediate pore blocking) because of their similar size to the membrane pores, or they are completely capping the pores after sitting on the pores because of their bigger size compared to the membrane pores (complete pore blocking). Hence, prediction on the outcome of the membrane separation process through mapping using ANN only becomes vulnerable because of possible experimental errors. Hence, including qualitative statements as in the case with ANFIS might reduce the propensity of gross prediction error that may arise due to experimental errors. Fuzzy interface neural network model extends the domain of prediction by incorporating some qualitative remarks in addition with pattern mapping. Moreover, if only pattern mapping network is used, it becomes quite hard to quantify those issues that are absolutely qualitative in nature. Consider an example where we want to analyze a membrane separation process for the treatment of oily wastewater. Operational parameters such as TMP, temperature and volume concentration factor (VCF) were selected as the inputs to train the network. From Darcy's law, it can be well predicted that with an increase in TMP, permeation flux will enhance in the pressure driven region. However, because of higher TMP, formation of a high density cake or gel on the membrane surface will take place accelerating the degree of membrane fouling leading the membrane separation process to a mass transfer controlled region. It is well understood from both the Arrhenius equation and Darcy's law that with the increase in temperature, steady state permeation flux increases. At elevated temperatures, viscosity gradually declines along with an increase in the diffusivity that increases permeation flux. In addition, an increase in CFV results an increase in Reynold's number, and thus lowering the viscous force attributing to a higher permeation flux. With higher Reynold's number, turbulence becomes pronounced that enhances mixing with lowering in the high shear rate and manifests less accumulation of cake/gel layer on the membrane surface. Further, with the increase in CFV, the mass transfer coefficient within the concentration boundary layer increases. This reduces the concentration polarization leading to a higher permeate flux. Now all of these above described conditions are quantitative in nature and needs to be included in the prediction model for membrane separation process. The performance of membrane separation process for oily wastewater also requires a complete understanding of the nature of membrane material. Suppose in this case, three types of membranes PES, PSf and PVDF, are applied for the treatment. Hence, one can make the statement like "*If the membrane is PES and if temperature is 50 °C and if the TMP is 4.5 bar and if the VCF is 1.5, then the performance will be 63 %...*" The below diagram (Figure 11.4) shows the above memberships and the result comes out of this memberships.

Now let us define some statements (rules) for the above mentioned process.

Rule 1: If the membrane is PES and if the TMP is 4.5 bar and if temperature is 50 °C and if the VCF is approximately 1.5, then the performance will be 63 %.

Rule 2: If the membrane is PVDF and if the TMP is 3 bar and if temperature is 50 °C and if the VCF is approximately 1.5, then the performance will be 92 %.

Rule 3: If the membrane is PVDF and if the TMP is 3 bar and if temperature is 30 °C and if the VCF is approximately 2, then the performance will be 34 %.

Rule 4: If the membrane is PSf and if the TMP is 1.5 bar and if temperature is 20 °C and if the VCF is approximately 3, then the performance will be 25 %.

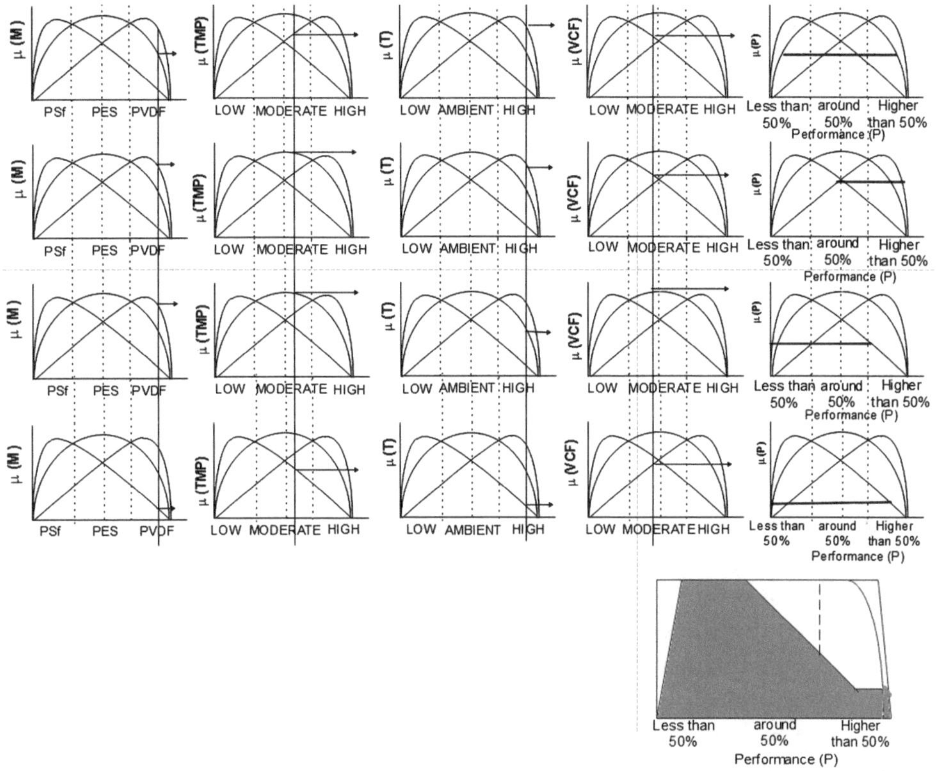

Figure 11.4: Final domain obtained after applying membership functions of the fuzzy rules.

Now say in an experiment, *if the membrane is PVDF and if the TMP is 2 bar and if temperature is 25 °C* and *if the VCF is approximately 2*, then one will try to understand permeate flux and the extent of the separation. Results against the above mentioned question can be determined using a weighted sum of the above predefined rules. The results can be evaluated using the centroid calculation, which says about the ratios in between the statistical moment for the whole domain and the area covered under the curve (refer to equation (11.2)). At the time of marking of the rules on the membership function graph manifesting the conclusion, operator "AND" (intersection operator) signifies the minimum value of the memberships for a specified rule among the rules, whereas operator "OR" (union operator) provides the maximum value (Figure 11.5).

$$x^* = \frac{\int \mu(x).xdx}{\int \mu(x)dx} \tag{11.2}$$

Thus ANFIS is a tool that can be used in order to have a predictive model for the membrane separation process, where the process performance is highly uncertain in

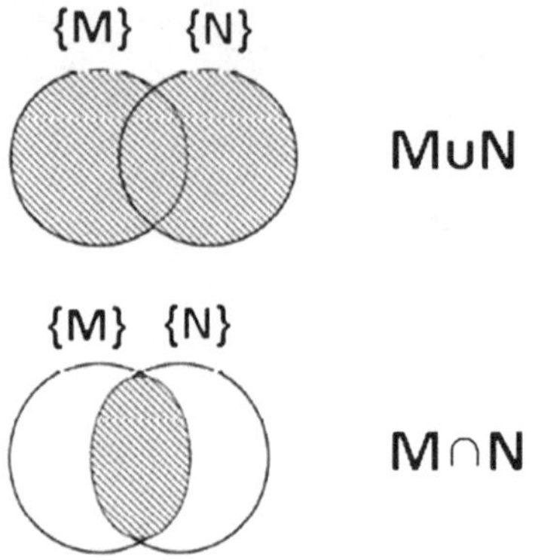

Figure 11.5: Pictorial representation of operator union and intersection for two given sets {*M*} and {*N*}.

itself. Especially, with the application of the process for the treatment of oily wastewater it is much difficult to judge the performance of the membrane separation process because of the presence of lower chains along with other impurities that either might get adsorbed on the surface or might be inserted into the pores. Therefore, substantial unforeseen scenarios make it difficult to apply any classical model to describe the process. At the same time, due to limitations with the ANN, it is difficult to extrapolate the process performance which is vital in case with the oil-water separation due to sudden unpredictable shift of the process from pressure driven to mass driven. ANFIS is thus providing an option to have a statement for the process (such as membrane material) that may have a great importance in making such shift.

11.8 Summary

ANFIS is a tool that can be used in order to have a predictive model for the membrane separation process, where the process performance is highly uncertain in itself. Especially, with the application of the process for the treatment of oily wastewater it is much difficult to judge the performance of the membrane separation process because of the presence of lower chains along with other impurities that either might get adsorbed on the surface or might be inserted into the pores. Therefore, substantial unforeseen scenarios make it difficult to apply any classical model to describe the process. At the same time due to limitations with the ANN, it is difficult to extrapolate the process performance which is vital in case with the oil-water separation due to sudden unpredictable shift of the process from pressure driven to mass driven. ANFIS is thus providing an option to have a statement for the process (such as membrane material) that may have a great importance in making such shift.

Compliance with ethical standards

Conflict of interest

The authors declare that they have no conflict of interest.

Informed consent

Informed consent was obtained from all individual participants included in the study.

Acknowledgment

The authors acknowledge DST sponsored TSDP funded project (vide sanction no. DST/TSG/AMT/2015/276 (General) dated 11.07.2016) for providing necessary support and facilities during the preparation of this book chapter.

Bibliography

[1] Janknecht P, Lopes AD, Mendes AM. Removal of industrial cutting oil from oil emulsions by polymeric ultra-and microfiltration membranes. Environ Sci Technol. 2004;38:4878–83.
[2] El-Naas MH, Al-Zuhair S, Al-Lobaney A, Makhlouf S. Assessment of electrocoagulation for the treatment of petroleum refinery wastewater. J Environ Manag. 2009;91:180–5.
[3] Cheryan M, Rajagopalan N. Membrane processing of oily streams. Wastewater treatment and waste reduction. J Membr Sci. 1998;151:13–28.
[4] Rezakazemi M, Khajeh A, Mesbah M. Membrane filtration of wastewater from gas and oil production. Environ Chem Lett. 2018;16:367–88.
[5] Marchese J, Ochoa N, Pagliero C, Almandoz C. Pilot-scale ultrafiltration of an emulsified oil wastewater. Environ Sci Technol. 2000;34:2990–6.
[6] Huang S, Ras RH, Tian X. Antifouling membranes for oily wastewater treatment: interplay between wetting and membrane fouling. Curr Opin Colloid Interface Sci. 2018;36:90–109.
[7] Zhu Y, Wang D, Jiang L, Jin J. Recent progress in developing advanced membranes for emulsified oil/water separation. NPG Asia Mater. 2014;6:e101.
[8] Noshadi I, Salahi A, Hemmati M, Rekabdar F, Mohammadi T. Experimental and ANFIS modeling for fouling analysis of oily wastewater treatment using ultrafiltration. Asia-Pac J Chem Eng. 2013;8:527–38.
[9] Asghari M, Dashti A, Rezakazemi M, Jokar E, Halakoei H. Application of neural networks in membrane separation. Rev Chem Eng. 2018;36(2):265–310.
[10] Zadeh LA. Fuzzy sets. Inf Control. 1965;8:338–53.
[11] Mamdani EH, Assilian S. An experiment in linguistic synthesis with a fuzzy logic controller. Int J Man-Mach Stud. 1975;7:1–13.
[12] Bai Y, Wang D. Fundamentals of fuzzy logic control – fuzzy sets, fuzzy rules and defuzzifications. In: Advanced fuzzy logic technologies in industrial applications. London: Springer; 2006. p. 17–36.
[13] Zadeh LA. The concept of a linguistic variable and its application to approximate reasoning – I. Inf Sci. 1975;8:199–249.
[14] Yager R. A characterization of the extension principle. Fuzzy Sets Syst. 1986;18:205–17.

[15] McCulloch WS, Pitts W. A logical calculus of the ideas immanent in nervous activity. Bull Math Biophys. 1943;5:115–33.
[16] Hebb DO. The organization of behavior: a neurophysical theory. New York: Wiley; 1962.
[17] Rosenblatt F. The perceptron: a probabilistic model for information storage and organization in the brain. Psychol Rev. 1958;65:386.
[18] Widrow B, Hoff ME. Adaptive switching circuits. Stanford Univ CA, Stanford Electronics Labs; 1960.
[19] Hopfield JJ. Neural networks and physical systems with emergent collective computational abilities. Proc Natl Acad Sci USA. 1982;79:2554–8.
[20] Sen D, Roy A, Bhattacharya A, Banerjee D, Bhattacharjee C. Development of a knowledge based hybrid neural network (KBHNN) for studying the effect of diafiltration during ultrafiltration of whey. Desalination. 2011;273:168–78.
[21] Jang JSR. Input selection for ANFIS learning. In: Proceedings of the fifth IEEE international conference on fuzzy systems. vol. 2. IEEE; 1996. p. 1493–9.
[22] Sargolzaei J, Asl MH, Moghaddam AH. Membrane permeate flux and rejection factor prediction using intelligent systems. Desalination. 2012;284:92–9.
[23] Salahi A, Abbasi M, Mohammadi T. Permeate flux decline during UF of oily wastewater: experimental and modeling. Desalination. 2010;251:153–60.
[24] Rahmanian B, Pakizeh M, Mansoori SAA, Esfandyari M, Jafari D, Maddah H, Maskooki A. Prediction of MEUF process performance using artificial neural networks and ANFIS approaches. J Taiwan Inst Chem Eng. 2012;43:558–65.
[25] Suparta W, Alhasa KM. A comparison of ANFIS and MLP models for the prediction of precipitable water vapor. In: 2013 IEEE international conference on space science and communication (IconSpace). IEEE; 2013. p. 243–8.
[26] Suparta W, Alhasa KM. Adaptive neuro-fuzzy interference system. In: Modeling of tropospheric delays using ANFIS. Switzerland: Springer; 2016. p. 5–18.
[27] Kumar SM RS. Recovery of water from sewage effluents using alumina ceramic microfiltration membranes. Sep Sci Technol. 2008;43:1034–64.

M. Mary Shanthi Rani and P. Shanmugavadivu

12 Deep learning based food image classification

Abstract: Nowadays, people are increasingly conscious about their health and well-being due to the alarming rise in the death rates due to heart attacks, cancer, etc. Physicians' foremost advice for an obese person is to keep track and control the intake of calories to reduce weight. Though there are quite a number of algorithms to identify images and provide a calorie based calculation for the requested image, there are limited data sets specifically on food items. This is a big challenge for deep learning systems in which learning rate and accuracy depends on the number of images in the training set. The main objective is to develop a novel deep learning-based system for classification of food images of South India.

Keywords: Deep learning, food image classification, food recognition

12.1 Introduction

Obesity is a major risk in adults and children which in turn invites a host of diseases like diabetes, cholesterol, enlargement of lungs, etc., leading to life-threatening disorders. It is obvious that irregular and uncontrolled diet patterns without any physical activity are the primary causes. Moreover, food intake without the knowledge of its composition and nutritional contents may lead to unpredictable health hazards. Body mass index which is the ratio of weight and height of a person can be used as a yardstick for measuring obesity. Normally, people search Google or other applications like calorie counter, coach, etc. to get calorie information about the food including fruits. There are many diet and nutrition apps like Fooducate, My Fitness Pal, My Diet Coach, Calorie Mama and Life Sum in the market. Calorie Mama is powered by Food AI API based on deep learning techniques. Furthermore, the recent and rapid development in computing power; robust software and voluminous data generation have motivated the researchers to explore the promising applications of deep learning. This chapter presents a novel approach of classification of food items specifically in food images of South India.

12.1.1 Deep learning

The traditional computational methods have lost their charms and are proved to be noncontextual and irrelevant in the recent years to handle and process volumes

M. Mary Shanthi Rani, P. Shanmugavadivu, Dept. of Computer Science & Applications, Gandhigram Rural Institute (Deemed to be University), Gandhigram, Tamil Nadu, India, e-mails: drmaryshanthi@gmail.com, psvadivu67@gmail.com

https://doi.org/10.1515/9783110671353-012

of data. The concept of Artificial Intelligence (AI) and Machine Learning (ML) had evolved in well in 1950s and was explored to design and develop the decision support system. Being interdisciplinary in nature, it percolates into several fields such as health care [8], agriculture, law enforcement, education, manufacturing, engineering, remote sensing, robotics, information security and finance. The outgrowth of AI into Machine learning (ML) and Deep Learning (DL) have introduced new promises in image processing.

Deep Learning is a variant of machine learning, which provides a deeper insight into data through multiple hierarchical abstractions of data. It ideally provides optimal solutions to the problems which cannot be solved by deterministic algorithms as the goal posts keep changing. They are well suited for prediction, detection and generation based on past and existing patterns. Deep learning has become a frontline research area with vast number of applications like learner profiling, image/video captioning, cancer detection, precision agriculture, pattern recognition in images, behavior analysis and prediction, recommendations, identification of specific markers in genomes, etc.

The availability and use of new ready-to-use models such as Artificial Neural Network (ANN) models, Convolutional Neural Networks (CNN) [7] and Recurrent Neural Networks (RNN) has attracted the focus of researchers to explore this promising domain. These ready-to-use models are proved to produce results with the maximum envisioned and also provide ample scope to customize/redesign, in order to augment precision of processing.

With the increase in computational power due to advent of GPUs and efficient ANN nets, several real life applications with societal benefit are developed powered by machine intelligence and deep learning techniques.

12.1.1.1 Deep neural architectures

An artificial neural network (ANN) is either a hardware or software implementation that simulates the information processing capabilities of its biological exemplar. It is typically composed of a great number of interconnected artificial neurons that are simplified models of their biological counterparts. Usually, a biological neuron receives its information from other neurons in a confined area called receptive field. Once the input exceeds a critical level, the neuron discharges a spike – an electrical pulse that travels from the body, down the axon, to the next neuron(s).

Basically, an ANN has three important layers; input layer, hidden layer and output layer. Input layer accepts input features and just pass on the information (features) to the hidden layer. Hidden layer is the layer of abstraction that performs all sort of computation and transfer the result to the output layer. Furthermore, ANN uses an activation function which is a decision making function to activate a neuron based on the presence of a particular neural feature. It is needed to introduce nonlinearity into the network to learn complex functions. Some popular activation functions are Sigmoid, Tan H, Softmax and Rectified Linear Unit (ReLu). Sigmoid function is well suited for binary classification. The softmax function is also a type of sigmoid function used to handle classification problems with multiple classes. ReLu learns much faster than Sigmoid and Tan H function due to nonsaturation form and it is popularly used in convolution neural networks. Linear regression is used for applications where output will be a real value like predicting housing price, or predicting price of a share in stock market. If the application involves multiple dependent variables, we call it as multiple linear regression. Logistic regression will be used if the output is a binary value.

The goal of ANN is to learn the weights of the network automatically from data such that the predicted output is close to the target for all inputs. Optimization functions are used to update and tune the parameters so as to minimize the error. Gradient Descent (GD) is a popular optimizing function. Traditional batch gradient descent and stochastic gradient descent are some variants of GD used to improve the performance of NN.

Adam is an optimization algorithm used in many applications instead of the classical stochastic gradient descent procedure to update network weights. It has become a very popular optimizer in which a learning rate is maintained for each network weight (parameter) and individually adapted as learning unfolds during the iterative process.

All previous ANN models rely on handcrafted features like color, shape and texture that are difficult to design. With the dawn of deep learning, specifically Convolutional Neural Networks (CNN) have automated the feature extraction process while significantly boosting performance

CNN is considered as the state of the art choice for object classification and detection. Its development has witnessed several milestones starting with introduction of AlexNet, LeNet, ZFFnet and VGGNet, etc. GoogleNet also known as Inception V1, being the winner of ILSVRC 2014 has brought the top-5 error 16.4 % obtained by AlexNet down to 6.7 %. It reduced the parameters by applying average pooling instead of fully connected layers. Following V1, several versions of the inception model have been introduced with enhanced features. The latest version inception-resnet design makes full use of residual connections which accelerates the learning process. Recent approaches like R-CNN use region proposal methods to generate potential bounding boxes in an image and run a classifier on these proposed boxes. But these complex pipelines are slow and hard to optimize as each individual component must be trained separately. You Only Look Once (YOLO) is a fast single convolutional network that simultaneously predicts multiple bounding boxes and class probabilities for those boxes. YOLO trains on full images and directly optimizes detection performance thus outperforming traditional methods of object detection.

12.2 Related works

Literature survey shows that there are good numbers of research publications on food calorie estimation.

Parisa Pouladzadeh et al. [1] proposed a new system for measuring the calorie present in the food by capturing images in the smartphone. This method helps the doctor to assess the intake of calorie of patient. This method is developed based on the combination of graph cut segmentation and deep neural network for the classification. Nearly 10,000 training samples have been used in this method which achieved an accuracy of 99 %.

Dae Han Ahn et al. proposed a novel approach which estimates the composition of carbohydrates, proteins and fats from hyper spectral food images obtained by using low-cost spectrometers. In this proposed system, feature extraction is done with hyperspectral signals and error avoidance is done with auto encoder. The proposed system detects and discards the estimation anomalies automatically [2].

Sulfayanti F. Situju et al. proposed an automatic ingredient estimation in food images. To reduce the risk of new diseases caused by unbalanced nutrition, the proposed system has been created. This method of estimation assess healthy food intake using multi-task CNN for classification and estimation by finding the relationship between the food category and salinity [3].

Isaksen et al. developed a network to estimate the weight of food from a single image using the convolutional neural network. The proposed method classifies the image to identify the food from different food images and estimation of weight is done. Comparison is done using standard food databases [4].

Liang, Yanchao and Jianhua Li proposed a method for estimation of calories automatically from the food images using Faster R-CNN. In this method, food's contour is found out using the Grabcut algorithm and then the volume of food is estimated using volume estimation formulae [5].

Takumi Ege et al. has introduced the simultaneous estimation of food categories and calories from a novel food image. In this method, single task CNN and multitask CNN are used for training where multitask CNN achieves a better performance on both food category classification and food calorie estimation than single-task CNNs [6].

12.3 Proposed work

Despite a good number of applications on food identification and calorie estimation, accurate classification of food image still remains as a challenge as there are various critical parameters like visually similar foods, quality of images, illumination condition and number of food items, etc. In the proposed design, we use pretrained VGG-16

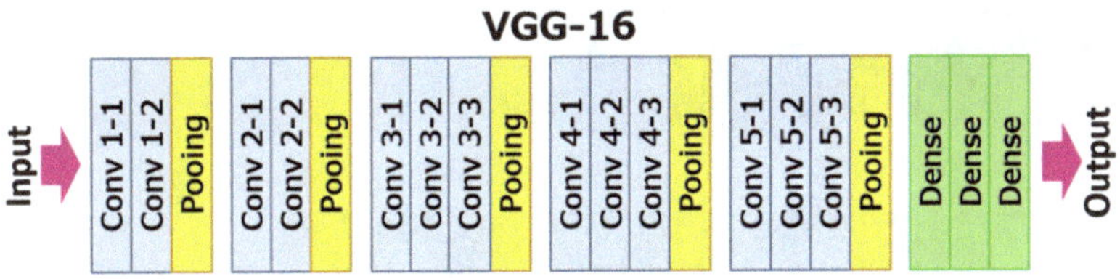

Figure 12.1: Architecture of the VGG16 model.

architecture for food classification. The novelty of the proposed method is creation and annotation of new data set for South Indian authentic food items for classification.

VGG16 is a convolutional neural network model proposed by K. Simonyan and A. Zisserman. The model is an improvement over AlexNet by replacing large kernels with 3×3 filters and achieves 92.7 % test accuracy when tested using ImageNet, which is a large data set of over 14 million images belonging to 1,000 classes. Figure 12.1 shows the architecture of VGG 16 model.

Max pooling layers are introduced between convolution layers in order to reduce the dimensions of the image. Batch normalization is a strategy used to eliminate the issue of internal covariate shift. It helps to learn faster and achieve higher overall accuracy. Both ReLU activation function and batch normalization are applied in our experiments.

12.3.1 Methodology

The proposed method involves the following modules and tasks:

Module 1 Collection of food images from various sources and preparation of data sets.

Step 1: This will involve collection of various food item images specifically authentic food items of South India from different sources.

Step 2: Preprocessing the images for effective classification. It is essential to preprocess the images that were taken in different environment and background to speed up learning pace and to improve the accuracy.

Module 2 Annotation of newly created data set for food item classification.

Module 3 Training the model for classification of food items from food images using the VGG16 model.

Module 4 Testing and validation of the model with test data set based on evaluation measures.

Module 5 Food item classification using the trained model.

The work flow diagram of the proposed methodology is depicted in Figure 12.2.

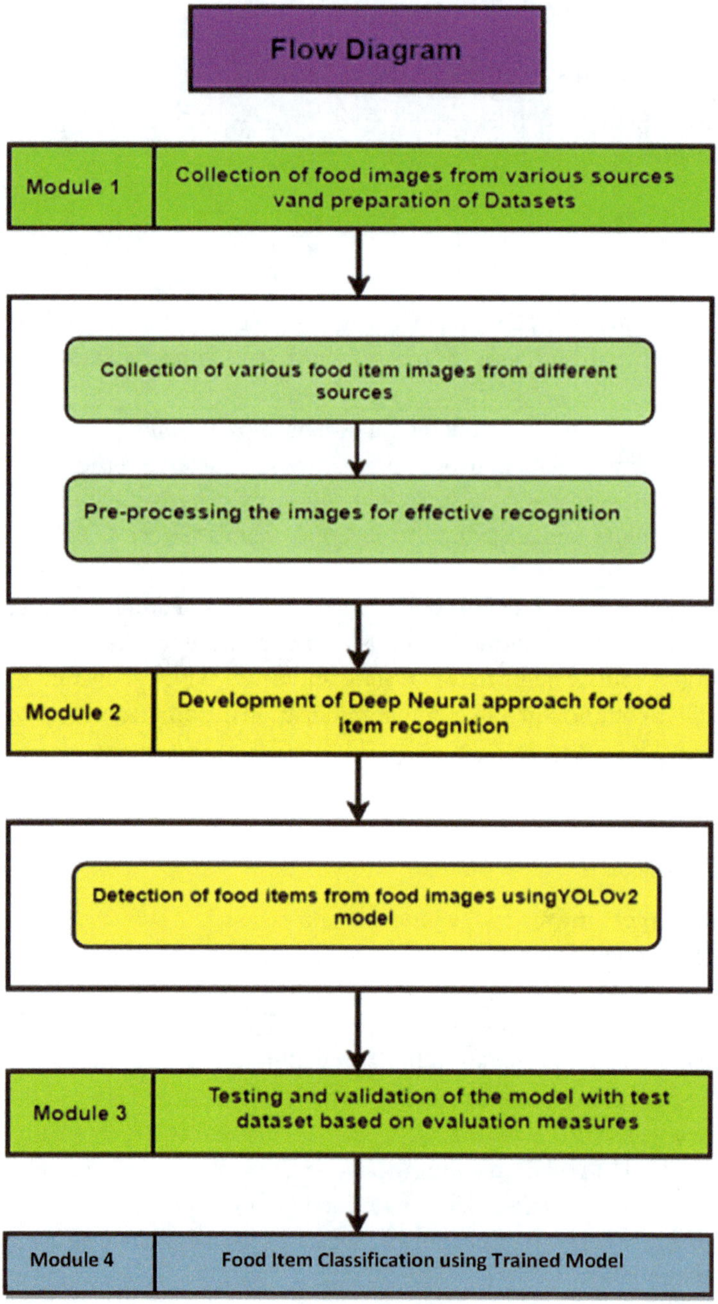

Figure 12.2: Flow diagram of the proposed methodology.

12.3.2 Materials and methods

Dataset

As there is no standard data set for Indian dishes, a new data set has been created and annotated. The data set created is preprocessed and used for training the model.

Model

We use a pretrained model VGG16 trained with 1,000 class images by Google's imagenet and retrain the system to identify food items of authentic Indian dishes as well with more accuracy and precision.

Evaluation measures

The performance of the proposed model is tested for classification accuracy using the test data set.

Software and hardware requirements

Deep learning libraries such as OpenCV, TensorFlow and Keras will be exploited for the development and experimentation of the project.

12.4 Results and discussion

This chapter presents a new approach for classification of South Indian dishes like Puttu, Vada and Dosai with greater accuracy. A data set with a total of 1,981 images of Indian dishes is used for training and testing. The ratio of images taken for training and testing are 80 % and 20 %, respectively.

Table 12.1 lists the materials and functions used in our experiment for training. The proposed approach used VGG16 and VGG19 neural architecture models for training. The model is trained with Adam optimizer, categorical cross entropy loss function and softmax activation function. Cross-entropy loss, or log loss, measures the performance of a classification model whose output is a probability value between 0 and Cross-entropy loss increases as the predicted probability diverges from the actual label.

The batch size is the number of samples processed before the model is updated. The number of epochs is the number of complete passes through the training data set. The size of a batch must be more than or equal to one and less than or equal to the number of samples in the training data set.

Table 12.1: Parameters used for training the model.

Batch size	32, 64
No. of epochs	200, 60, 50, 25
Optimizer	Adam
Entropy	Categorical_cross entropy
Architecture	VGG16 & VGG19
Activation function	Soft max
Data set	Google images
Types of images	Beef chilly Dosai Panniyaram Puttu Vada
Total Images	1941 Training – 1553 Testing – 388 Training – 80 % Testing – 20 %

Accuracy is one of the metrics for evaluating classification models which is defined as the fraction of number of correct predictions to the total number of predictions as shown in equation (5.1).

$$\text{Accuracy} = \frac{\text{Number of correct predictions}}{\text{Total number of predictions}} \tag{12.1}$$

For binary classification, accuracy is calculated in terms of positives and negatives as follows:

$$\text{Accuracy} = \frac{TP + TN}{TP + TN + FP + FN}, \tag{12.2}$$

where TP = True Positives, TN = True Negatives, FP = False Positives and FN = False Negatives.

Table 12.2 presents the performance of the proposed approach in terms of accuracy for different epochs and batch sizes.

Table 12.2: Performance analysis of the proposed method using VGG16.

Epoch	**60**		**50**		**25**	
Batch Size	64	32	64	32	64	32
Accuracy	0.8075	0.9392	0.7734	0.9188	0.6829	0.7062

It is obvious from Table 12.2 that the proposed model has achieved greater accuracy of for 60 epochs. Moreover, it is worth noting that the model achieves more than 90 % accuracy for a batch size of 32.

Figure 12.3 presents the graphical representation of the comparative analysis of the performance the proposed model using VGG16 for different epochs and batch sizes.

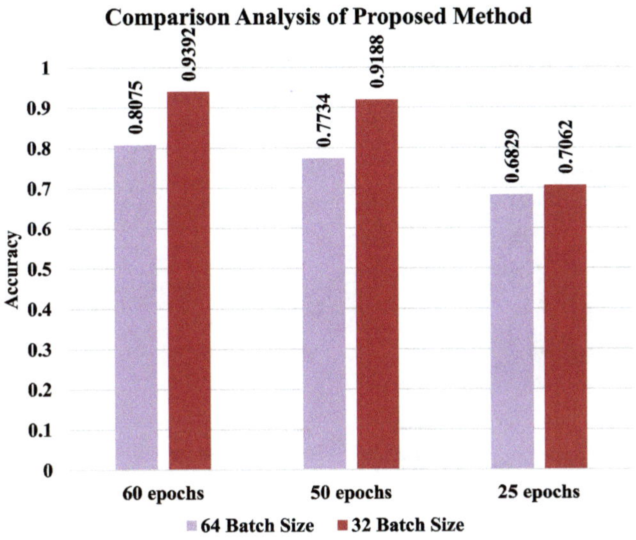

Figure 12.3: Performance analysis in terms of accuracy.

Table 12.3: Performance analysis of the proposed method using VGG19.

Epoch	200		60		50	
Batch Size	64	32	64	32	64	32
Accuracy	0.9052	0.9373	0.7573	0.7657	0.7293	0.7271

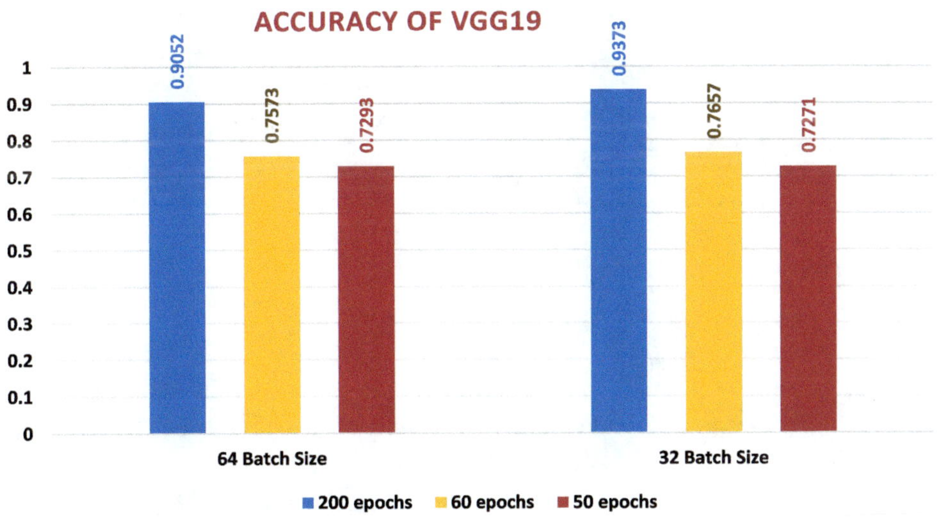

Figure 12.4: Performance analysis in terms of accuracy.

Figure 12.5: Test images.

Table 12.3 presents the comparative analysis of the results in terms of accuracy for different batch sizes (64, 32) and for different number of epochs (200, 60, 32) using the VGG19 model.

Table 12.3 reveals that accuracy is higher for 200 epochs and for 32 batch size. Figure 12.4 demonstrates the graphical representation of the performance of the VGG19 model.

Figure 12.5 demonstrates the visual presentation of the outputs of the proposed model for test images Beef chilly, Vada, Dosai and Puttu using VGG16 and VGG19, respectively.

It is obvious from Figure 12.5 that our trained model has achieved 100 % accuracy in classifying beef chilly and Vada, 99 % accuracy in classifying Puttu.

12.5 Conclusions

In this chapter, a new method of classification of South Indian food items using deep learning model VGG 16 and VGG 19 has been proposed. The model is trained with newly created and annotated south Indian food image data set. Experimental results prove the superior performance of the model with 99 % accuracy . Further, the model is able to achieve 100 % accuracy for classification of certain food items like beef chilly during testing. Our future work would be to train the model with more number of food items, along with volume estimation to compute the calorie content as well.

Bibliography

[1] Pouladzadeh P, Kuhad P, Peddi SV, Yassine A, Shirmohammadi S. Food calorie measurement using deep learning neural network. In: 2016 IEEE international instrumentation and measurement technology conference proceedings, 2016 May 23. IEEE; 2016. p. 1–6. 2016.

[2] Ahn D, Choi JY, Kim HC, Cho JS, Moon KD, Park T. Estimating the composition of food nutrients from hyperspectral signals based on deep neural networks. Sensors. 2019;19(7):1560.

[3] Situju SF, Takimoto H, Sato S, Yamauchi H, Kanagawa A, Lawi A. Food constituent estimation for lifestyle disease prevention by multi-task CNN. Appl Artif Intell. 2019;33(8):732–46.

[4] Isaksen R, Knudsen EB, Walde AI. A deep learning segmentation approach to calories and weight estimation of food images. Master's thesis. Universitetet i Agder; University of Agder.

[5] Liang Y, Li J. Deep learning-based food calorie estimation method in dietary assessment. arXiv preprint arXiv:1706.04062 (2017 Jun 10).

[6] Ege T, Yanai K. Simultaneous estimation of food categories and calories with multi-task CNN. In: 2017 fifteenth IAPR international conference on machine vision applications (MVA), 2017 May 8. IEEE; 2017. p. 198–201.

[7] Sangeetha R, Mary Shanthi Rani M. Tomato leaf disease prediction using convolutional neural network. Int J Innov Technol Explor Eng. 2019;9(1):1348–52.

[8] Kalpana Devi M, Mary Shanthi Rani M. A review on detection of diabetic rectinopathy. Int J Sci Technol Res. 2020;9(2):1922–4.

Vandana Khanna, B. K. Das, and Dinesh Bisht

13 Particle swarm optimization and differential evolution algorithms: application to solar photovoltaic cells

Abstract: This work has been done on the large area (area ~ $154.8\,\text{cm}^2$) single crystalline silicon solar cells. The most common equivalent circuit models for solar cells are one-diode and two-diode models. Two-diode model is more accurate though little complicated than one-diode model. The seven parameters of two-diode model namely photon current, reverse saturation currents and ideality factors of two diodes, series resistance and shunt resistance have been estimated using population-based algorithms, Particle Swarm Optimization (PSO) and Differential Evolution (DE). The mean absolute error has been considered as fitness function to be minimized for both the algorithms. Very good results with good fit of calculated $I-V$ characteristics to the measured $I-V$ characteristics have been achieved for all the solar cell samples. Average values of ideality factors n_1 and n_2 were estimated as 1.2 and 2.7, respectively, as compared to n_1 and n_2 values as 1 and 2, respectively, for an ideal two-diode model. Practical two-diode model best represents these large area solar cells. Results of the slope method (used by solar simulator), PSO and DE algorithms have been compared.

Keywords: Crystalline silicon solar cells, particle swarm optimization, differential evolution, solar cell models, parameters estimation

13.1 Introduction

Optimization problems have been targeted in either analytical or population based evolutionary approaches in literature. Analytical approaches are simple in nature and are mostly based on some measured key points. These key points may be measured manually or with the help of any software which may work on few assumptions; these measured points may not be very precise. To overcome this issue, the researchers have applied various types of population-based optimization algorithms such as Pattern Search (PS), Genetic Algorithms (GA), Particle Swarm Optimization (PSO), Simulated Annealing (SA), Differential Evolution (DE), Artificial Bee Colony (ABC), Artificial Bee Swarm Optimization (ABSO), Artificial Fish Swarm Algorithm (AFSA) and many other algorithms for the optimization problems in different fields of engineering. These algorithms have been found to provide good results in numerous applications.

Vandana Khanna, B. K. Das, Department of EECE, The NorthCap University, Gurugram, India, e-mails: vandanakhanna@ncuindia.edu, bkdas45@gmail.com
Dinesh Bisht, Department of Mathematics, Jaypee Institute of Information Technology, Noida, India, e-mail: drbisht.math@gmail.com

https://doi.org/10.1515/9783110671353-013

13.2 Particle Swarm Optimization (PSO) and Differential Evolution (DE) algorithms

Around the 1970s, Holland [1] developed the Genetic Algorithm (GA) for optimization problems. GA was inspired by natural biological evolution and was based on biological mutation, crossover, etc. In the mid-1990s, Kennedy and Eberhart [2] introduced particle swarm optimization (PSO) – a new algorithm for complex non-linear optimization problems. It was based on the phenomenon of collecting food by a group of birds. At about the same time, Price and Storn [3] replaced basic operators of mutation and crossover in GA by differential operators and introduced a new algorithm called differential evolution (DE). Panduro et al. [4] have compared three algorithms, namely GA, PSO and DE for the design of scannable circular antenna arrays. In the antenna array design problem, authors concluded that PSO and DE gave similar results and better than obtained through GA optimization. Das et al. [5] have analyzed both PSO and DE algorithms in detail, and applications of these algorithms in various engineering domains have been surveyed. Moreover, the authors have worked on the ways of selection of parameters for convergence of both algorithms. PSO and DE optimization methods are conceptually very simple and have few parameters to be tuned for specific optimization problem; this makes these two algorithms useful for many engineering problems.

13.2.1 Particle swarm optimization

Particle Swarm Optimization (PSO) algorithm is developed on the phenomenon of search for food by a group of birds. Birds move to find the food as per their knowledge of how far the food is. To start with, a number of birds move in arbitrary directions and with different speeds, but after a while, one of the bird takes the lead and locates the food, other birds follow the leader, by observing their own and other birds' flight.

PSO has been applied to various optimization problems by researchers. The problem to be optimized is defined as a fitness function. The search space for this particular problem is initially defined based on the past experience. For the optimization problem, the first step is to generate a number (n) of random solutions or particles, where each "particle" (identified as a "bird") has velocity and position as $v[t]$ and $x[t]$, respectively, at time t, in the search space. The fitness values for each particle are calculated from the fitness function in each iteration. The best fitness value attained by each particle is considered as "pbest" (particle best); the best fitness value in the whole population is taken as "gbest" (global best).

After this, velocities of all particles are updated as per equation (13.1).

$$v_{t+1} = w * v_t + c_1 * rand * (\text{pbest}_t - x_t) + c_2 * rand * (\text{gbest}_t - x_t) \tag{13.1}$$

v_{t+1} is the velocity of each particle at time $t + 1$. w is called inertia weight which can take values between 0 and 1 and its value is made to decrease in subsequent iterations. pbest_t and gbest_t are defined above. "rand" is a random number generated during runtime and its value lies between 0 and 1. The second term of equation (13.1) represents a "cognitive" model and c_1 is given a constant value and is called cognitive or local weight; cognitive model represents each particle's own experience. The third term in equation (13.1) represents a "social" model and c_2 is also given a constant value and is called social or global weight; the social model represents the cooperation between particles. Usually, c_1 and c_2 are taken equal to 2.

A good optimization algorithm balances between exploration and exploitation efficiently. PSO uses the concepts of inertia weight and velocity clamping [6, 7] for this purpose.

Inertia weight w controls the momentum of the particles' velocity during the iterations in the algorithm. The variation of the value of w during iterations plays an important role in convergence of the PSO algorithm. Large values of w means more momentum of the particles, and thus better exploration, whereas the low values facilitate local exploitation [7]. Initially, more exploration is required which is ensured by taking large values of w during initial time of the algorithm. Over time, the solutions need to be converged by good local exploitation, which is ensured by taking lower values of w. Value of the inertia weight is dynamically changed for a better trade-off between exploration and exploitation. In this work, w was initialized to 1 and it was made to decrease linearly with iterations to make value of w as 0 in the final iteration.

In PSO algorithm, the values of the velocities of particles can easily go high, and thus, positions of particles when updated may take values which may extend outside the lower and upper boundaries of the search space. *Velocity clamping* [6, 7] is the process to clamp the velocities of the particles to a particular boundary value of velocities. Velocity clamping was used in this work as per equation (13.2).

$$v_{t+1} = \left\{ \begin{array}{ll} v_{t+1}, & \text{if } v_{t+1} < v_{\max} \\ v_{\max}, & \text{otherwise} \end{array} \right\} \tag{13.2}$$

$v_{\max}$ denotes the particle's maximum allowed velocity in the solution space. Equation (13.2) is applied to clamp the velocity of the particles. Usually, $v_{\max}$ value equals a fraction of the domain of each dimension of the search space as given below.

$$v_{\max} = \delta(x_{\max} - x_{\min}) \tag{13.3a}$$

$$v_{\min} = \delta(x_{\min} - x_{\max}) \tag{13.3b}$$

where, $x_{\min}$ and $x_{\max}$ and are the minimum and maximum positions of the particles respectively. Value of δ is taken as 0.1 in this work. Particles' velocities on each dimension are clamped to maximum and minimum velocities: $v_{\max}$ and $v_{\min}$ respectively,

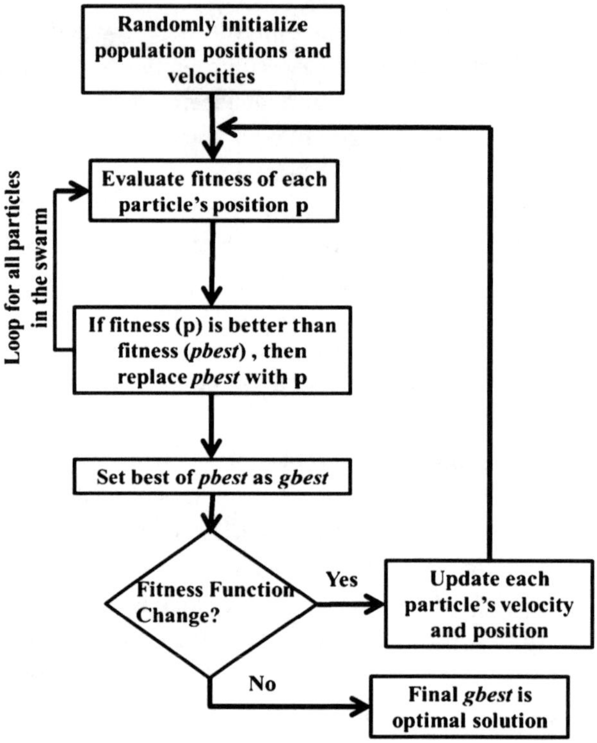

Figure 13.1: Flow chart of the PSO algorithm [38].

through equation (13.3b). Next, positions of all particles are updated as per equation (13.4).

$$x_{t+1} = x_t + v_{t+1} \tag{13.4}$$

v_{t+1} and x_{t+1} are the velocity and position of each particle at time $t + 1$.

The particle's best results and overall best solution can be achieved through particle swarm optimizer by changing each particle's velocity and position for better results in each iteration. The flowchart of PSO algorithm is shown in Figure 13.1.

13.2.2 Differential evolution algorithm

Differential Evolution (DE) is also a population-based metaheuristic optimization algorithm. In this algorithm also, initial random population is generated, which is then improved through generations by applying mutation, crossover and selection operations. This continues until a stopping criteria is reached, which is either predefined maximum number of generations or a minimum value of the fitness function.

The DE algorithm can be implemented as per following steps:

1) Initialization: Population size (N), maximum number of generations/iterations ($it_{\max}$) and control parameters of DE algorithm: the scaling factor (F) and the crossover rate (CR) are defined initially. Random population of N, D-dimensional parameter vectors is generated, given by

$$X_{i,g} = [x_{i,1,g}, \ldots, x_{i,j,g}, \ldots, x_{i,D,g}]$$

where $i = 1, 2, \ldots, N$, and $j = 1, 2, \ldots, D$, g is the generation number, for initial population, $g = 1$.
This random population is generated uniformly within the lower ($X_L = [X_{1,L}, X_{2,L}, \ldots, X_{D,L}]$) and upper ($X_H = [X_{1,H}, X_{2,H}, \ldots, X_{D,H}]$) limits of the search space interval as follows:

$$X_{i,1} = X_L + (X_H - X_L) * rand[0, 1] \tag{13.5}$$

2) Mutation: Mutant vector $V_{i,g}$ corresponding to each $X_{i,g}$, (called target vector) is produced in the current generation during mutation operation. In this work, the mutation strategy of "DE/best/1" (a nomenclature is used to refer to different DE mutation strategies [8]) is used, where "DE" denotes differential evolution, 'best" denotes the target vector selected for mutation to be the individual with the best fitness value in the current generation of population, "1" represents the number of difference vectors to be generated. Mutation vector is generated as follows:

$$V_{i,g} = X_{\text{best},g} + F(X_{r_1,g} - X_{r_2,g}), \tag{13.6}$$

where r_1 and r_2 are randomly generated integers in the range $[1, N]$ for each mutant vector, $X_{r_1,g}$ and $X_{r_2,g}$ are the two randomly selected vectors from the population at generation g, F is one of the control parameter called scaling factor to scale the difference vector, $X_{\text{best},g}$ is the vector with the lowest/highest fitness value for a minimization/maximization problem respectively among all vectors in current population at generation g and

$$V_{i,g} = [v_{i,1,g}, v_{i,2,g}, \ldots, v_{i,j,g}, \ldots, v_{i,D,g}]$$

3) Crossover: After the mutation phase, crossover operation is introduced to generate a trial vector $U_{i,g}$ by mixing the components of each pair of target vector $X_{i,g}$ and corresponding mutant vector $V_{i,g}$.

$$U_{i,g} = [u_{i,1,g}, u_{i,2,g}, \ldots, u_{i,j,g}, \ldots, u_{i,D,g}]$$

Two types of crossover methods such as exponential and binomial (uniform) can be applied in a DE algorithm. In this work, binomial crossover strategy is used which is defined as follows:

$$u_{i,j,g} = \begin{cases} v_{i,j,g}, & \text{if } (rand[0, 1] \le CR) \text{ or } (j = rand[1, D]) \\ x_{i,j,g}, & \text{otherwise} \end{cases} \tag{13.7}$$

CR is another control parameter of DE algorithm called crossover rate. $rand[1,D]$ is a random integer from the range $[1,D]$. The condition $j = rand[1,D]$ is introduced to ensure that $U_{i,g}$ attains at least one element from $V_{i,g}$. During binomial crossover operation, the jth component of the mutant vector $V_{i,g}$ is copied to the trial vector $U_{i,g}$, if $(rand[0,1] \leq CR)$ or $(j = rand(1,D))$, else target vector $X_{i,g}$ is copied. During the operations of mutation and crossover, the parameter values might violate the boundary constraints $[X_L, X_H]$ of the search space of parameters. The following operation is performed to ensure that the parameter values lie within the specified limits after recombination:

$$U_{i,g} = \begin{cases} U_{i,g} + rand[0,1] * (X_H - X_L), & \text{if } U_{i,g} \leq X_L \\ U_{i,g} - rand[0,1] * (X_H - X_L), & \text{if } U_{i,g} \geq X_H \end{cases} \tag{13.8}$$

4) Evaluation and selection: The fitness of the trial vector is evaluated, and this trial vector becomes member of the $(g+1)$ generation, if following condition is met:

$$X_{i,g+1} = \begin{cases} U_{i,g}, & \text{if fitnessvalue}(U_{i,g}) < \text{fitnessvalue}(X_{i,g}) \\ X_{i,g}, & \text{otherwise} \end{cases} \tag{13.9}$$

For the minimization problem, if the fitness or objective function value of the trial vector is less than the original target vector, then this trial vector $U_{i,g}$ is selected as a vector of the generation $(g+1)$ replacing the original target vector $X_{i,g}$. Otherwise the trial vector is rejected and original vector retained, so the population gets better or remains same with generations, but never gets worse.

Steps 2–4 are repeated until a stopping criteria of either the maximum number of iterations (generations) is achieved or a minimum acceptable value of the fitness value is achieved for a minimization problem.

Control parameters of DE algorithm

The scaling factor F and crossover rate CR are the control parameters for the DE algorithm, literature on this algorithm [8, 9, 10, 11, 12, 13, 14] suggests using different values and ranges for these parameters. The values and ranges of these parameters depend on the problem on which this algorithm is applied.

Das et al. [13] applied DE on a problem of automatic clustering of large unlabeled data sets. They suggested varying the values of F and CR in a particular manner. The scaling factor F was varied randomly in the range $(0.5, 1)$ using the equation (13.10):

$$F = 0.5 * (1 + rand(0,1)) \tag{13.10}$$

Also, the crossover rate CR was varied with each iteration or generation as in equation (13.11):

$$CR = (CR_{\max} - CR_{\min}) * (it_{\max} - it)/it_{\max} \tag{13.11}$$

where the CR_{max} and CR_{min} are the maximum and minimum value of crossover which are taken as 1.0 and 0.5, respectively, it_{max} is the maximum number of iterations to be done, *it* is the current iteration number. So, the value of *CR* decreases from 0.5 to 0 with increasing iterations, a high value of *CR* in the beginning helps in the exploration of the search space and at the later stages of search, lower value of *CR* helps in local exploitation.

13.3 Optimization problem: solar photovoltaics

With the advancement in technologies, the energy requirement for an average human being is increasing day by day. The fossil fuels are no longer preferred to meet the energy requirement as there are many problems related to these fuels, such as environmental pollution and global warming. Apart from these problems, their availability on the planet is limited. These problems have been recognized by the researchers since last few decades and they have been giving a lot of importance to the renewable energy in the 21st century. The sources that can be naturally replenished such as sunlight, wind, rain, tides, waves, biomass, etc. are being exploited and researched throughout the world. Solar energy is being encouraged the most as renewable alternatives for nonrenewable energy sources, due to sunlight being easily and freely available throughout the world. Moreover, the conversion of solar energy to electricity using solar photovoltaics (PV) has advantages such as there is no pollution or noise in energy generation.

13.3.1 Solar cell technology

Solar cells based on crystalline silicon wafers are mostly used, and their efficiencies lie in the range of 15–20 %. These solar cells are commercialized, and these are the ones mostly seen on rooftops. Two types of crystalline silicon solar cells (single or monocrystalline and poly or multicrystalline) are commonly used for deployment for the generation of electricity. Solar panels of crystalline silicon constitute more than 80 % of all the solar panels sold around the world.

Both monocrystalline and multicrystalline silicon PV technologies are proven and are being used around the world commercially and in households for power generation. It has been found that silicon based solar cells have more efficiency and much better lifetime as compared to nonsilicon based cells. But at the same time, they are at a disadvantage when working at higher temperatures during hot sunny days due to loss of their efficiency, wherever nonsilicon based thin-film solar cells show up better in these conditions

13.3.2 Solar cell models

Solar cells deployed in fields cannot be tested for prediction of output power all the time, as this uses expensive equipment and lots of resources. Simulation of the equivalent circuit models developed for solar cells can give a very good approximation to the actual power out that would be received in different environmental conditions. Researchers have developed and simulated many models for solar PV cells and PV modules at various environmental conditions of solar irradiance and temperature [15, 16, 17, 18, 19, 20, 21, 22, 23, 24, 25, 26, 27, 28]. In practice, one-diode and two-diode models [15, 16, 17, 18, 19, 20, 21, 22, 23, 24, 25, 26, 27, 28] are mostly used for a p–n junction silicon solar cell.

One-diode model

An ideal solar cell is equivalent to a constant current source, which is represented by I_{ph} – photon generated current in the solar cell by the solar irradiation falling on the solar cell. Due to some losses (both optical and electrical) that occur in a practical solar cell, the solar cell model deviates from its ideal behavior, thus a diode is connected in parallel to I_{ph}, which forms the one-diode model shown in Figure 13.2(a). One series resistance (R_s) and one shunt resistance (R_{sh}) are also added to represent the resistive losses in the solar cell. The load current (I), which depends on the diode parameters, and series and shunt resistances, varies with the load voltage (V) as follows:

$$I = I_{\mathrm{ph}} - I_d - I_{\mathrm{sh}} = I_{\mathrm{ph}} - I_0\left[\exp\left(\frac{V_d}{nV_t}\right) - 1\right] - \frac{V_d}{R_{\mathrm{sh}}} = I(V, I, \text{parameters1}) \tag{13.12a}$$

$$\text{where parameters1} = \{I_{\mathrm{ph}}, I_0, n, R_s, R_{\mathrm{sh}}\} \tag{13.12b}$$

where

I_{ph} = the photon generated current,

I_d = the current flowing through the diode,

I_0 = the reverse saturation current of the diode,

V_d = voltage across the diode = $V + IR_s$,

n = the ideality factor of the diode,

R_s = the series resistance is the sum of the contact resistances and resistances in the bulk and emitter regions,

R_{sh} = the shunt resistance to take care of the resistive loss due to leakage of the current across the p–n junction of the solar cell,

I_{sh} = the current flowing through the shunt resistance, R_{sh}

V_t = the thermal voltage = $k_B T/q$,

where k_B is Boltzmann constant, T is temperature of solar cell in degrees Kelvin and q is electronic charge.

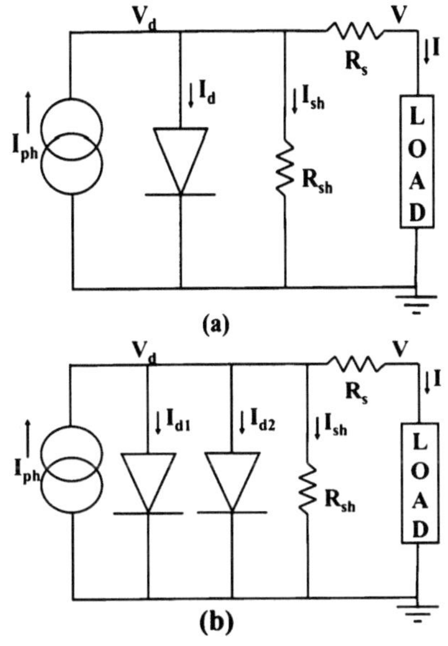

Figure 13.2: (a): One-diode model of a solar cell. (b): Two-diode model of a solar cell.

Ideal one-diode model with $n = 1$ is used, if only diffusion and recombination of carriers in the bulk and emitter regions is considered and it is assumed that there is no recombination in the depletion regions of the p–n junction of the diode. As per experiments by researchers in [29, 30, 31], practically n can vary between 1 and 2 [29, 30, 31].

Two-diode model

Figure 13.2(b) shows the two-diode model where two diodes are connected in parallel to the current source; the series resistance R_s, and the shunt resistance R_{sh} are connected similar to one-diode model. The current through the first diode represents the current loss due to the diffusion and recombination of carriers in the bulk and emitter regions. The current through the second diode represents the current loss due to recombination in the depletion regions near the p–n junction of the solar cell. As the currents through diodes represent the losses in current, direction of these currents is considered opposite to the photon current, I_{ph}. Load current for this model [29, 31] is given by equation (13.13a).

$$\begin{aligned} I &= I_{ph} - I_{d_1} - I_{d_2} - I_{sh} \\ &= I_{ph} - I_{01}\left[\exp\left(\frac{V_d}{n_1 V_t}\right) - 1\right] - I_{02}\left[\exp\left(\frac{V_d}{n_2 V_t}\right) - 1\right] - \frac{V_d}{R_{sh}} \\ &= I(V, I, \text{parameters2}) \end{aligned} \tag{13.13a}$$

where I_{d_1} and I_{d_2} are the currents flowing through the two diodes, I_{01} and I_{02} are the reverse saturation currents of the two diodes and n_1 and n_2 are the ideality factors of the two diodes. In an ideal two-diode model, the recombination in the depletion region of the p–n junction is due to mid-gap states and $n_1 = 1$ and $n_2 = 2$. Thus, five parameters come into picture in the ideal two-diode model, which are given in equation (13.13b).

$$\text{parameters2(ideal two-diode model)} = \{I_{ph}, I_{01}, I_{02}, R_s \text{ and } R_{sh}\} \tag{13.13b}$$

For ideal model, the ideality factors are fixed as 1 and 2, respectively; whereas these are taken as variables in practical model. So, the count of parameters to be estimated for the practical two-diode model (equation (13.13c)) became seven as given below:

$$\text{parameters2(practical two-diode model)} = \{I_{ph}, I_{01}, n_1, I_{02}, n_2, R_s \text{ and } R_{sh}\} \tag{13.13c}$$

Although the ideal two-diode model is theoretically correct and well understood, it does not work well with the experimental $I–V$ data for silicon solar cells [30]. In literature, parameters of two-diode model (equation (13.13c)) for the solar cells have been extracted using population-based metaheuristic optimization algorithms, such as PS [32, 33], GA [34], PSO [35], SA [36], DE [11] and ABSO [37]. This is done to achieve better curve-fitting of the $I–V$ data of cells using evolutionary method like PSO with minimum value of error.

The advantages and disadvantages of one-diode and two-diode models have been discussed in Table 13.1.

13.4 Implementation of PSO algorithm

Single crystalline silicon solar cells of industrial size (area ~ 154.8 cm^2) were used as samples in the present work. These cells had average efficiency, open circuit voltage (V_{oc}) and short circuit current (I_{sc}) values of ~16 %, 0.6254 V and 5.61 A, respectively.

PSO and DE algorithms were applied for the estimation of parameters of two-diode model of these solar cell samples [38]. In each iteration, the fitness function (which was to be minimized) was calculated as the mean of the absolute values of the differences between the calculated value of output current, $I_{calculated}$, and the measured output current, $I_{measured}$ (at given output voltages, V) for ~2500 measured points in the $I–V$ curve. The lesser was the value of the fitness function achieved, the better was the estimation of parameters. Defined number of particles (sets of seven parameters to be estimated) were randomly generated in the PSO algorithm. The velocities and positions of the particles (or the parameters) were updated through a specified number of iterations; 500 iterations worked fine for samples in this work (will be discussed in Section 13.5).

Table 13.1: Comparison of one-diode model and two-diode model.

	Solar Cell Models	
	One-diode model	**Two-diode model**
Advantages	It is simple and has less parameters to work upon. Most of the researchers have found that the simulated results of this model match very well to the experimental results	The two-diode model clearly separates out the two recombination current loss components of the carriers in the solar cell through the parameters n_1 and n_2. The recombination current in the space charge regions dominates near the low diode current values represented by the second diode current with ideality factor n_2 ($n_2 = 2$ in the ideal mid-band recombination case) and recombination current in the quasi neutral region dominates near the high diode current values, which is represented by the first diode through parameter n_1 ($n_1 = 1$ in the ideal case)
	Due to simpler current equation, it saves time in simulations and in curve-fitting for estimation of parameters	This is more accurate model and hence more reliable for modeling and simulations
Disadvantages	Though a simple model, it does not represent the two current components very clearly, it simply combines the effect of two recombination currents in a quasi-neutral region and space charge region in the parameter n	This model has more parameters as compared to one-diode model and hence it is more complex. Simulations and curve-fitting using this model needs more time

The PSO algorithm was implemented using MATLAB coding as per following steps:

(i) The initial ranges of all the parameters were defined.

(ii) Values of all the parameters were randomly picked from the predefined ranges for a defined number of particles. The velocities and positions of all the particles were defined for time t.

(iii) $I_{\text{calculated}}$ was found for a given V at each point of the I–V characteristics; this was done by finding the roots of the following current equation using the Newton–Raphson method:

$$I_{\text{calculated}} - I(V, I_{\text{calculated}}, \text{parameters2}) = 0 \tag{13.14}$$

(iv) Mean Absolute Error (MAE) (defined in Section 13.4.3) was considered as the objective (fitness) function which needed to be minimized in this work.

(v) Velocity and position of each particle was updated for time $t + 1$ as per PSO algorithm equations.

(vi) Steps (iii) to (v) were repeated for a defined number of iterations (which were defined based on the results of MAE values for few samples; it will be discussed in Section 13.5).

13.4.1 Initial ranges of different parameters

The custom PSO algorithm for the estimation of parameters was implemented using MATLAB coding. Equation (13.13a) is the current equation for two-diode model, and the parameters to be estimated were 7 in number, which were I_{ph}, I_{01}, n_1, I_{02}, n_2, R_s and R_{sh} as per equation (13.13c). Initial ranges of all the parameters were set so that the algorithm picked the guess random values within those ranges of individual parameters.

For initial estimates of different parameters, the following methodologies were used:

- Initial estimates of I_{01}, n_1, I_{02}, and n_2 were found from the dark forward I–V characteristics (using Log I_d vs. V_d curve) of the solar cell.
- Initial estimates of R_s and R_{sh} were taken from the solar simulator at National Physical Laboratory (NPL).
- Initial estimate of I_{ph} was taken from the value of I_{sc} given by the solar simulator at NPL.

Estimate of I_{01}, I_{02}, n_1 and n_2

Initial estimates of I_{01}, I_{02}, n_1 and n_2 were taken by analyzing dark characteristics of 4–5 samples of solar cells. The diode current was calculated from the dark current as follows:

Diode voltage (V_d) = voltage applied – voltage drop across the R_s
Diode current (I_d) = Dark current (I) – current through R_{sh}

$$V_d = V - IR_s \quad \text{and} \quad I_d = I - \frac{V_d}{R_{sh}} \tag{13.15}$$

where I_d is the diode current passing through the two diodes and is a sum of I_{d_1} and I_{d_2}. Log(I_d) vs. diode voltage (V_d) curve was drawn (Figure 13.3) to find the estimate for I_{01}, I_{02}, n_1 and n_2 in different voltage ranges.

I_{01}, I_{02}, n_1 and n_2 values can be approximated from the Log(I_d) vs. V_d curve. The curve in Figure 13.3 was fitted in two different regions, first at high voltage values between 0.5 to 0.56 V and then at low voltage values between 0.35–0.45 V. At high and low voltages, the curves exponentially fitted the equations (13.16a) and (13.16b), respectively. The curve-fitting was really good in both cases, with very good values of coefficient of determination (R^2) [39] as 0.996 and 0.999. The coefficient of determination is a statistical measure to assess how well a model fits for the particular data

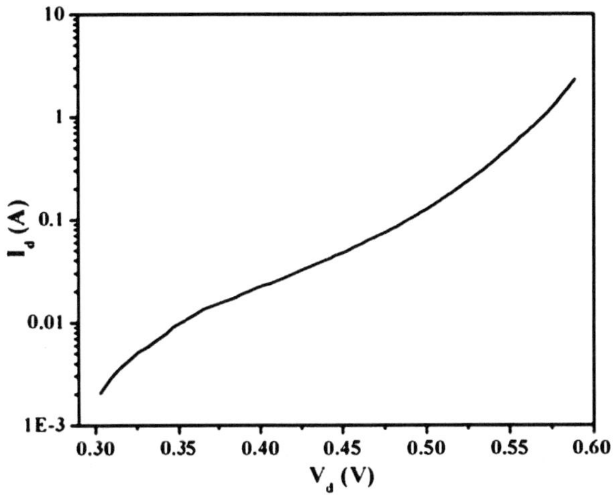

Figure 13.3: Log(I_d) vs. V_d curve generated from the dark characteristics of a typical solar cell.

set. In simple terms, more the value of R^2 toward 1, more confident we can be with the model equation obtained as the outcome of a curve-fit.

The equations came out to be the following:

$$I_{d_1} \text{ (in A)} = 5\text{E}{-}08\text{e}^{29.48V_d} \tag{13.16a}$$

$$I_{d_2} \text{ (in A)} = 5\text{E}{-}05\text{e}^{15.33V_d} \tag{13.16b}$$

Thus,

$$I_{01} = 5\text{E}{-}08\text{ A} \quad \text{and} \quad \frac{1}{n_1 V_t} = 29.48\text{ V} \tag{13.17a}$$

$$I_{02} = 5\text{E}{-}05\text{ A} \quad \text{and} \quad \frac{1}{n_2 V_t} = 15.33\text{ V} \tag{13.17b}$$

where

$$V_t = \frac{kT}{q} = 25.9\text{ mV} = 0.0259\text{ V} \quad \text{at 300 K} \tag{13.18}$$

Hence, from equations (13.17a) and (13.17b), we can find values of n_1 and n_2, as follows:

$$n_1 = 1/(29.48 * 0.0259) = 1.310 \tag{13.19a}$$

$$n_2 = 1/(15.33 * 0.0259) = 2.519 \tag{13.19b}$$

The calculated values of I_{01}, I_{02}, n_1 and n_2 in equations (13.16a) to (13.19b) above, were typical values for one of the solar cell samples, whose dark forward characteristics were studied to get the initial estimate of these parameters.

Estimate of R_s and R_{sh}

A solar simulator at NPL gave $I-V$ data points and curves of the solar cells and the attached LabView software calculated certain parameters; R_s and R_{sh} were among those parameters. The LabView software in the solar simulator estimated R_s through the slope of $I-V$ characteristics at open-circuit voltage and R_{sh} were estimated through the slope of $I-V$ characteristics at short-circuit current. These values were taken as initial guess values for these parameters.

Estimate of I_{ph}

LabView software in the NPL solar simulator system also gave the values of V_{oc} and I_{sc} of the solar cell, whose $I-V$ characteristics were taken. The initial estimate of the photon current I_{ph} was found using the short circuit current I_{sc} as per equation (13.20):

$$I_{ph} = I_{sc}\left(1 + \frac{R_s}{R_{sh}}\right) \tag{13.20}$$

13.4.2 Newton–Raphson method

The $I-V$ characteristics obtained from the solar simulator, gave about 2500 $I-V$ points for each solar cell. After randomly picking the values of all parameters from the initial defined ranges, the algorithm proceeds to calculate the roots of the function given below. The *Newton–Raphson method* is used for this purpose to find $I_{calculated}$ at corresponding voltage value at all points of the $I-V$ curve.

$$\begin{aligned} f(I_{calculated}) = I_{calculated} - \Bigg\{ I_{ph} - I_{01}\left[\exp\left(\frac{V + I_{calculated}R_S}{n_1 V_t}\right) - 1\right] \\ - I_{02}\left[\exp\left(\frac{V + I_{calculated}R_S}{n_2 V_t}\right) - 1\right] - \frac{V + I_{calculated}R_S}{R_{sh}} \Bigg\} = 0 \end{aligned} \tag{13.21}$$

13.4.3 Individual absolute error and mean absolute error

Individual Absolute Error (IAE) was found as the absolute error in the measured and calculated current value at a particular $I-V$ data point.

Mean Absolute Error (MAE) is an error defined as the mean of the individual errors (IAEs) calculated at n number of $I-V$ data points (n was ~2500 in this work). MAE was taken as the fitness function during PSO optimization. MAE and IAE were found as per equation (13.22):

$$\text{MAE} = \frac{\sum_{i=1}^{n} |(I_{calculated} - I_{measured})|}{n} = \frac{\sum_{i=1}^{n} \text{IAE}}{n} \tag{13.22}$$

MAE value demonstrates the (mis)match between the calculated $I-V$ curve and the measured $I-V$ curve. In the ideal case, MAE value should be equal to zero, which means calculated curve must completely overlap measured curve. In this work, MAE value was reduced up to about 0.00645 (average value) in 500 iterations.

13.4.4 Initial ranges of all parameters and values of PSO parameters

- From the dark $I-V$ characteristics, estimated values of ideality factors were 1.3 and 2.6. For the estimation of n_1 and n_2 variables through PSO algorithm, the initial ranges were set as 1 to 1.5 for n_1 and 2 to 3 for n_2.
- Ranges of I_{01} and I_{02} were taken around the initial estimate found through dark characteristics (Section 13.4.1). Ranges of R_s and R_{sh} were defined as ±some% around the initial estimate (say, R_{s_npl} and R_{sh_npl}) got from the NPL solar simulator.
- The value of R_{s_npl}/R_{sh_npl} for all solar cells was in the range of 1E-03 to 1E-04, as per the data of all solar cells obtained from the NPL solar simulator. The value of I_{ph} was not calculated from equation (13.21), as I_{ph} would have been approximately equal to I_{sc} value, R_s/R_{sh} being very small (~0.001) in value. Thus, I_{ph} of each sample was considered equal to its I_{sc} value obtained from the NPL solar simulator; and it was not estimated through the PSO.
- After several trial runs of PSO, the initial ranges of parameters to be defined in PSO were finalized. These ranges were the search space for the PSO algorithm; the positions and velocities of particles were updated in each iteration and then were clamped to remain in the defined search space only. The ranges of parameters that were initially defined in the MATLAB code in this work are shown in Table 13.2. The PSO parameters chosen for the present work are presented in Table 13.3.

Table 13.2: The initial ranges of all parameters defined in PSO.

Parameter	Lower Limit	Higher Limit
I_{01E}	0.01 nA	10 nA
I_{02E}	1 μA	100 μA
n_{1E}	1	1.5
n_{2E}	2	3
R_s	* $R_{s\_npl} - 70\,\%\, R_{s\_npl}$	$R_{s\_npl} + 70\,\%\, R_{s\_npl}$
R_{sh}	* $R_{sh\_npl} - 5\,\%\, R_{sh\_npl}$	$R_{sh\_npl} + 5\,\%\, R_{sh\_npl}$

* R_{s_npl} and R_{sh_npl} were the initial estimates of series and shunt resistances that were obtained from the solar simulator at NPL.

Table 13.3: PSO parameters chosen for this work.

PSO Parameters	Value
Population size	50
Number of iterations	500
Cognitive or local weight c_1	2
Social or global weight c_2	2
Inertia weight w	initial value = 1, final value = 0; the value linearly decreased in each iteration

Table 13.4: Results of estimated parameters for Sample 1 using PSO; 10 runs were done with same setup of PSO.

Sample 1							
PSO Run	I_{01} **(×10^{-9} A)**	n_1	I_{02} **(×10^{-6} A)**	n_2	R_s **(×10^{-3} Ω)**	R_{sh} **(Ω)**	**MAE (A)**
1	6.57	1.22	35.82	2.72	11.66	66.76	0.00626
2	6.35	1.22	10.67	2.33	11.59	62.41	0.00631
3	4.88	1.20	16.13	2.43	11.68	65.35	0.00628
4	4.04	1.19	32.45	2.64	11.79	64.67	0.00636
5	3.59	1.18	27.87	2.58	11.81	65.76	0.00637
6	6.45	1.22	36.20	2.73	11.67	63.46	0.00626
7	5.80	1.21	23.99	2.57	11.67	64.81	0.00626
8	6.79	1.22	15.59	2.44	11.59	63.79	0.00629
9	3.71	1.19	27.88	2.58	11.81	66.96	0.00635
10	3.13	1.18	13.07	2.34	11.78	62.49	0.00632
std dev	1.42	0.017	9.54	0.145	0.086	1.61	4.27E−05

13.5 Results and discussion

Parameters of the two-diode model for 12 solar cell samples were estimated using the PSO algorithm. PSO being a random and heuristic method, the parameters estimated were not exactly same after every run of PSO, even with the entirely same setup. So, 10–12 runs (each with 500 iterations) of PSO was done for each of the solar cell sample to check the spread in results. Table 13.4 shows the results of multiple PSO runs for Sample 1.

Standard deviations of all the estimated parameters were found; the observation of these values in Table 13.4 shows that the results attained in various PSO runs were consistent, despite the random nature of this algorithm.

For Sample 1, the MAE values found in each iteration up to 500 iterations are shown in Figure 13.4 for five different runs. The mean absolute errors decreased at a fast rate during initial 100 iterations, after that the MAE values decreased slowly and achieved the minimum values at around 300 iterations. For five runs shown in Figure 13.4, the minimum MAE values reached at 295, 312, 236, 425 and 328 iterations;

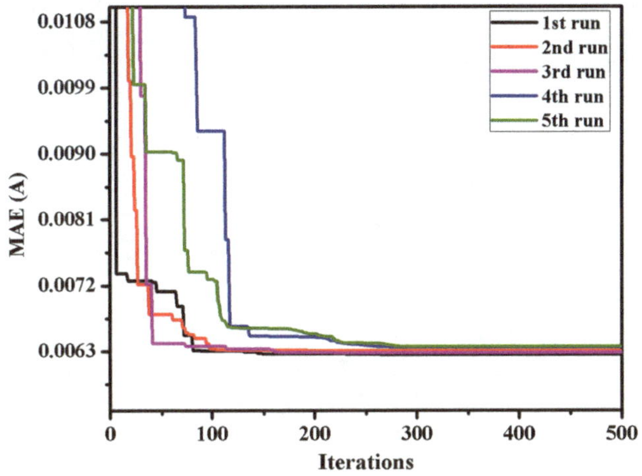

Figure 13.4: MAE values of Sample 1 achieved with iterations (with PSO in progress) for five different runs.

after which the values of MAE remained almost the same until 500 iterations for all the runs. So, the MAE values converged between 295 to 425 iterations in different runs for Sample 1. After observing this pattern of MAE values reduction with iterations, the number of iterations for each PSO run of all solar cell samples were kept as 500.

Out of multiple runs done for a cell, the set of parameters were chosen corresponding to the run that had attained the least MAE value. MAE values were less than 0.01 A for all the solar cells. These values indicated that the estimation of the parameters was good, as the $I-V$ curves calculated using Newton–Raphson method fitted well on $I-V$ curves measured by solar simulator; this happened for all the cells.

Figure 13.5 shows the matched $I_{\text{calculated}}-V$, $I_{\text{measured}}-V$ characteristics and IAE values for two solar cells (Samples 1 and Sample 3). These were obtained after 500 iterations using the PSO algorithm. At this point, the values of parameters along with MAE values were noted down. MAE values obtained were 0.00626 A (0.11 % of I_{sc}) and 0.00516 A (0.092 % of I_{sc}) for Sample 1 and Sample 3, respectively. As per observation from the figure, the IAE values were smaller near the maximum power points and higher near open-circuit voltages (V_{oc}) in all cases; as the Newton–Raphson method could not converge properly near V_{oc}.

Table 13.5 shows the extracted parameters of 12 solar cell samples. The table also shows the mean absolute error attained by PSO algorithm while extraction of parameters of each of the solar cell and the percentage error found by using equation (13.23). The average values of n_1 and n_2 were approximately 1.2 and 2.7, as compared to n_1 as 1 and n_2 as 2 for ideal two-diode model.

$$\text{Percentage error (in \%)} = \frac{\text{MAE}}{I_{sc}} \times 100\,\% \qquad (13.23)$$

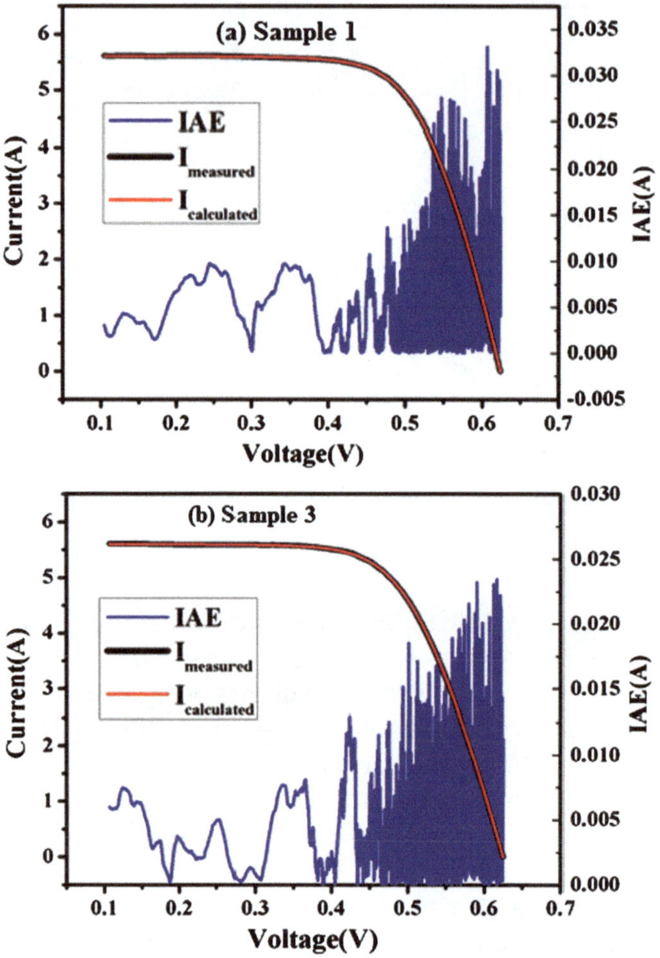

Figure 13.5: $I_{measured}$, $I_{calculated}$ and IAE values vs. output voltage for (a) Sample 1 and (b) Sample 3.

13.5.1 Comparison of results of PSO, DE and slope methods

While taking data from the solar simulator at NPL, the LabView setup attached with the simulator estimated R_s value by the slope of $I-V$ curve at the open circuit condition. Similarly, R_{sh} value was estimated by the slope of $I-V$ curve at short-circuit condition. It was called slope method and these values were used as initial guess value for PSO algorithm and later for DE algorithm as well.

Differential Evolution (DE) algorithm [9, 10, 11] was also run to estimate the parameters of two-diode model for solar cell Sample 1 and Sample 2. The parameters for DE algorithm are the scaling factor F and crossover rate CR, apart from population size and number of iterations. DE algorithm was run many times by using values and ranges of the DE parameters as suggested by various researchers

Table 13.5: Results of estimated parameters for solar cell samples 1 to 12.

Sample	I_{sc} (A)	I_{01} ($\times10^{-9}$A)	n_1	I_{02} ($\times10^{-6}$A)	n_2	R_s ($\times10^{-3}\Omega$)	R_{sh} (Ω)	MAE (A)	Percentage Error (%)
1	5.61	6.45	1.22	36.20	2.73	11.67	63.46	0.00626	0.112
2	5.63	4.31	1.20	63.32	2.79	13.29	16.18	0.00604	0.107
3	5.61	8.40	1.23	15.29	2.57	15.15	33.70	0.00516	0.092
4	5.64	11.15	1.25	18.78	2.67	11.86	9.64	0.00602	0.107
5	5.55	7.37	1.22	16.14	2.60	8.79	21.31	0.00589	0.106
6	5.59	2.26	1.16	36.31	2.85	12.42	13.99	0.00648	0.116
7	5.6	4.45	1.20	10.96	2.49	11.13	15.19	0.00552	0.099
8	5.56	6.50	1.22	2.71	2.25	9.96	105.77	0.00828	0.149
9	5.62	2.73	1.17	81.35	2.97	11.24	20.44	0.00603	0.107
10	5.62	5.70	1.21	28.15	2.84	8.40	12.50	0.00709	0.126
11	5.63	6.83	1.22	36.69	2.74	8.21	11.48	0.00766	0.136
12	5.65	5.18	1.20	30.84	2.67	9.52	12.21	0.00694	0.123

[9, 10, 11, 12, 13, 14]. The values of the parameters were finally chosen as per equations (13.10) and (13.11) (discussed in Section 13.2.2). The number of particles and number of iterations were taken as 50 and 500 respectively, similar to the PSO algorithm. Moreover, the initial search space of two-diode model parameters was defined exactly same as in PSO (discussed in Section 13.4.4). The method suggested by Das et al. [13] was used for this work of parameter extraction of two-diode model of the industrial solar cells and it helped in the proper convergence of the optimization problem.

The results of PSO and DE algorithms in terms of the estimated parameters of Sample 1 and Sample 2 for two-diode model are tabulated in Table 13.6. As can be seen from Table 13.6 (for Sample 1 and Sample 2), the R_s values found by the solar simulator at NPL using slope method, were higher than the R_s values estimated by PSO method; this is expected as the slope method uses the simplified single-diode model for its modeling. At the same time, R_{sh} values estimated by PSO method were very similar to the values found by the "Slope" method. The values of different solar cell parameters estimated by the PSO and DE algorithms were quite comparable.

13.6 Summary and conclusions

Solar cell samples of large area crystalline silicon were considered for this work, the parameters of two-diode model for these samples were estimated using PSO algorithm. While implementing the PSO algorithm, many challenges were encountered, which were resolved by detailed literature study related to PSO, and many trial runs of PSO algorithm done through MATLAB coding. Multiple runs were done for each so-

Table 13.6: Comparison of estimated parameters for Sample 1 and Sample 2, using the slope method (values taken from solar simulator at NPL), PSO and DE algorithms.

	Method	I_{01} ($\times10^{-9}$A)	n_1	I_{02} ($\times10^{-6}$A)	n_2	R_s ($\times10^{-3}\Omega$)	R_{sh} (Ω)	MAE (A)
Sample 1	Values taken from NPL Solar Simulator – the Slope Method					17.2	64.3	
	PSO	6.45	1.22	36.20	2.73	11.67	63.46	0.00626
	DE	8.72	1.23	51.04	2.91	11.61	63.52	0.00625
Sample 2	Values taken from NPL Solar Simulator – the Slope Method					19.0	15.8	
	PSO	4.31	1.19	63.32	2.79	13.29	16.18	0.00604
	DE	5.341	1.21	69.99	2.85	13.23	16.48	0.00605

lar cell sample, to get confidence on the results for that particular sample by checking the standard deviation of each of the parameter estimated in different runs. The following conclusions could be drawn:

- During implementation of the PSO algorithm for all the solar cells to reduce the mean absolute error, MAE, it was found that 500 iterations were sufficient to reduce the MAE up to third decimal place and further significant reduction in MAE was not observed for iterations more than 500.
- Minimum MAE achieved was 0.00516 A (which was equal to 0.092 % of I_{sc}) for the Sample 3. The maximum MAE value obtained was 0.00828 A (equal to 0.149 % of I_{sc}) for the sample 8, while rest of the MAE values were much lower than this, making the average MAE as 0.0065 A, which was approximately 0.12 % of I_{sc} value. These values were significantly low for such precise I–V data of approximately 2500 points for each solar cell. The well matched $I_{\text{calculated}}-V$ and $I_{\text{measured}}-V$ curves for the solar cells is a proof of the high accuracy of the results. Implementation of the DE algorithm for estimation of the cell parameters were also undertaken for a few solar cells. The parameters estimated for solar cells Sample 1 and Sample 2 through DE algorithm were very close to the ones estimated by the PSO algorithm, thus giving confidence to the PSO results for cell parameters of all the solar cells.
- The highlights of the results were that the n_1 values of all the solar cell samples were > 1 and the average value of n_1 was around 1.2. Similarly, the n_2 values of all solar cells were between 2.5 and 3 (means > 2), averaging it around 2.7. Thus, the industrial size crystalline silicon solar cells were not represented by the ideal two-diode model, where n_1 and n_2 values are 1 and 2, respectively. In fact, practical two-diode model best represents these types of solar cells.

Bibliography

[1] Holland JH. Adaptation in natural and artificial systems: an introductory analysis with applications to biology, control and artificial intelligence. Cambridge, MA, USA: MIT Press; 1992.
[2] Kennedy J, Eberhart R. Particle swarm optimization. In: Proceedings of IEEE international conference on neural networks. Perth, Australia. 1995. p. 1942–8.
[3] Storn R, Price K. Differential evolution – a simple and efficient heuristic for global optimization over continuous spaces. J Glob Optim. 1997;11(4):341–59.
[4] Panduro MA, Brizuela CA, Balderas LI, Acosta DA. A comparison of genetic algorithms, particle swarm optimization and the differential evolution method for the design of scannable circular antenna arrays. Prog Electromagn Res B. 2009;13:171–86.
[5] Das S, Abraham A, Konar A. Particle swarm optimization and differential evolution algorithms: technical analysis, applications and hybridization perspectives. In: Advances of computational intelligence in industrial systems. Studies in computational intelligence. vol. 116. 2008. p. 1–38.
[6] Shahzad F, Baig AR, Masood S, Kamran M, Naveed N. Opposition-based particle swarm optimization with velocity clamping (OVCPSO). Adv Intell Soft Comput. 2009;116:339–48.
[7] Engelbrecht AP. Computational intelligence: an introduction. 2nd ed. Wiley; 2007.
[8] Qin AK, Huang VL, Suganthan PN. Differential evolution algorithm with strategy adaptation for global numerical optimization. IEEE Trans Evol Comput. 2009;13(2):398–417.
[9] Ishaque K, Salam Z. An improved modeling method to determine the model parameters of photovoltaic (PV) modules using differential evolution (DE). Sol Energy. 2011;85(9):2349–59.
[10] Ishaque K, Salam Z, Mekhilef S, Shamsudin A. Parameter extraction of solar photovoltaic modules using penalty-based differential evolution. Appl Energy. 2012;99:297–308.
[11] Jiang LL, Maskell D, Patra JC. Parameter estimation of solar cells and modules using an improved adaptive differential evolution algorithm. Appl Energy. 2013;112:185–93.
[12] Qin AK, Suganthan PN. Self-adaptive differential evolution algorithm for numerical optimization. In: IEEE congress on evolutionary computation. vol. 2. 2005. p. 1785–91.
[13] Das S, Abraham A, Konar A. Automatic clustering using an improved differential evolution algorithm. IEEE Trans Syst Man Cybern, Part A, Syst Hum. 2008;38(1):218–37.
[14] Tamrakar R, Gupta A. Extraction of solar cell modelling parameters using differential evolution algorithm. Int J Innov Res Electr Electron Instrum Control Eng. 2015;3(11):78–82.
[15] Saloux E, Teyssedou A, Sorin M. Explicit model of photovoltaic panels to determine voltages and currents at the maximum power point. Sol Energy. 2011;85(5):713–22.
[16] Ishaque K, Salam Z, Taheri H. Simple, fast and accurate two-diode model for photovoltaic modules. Sol Energy Mater Sol Cells. 2011;95(2):586–94.
[17] Sheriff MA, Babagana B, Maina BT. A study of silicon solar cells and modules using PSPICE. World J Appl Sci Technol. 2011;3(1):124–30.
[18] Nema RK, Nema S, Agnihotri G. Computer simulation based study of photovoltaic cells/modules and their experimental verification. Int J Recent Trends Eng. 2009;1(3):151–6.
[19] Tsai HL. Insolation-oriented model of photovoltaic module using Matlab/Simulink. Sol Energy. 2010;84(7):1318–26.
[20] Altas IH, Sharaf AM. A photovoltaic array simulation model for Matlab-Simulink GUI environment. In: International conference on clean electrical power (ICCEP' 07). 2007. p. 341–5.
[21] Gow JA, Manning CD. Development of a photovoltaic array model for use in power-electronics simulation studies. IEE Proc, Electr Power Appl. 1999;146(2):193–200.
[22] Can H, Ickilli D, Parlak KS. A new numerical solution approach for the real-time modeling of photovoltaic panels. In: Asia-Pacific power and energy engineering conference. Shanghai.

2012. p. 1–4.
[23] Villalva MG, Gazoli JR, Filho ER. Modeling and circuit-based simulation of photovoltaic arrays. In: Brazilian power electronics conference. Bonito-Mato Grosso do Sul. 2009. p. 1244–54.
[24] Jiang Y, Qahouq JAA, Batarseh I. Improved solar PV cell Matlab simulation model and comparison. In: Proceedings of IEEE international symposium on circuits and systems. Paris: ISCAS; 2010. p. 2770–3.
[25] Lineykin S, Averbukh M, Kuperman A. Five-parameter model of photovoltaic cell based on STC data and dimensionless. In: IEEE 27th convention of electrical & electronics engineers in Israel (IEEEI). Eilat. 2012. p. 1–5.
[26] Ramos-Hernanz JA, Campayo JJ, Larranaga J, Zulueta E, Barambones O, Motrico J, Fernandez Gamiz U, Zamora I. Two photovoltaic cell simulation models in Matlab/Simulink. Int J Tech Phys Probl Eng. 2012;4(1):45–51.
[27] Nikhil PG, Subhakar D. An improved simulation model for photovoltaic cell. In: International conference on electrical and control engineering (ICECE). China. 2011. p. 1978–82.
[28] Jiang Y, Qahouq JAA, Orabi M. Matlab/Pspice hybrid simulation modeling of solar PV cell/module. In: Twenty-sixth annual IEEE applied power electronics conference and exposition (APEC). Fort Worth, TX. 2011. p. 1244–50.
[29] Green MA. Solar cells: operating principles, technology and system applications. Englewood Cliffs, NJ, USA: Prentice-Hall Inc.; 1982. p. 93–5.
[30] Ma T, Yang H, Lu L. Solar photovoltaic system modeling and performance prediction. Renew Sustain Energy Rev. 2014;36:304–15.
[31] Solanki C. Solar photovoltaics; fundamentals, technologies and applications. New Delhi, India: Prentice-Hall India Learning Private Ltd.; 2012. p. 102–3.
[32] AlHajri M, El-Naggar K, AlRashidi M, Al-Othman A. Optimal extraction of solar cell parameters using pattern search. Renew Energy. 2012;44:238–45.
[33] AlRashidi MR, AlHajri MF, El-Naggar KM, Al-Othman AK. A new estimation approach for determining the *I–V* characteristics of solar cells. Sol Energy. 2011;85(7):1543–50.
[34] Jervase J, Bourdoucen H, Al-Lawati A. Solar cell parameter extraction using genetic algorithms. Meas Sci Technol. 2001;12:1922–5.
[35] Sandrolini L, Artioli M, Reggiani U. Numerical method for the extraction of photovoltaic module double-diode model parameters through cluster analysis. Appl Energy. 2010;87(2):442–51.
[36] El-Naggar KM, AlRashidi MR, AlHajri MF, Al-Othman AK. Simulated annealing algorithm for photovoltaic parameters identification. Sol Energy. 2012;86(1):266–74.
[37] Askarzadeh A, Rezazadeh A. Artificial bee swarm optimization algorithm for parameters identification of solar cell models. Appl Energy. 2013;102:943–9.
[38] Khanna V, Das BK, Bisht Vandana D, Singh PK. A three diode model for industrial solar cells and estimation of solar cell parameters using PSO algorithm. Renew Energy. 2015;78(C):105–13.
[39] Hardle W, Simar L. Applied multivariate statistical analysis. 2nd ed. Springer; 2007.

S. N. Kumar, A. Lenin Fred, Parasuraman Padmanabhan, Balazs Gulyas, and Ajay H. Kumar

14 Multilevel thresholding using crow search optimization for medical images

Abstract: Segmentation is the process of delineation of the desired region of interest and in the case of medical images; the region of interest are anatomical organs, tumor or cyst. The multilevel thresholding gains much importance for the images with complex objects and the role of the optimization algorithm is vital in the selection of threshold values. Thresholding is a classical segmentation algorithm and for the estimation of threshold values, Otsu's or Kapur's technique is used. This research work employs various optimization techniques like electromagnetic optimization, harmony search optimization and crow search optimization for the optimum selection of threshold values. The electromagnetism like optimization algorithm (EMO) is based on the electromagnetism laws of physics. The harmony search algorithm key concept is that, when musicians compose the harmony, they usually try various possible combinations of the music pitches stored in the memory, based on this concept, this algorithm was evolved. The bio-inspired optimization techniques are gaining prominence in many applications and the crow search optimization is based on the biological traits of crows. The multilevel thresholding, when coupled with crow search optimization, was found to yield efficient results. The less number of parameter tuning and less complexity makes crow search optimization, an efficient one for solving the real-world problems. The algorithms are developed in Matlab 2010a tested on real-time CT abdomen DICOM data sets. The performance metrics evaluation reveals the efficiency of crow search optimization in the multilevel thresholding segmentation approach.

Keywords: Segmentation thresholding, crow search optimization, harmony search optimization, electromagnetism optimization

14.1 Introduction

The image processing role is inevitable in the medical field, remote sensing, computer vision and robotics. The computer-aided algorithms are used for the analysis of im-

S. N. Kumar, Amal Jyothi College of Engineering, Kanjirappally, Kerala, India, e-mail: appu123kumar@gmail.com
A. Lenin Fred, Ajay H. Kumar, Mar Ephraem College of Engineering and Technology, Marthandam, Tamilnadu, India, e-mails: leninfredip@gmail.com, ajayhakkumar@gmail.com
Parasuraman Padmanabhan, Balazs Gulyas, Nanyang Technological University, Jurong West, Singapore, e-mails: ppadmanabhan@ntu.edu.sg, balazs.gulyas@ntu.edu.sg

https://doi.org/10.1515/9783110671353-014

ages for specific applications. Segmentation is the technique of extraction of desired region of interest. The segmentation algorithms are classified into semiautomatic and automatic techniques [1].

Optimization algorithms role is inevitable in computer vision and image processing. The optimization techniques are coupled with classical algorithms for yielding better accuracy in computer-aided diseases diagnosis and for therapeutic applications [2]. Optimization techniques are used to properly tune the parameters, thereby eliminating manual intervention in parameter selection. Optimization algorithms are applicable in all domains of medical image processing including filtering, delineation of the region of interest, classification of abnormalities and compression. Apart from parameter tuning, computation speed is also crucial. The optimization techniques minimize the computation complexity, thereby parallel processing can yield faster results [3].

Metaheuristic techniques perform a robust search in the solution space, thereby escaping from the local optima [4]. The performances of classical algorithms in image processing are greatly enhanced by the incorporation of optimization techniques. Hybrid optimization techniques are also now widely used in many applications by inheriting the features of two or more algorithms. The combination of optimization techniques is also important to yield fruitful results [5, 6]. Though many optimization techniques are there, there is an urge for the development of novel optimization techniques for solving real-world problems.

The histogram thresholding with multiobjective criterion performs better for gray level image when compared with the Otsu method, Gaussian curve fitting-based method, valley emphasis-based method and two-dimensional Tsallis entropy-based method [7]. The Two-Stage Multithreshold Otsu (TSMO) method was proposed for reducing the time complexity of the multilevel thresholding, the accuracy of TSMO method was calculated using misclassification errors [8]. Maximum entropy-based honey bee mating optimization thresholding (MEHBMOT) method was proposed for the multilevel thresholding segmentation. Efficient results were produced when compared with PSO, the hybrid cooperative-comprehensive learning-based PSO algorithm (HCOCLPSO) and the fast Otsu's method [9]. The artificial bee colony (ABC) optimization when coupled with Kapur's entropy thresholding generates efficient results for and PSO was found to be efficient for bi-level thresholding [10]. M. A. E. Aziz et al. proposed two multilevel thresholding based on Otsu and Kapur's method for the image segmentation and analyzed the performance of the Whale Optimization Algorithm (WOA) and Moth-Flame Optimization (MFO) were compared with the SCA, HS, SSO, FASSO and FA optimization algorithm [11].

Thresholding techniques are broadly classified into two types; parametric and nonparametric [12, 13, 14, 15]. The parameter technique relies on the probability density estimation fort the modeling of each class and nonparameter techniques uses variance or entropy of class for segmentation [16, 17, 18]. Otsu thresholding maximizes the variance between the classes and Kapur's thresholding is based on the maximization of entropy [17, 18]. The genetic algorithm was coupled with the Gaussian model for the multilevel thresholding [19] and in [20] also genetic algorithm was used for the selection of optimum threshold value. The particle swarm optimization (PSO) and artificial bee colony (ABC) were employed for the optimum selection of threshold value in Kapur's thresholding technique [21]. In [22], the bacterial foraging algorithm was used for the optimum selection of threshold value. Section 14.2 describes the multilevel thresholding and various optimization techniques that are coupled with the thresholding algorithm. The results and discussion are described in Section 14.3.

14.2 Multilevel thresholding

Thresholding is a classical segmentation algorithm in which the grouping of pixels into distinct classes based on the gray values. The selection of threshold value is a vital one in this case and C_s expressed as follows:

$$\begin{aligned} S_1 &\to g \quad \text{if } 0 \le g \le t \\ S_2 &\to g \quad \text{if } t \le g \le N-1 \end{aligned}$$

The input is a grayscale image with $x \times y$ pixels and g represents one of the pixels. The image has N gray values $N = \{0, 1, 2, \ldots, N-1\}$ and will be in the range 0–255. The t represents the threshold value and S_1, S_2 represents the classes. Equation above represents the bilevel thresholding and works well for images with a simple and single region of interest. When multiple ROI's are there, bilevel thresholding values and the classical thresholding is sensitive to noise.

The multilevel thresholding is expressed as follows:

$$\begin{aligned} S_1 &\to g \quad \text{if } 0 \le g \le t_1 \\ S_2 &\to g \quad \text{if } t_1 \le g \le t_2 \\ S_3 &\to g \quad \text{if } t_i < g < t_{i+1} \\ S_4 &\to g \\ S_n &\to g_n \quad \text{if } t_n < g < N-1, \end{aligned}$$

where $\{t_1, t_2, \ldots, t_i, t_{i+1}, t_k\}$ represents different threshold values. The issue in bilevel thresholding and multilevel thresholding is the selection of appropriate threshold values. The Otsu and Kapur techniques are used for the estimation of threshold values. The objective function is maximized to estimate the optimum threshold values.

The classical thresholding is sensitive to noise, hence preceding to segmentation, preprocessing was done by nonlinear tensor diffusion filter.

The Otsu's method relies on the variance between the classes and it is a nonparametric technique. The criteria for segmentation used is maximum variance value of classes.

The probability distribution of gray value intensity is expressed as

$$P_i^c = \frac{H_i^c}{N}$$

$$\sum_{i=1}^{N} P_i^c = 1$$

$$c = \begin{cases} 1,2,3, & \text{if the input image is colour (RGB)} \\ 1, & \text{if the input image is gray scale,} \end{cases}$$

where "i" represents the specific intensity value $0 \leq i \leq N-1$, c is the component of the image and it relies on whether the input is grayscale or color (RGB), N is the total pixel count in the image, H_i^c is the count of pixels (histogram) that represents the i intensity level in c. The normalization of the histogram is done with the probability distribution P_i^c.

The two classes for bilevel thresholding is defined as follows:

$$C_1 = \frac{P_i^c}{W_0^c(t)} \cdots \frac{P_t^c}{W_0^c(t)}$$

$$C_2 = \frac{P_{t+1}^c}{W_1^c(t)} \cdots \frac{P_N^c}{W_1^c(t)},$$

where $W_0(t)$ and $W_1(t)$ represent the probability distributions for C_1 and C_2 and is represented as follows:

$$W_0^c(t) = \sum_{i=1}^{t} P_i^c$$

$$W_1^c(t) = \sum_{i=t+1}^{N} P_i^c$$

For the estimation of the variance of classes, the mean values have to be estimated. The mean values are represented as follows:

$$\mu_0^c = \sum_{i=1}^{t} \frac{iP_i^c}{W_0^c(t)}$$

$$\mu_1^c = \sum_{i=t+1}^{N} \frac{iP_i^c}{W_1^c(t)}$$

The variance of classes C_1 and C_2 as expressed as follows:

$$V_1^c = W_0^c(\mu_0^c + \mu_t^c)^2,$$

where $\mu_t^c = W_0^c\mu_0^c + W_1^c\mu_1^c$ and $W_0^c + W_1^c = 1$.

The objective function is denoted as follows:

$$\mathcal{O}(t) = \max(V^{2c}(t)); \quad 0 \le t \le N-1,$$

where $V^2(t)$ is the Otsu's variance. The objective of the optimization problem is to determine the intensity of the pixel level that maximizes the function in the equation.

For multilevel thresholding, the objective function is expressed as follows:

$$\mathcal{O}(T) = \max(V^{2c}(T)), \quad 0 \le t_i \le N-1, \quad i = 1, 2, \ldots, k$$

Here, i represents the class, W_i^c and μ_i^c represent probability value and mean value of the class.

The Kapur thresholding method relies on the entropy and it also a nonparametric method to estimate the optimal threshold values. The optimum threshold value is determined by the maximization of overall entropy. The entropy represents the compactness and separability among classes. The optimal threshold value separates the classes, where the entropy has a maximum value. The objective function of Kapur thresholding for bilevel technique is expressed as follows:

$$\mathcal{O}(t) = E_1^c + E_2^c,$$

$$c = \begin{cases} 1, 2, 3, & \text{if the input image is color (RGB)} \\ 1, & \text{if the input image is gray scale,} \end{cases}$$

where the entropy values E_1 and E_2 are represented as follows:

$$E_1^c = \sum_{i=1}^{t} \frac{P_i^c}{W_0^c} \ln\left(\frac{P_i^c}{W_0^c}\right)$$

$$E_2^c = \sum_{i=t+1}^{N} \frac{P_i^c}{W_1^c} \ln\left(\frac{P_i^c}{W_1^c}\right),$$

where P_i^c is the probability, $W_0(t)$ and $W_1(t)$ are the probability distribution for C_1 and C_2. Similar to Otsu's method, Kapur's method can be extended to the multiple threshold values, the image is divided into k classes.

The objective function is expressed as follows:

$$\mathcal{O}(\bar{t}) = \max\left(\sum_{i=1}^{k} E_i^c\right)$$

$$c = \begin{cases} 1, 2, 3, & \text{if the input image is color (RGB)} \\ 1, & \text{if the input image is gray scale,} \end{cases}$$

where $\bar{t} = \{t_1, t_2, \ldots, t_{k-1}\}$ represents a vector that comprises multiple threshold values. The entropy values are estimated separately and represented as follows:

$$E_1^c = \sum_{i=1}^{t_1} \frac{P_i^c}{W_0^c} \ln\left(\frac{P_i^c}{W_0^c}\right)$$

$$E_2^c = \sum_{i=t+1}^{t_2} \frac{P_i^c}{W_1^c} \ln\left(\frac{P_i^c}{W_1^c}\right)$$

$$E_k^c = \sum_{i=t_{k+1}}^{N} \frac{P_i^c}{W_{k-1}^c} \ln\left(\frac{P_i^c}{W_{k-1}^c}\right)$$

This work proposes optimization algorithms for the optimum selection of threshold values at each level. The crow search optimization based multilevel thresholding was proved to be efficient and for comparative analysis, electromagnetism optimization and harmony search optimization are employed.

14.2.1 Electromagnetism like optimization algorithm

The electromagnetism like optimization algorithm (EMO) is based on the electromagnetism laws of physics. The EMO algorithm was designed in such a manner that it can find a global solution for a nonlinear optimization problem. It is a population-based technique [23].

The initial population of the EMO algorithm is represented as follows:

$$S_t = \{q_{1,t}, q_{2,t}, \ldots, q_{N,t}\}$$

Based on the electromagnetism theory of physics, each point $q_{i,j} \in S_t$ is assumed as a charged particle. The attraction-repulsion mechanism is a process in which points with more charge attract other points, in search region and points with less charge repels other points. The net force exerted on j^{th} point $q_{j,t}$ is determined by summing the attraction-repulsion forces for each $q_{i,t} \in S_t$ moved in the direction of total force to the location $y_{i,t}$.

The members $q_{i,t+1} \in S_{t+1}$ is determined by using

$$q_{i,t+1} = \begin{cases} y_{i,t} & \text{if } f(y_{i,t}) < f(z_{i,t}) \\ z_{i,t} & \text{otherwise} \end{cases}$$

The flow chart of EMO optimization is depicted in Figure 14.1.

The steps in the EMO algorithm are summarised as follows:

1. Initialize the parameters maximum count of iterations $iteration_{\max}$, population size (N), local distance (δ), iteration center ($t = 1$).
2. Determine the best point in S_t:

$$S_t = \{x_1, x_2, \ldots, x_n\}$$

3. *while* $t < iteration_{\max}$, *do*

$$F_i^t \rightarrow \textit{calculate}\ F(S_t)$$

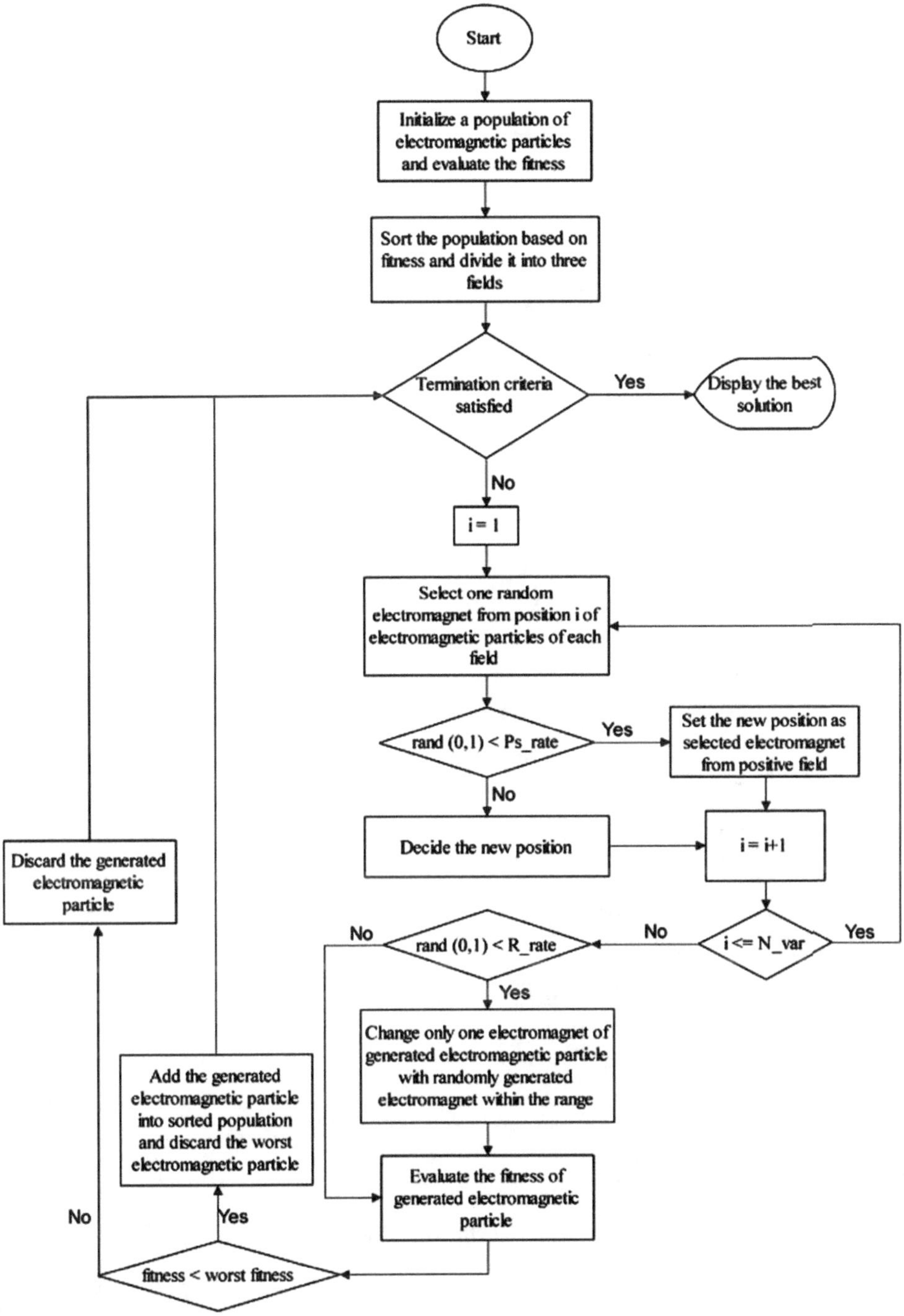

Figure 14.1: Flowchart of the EMO algorithm.

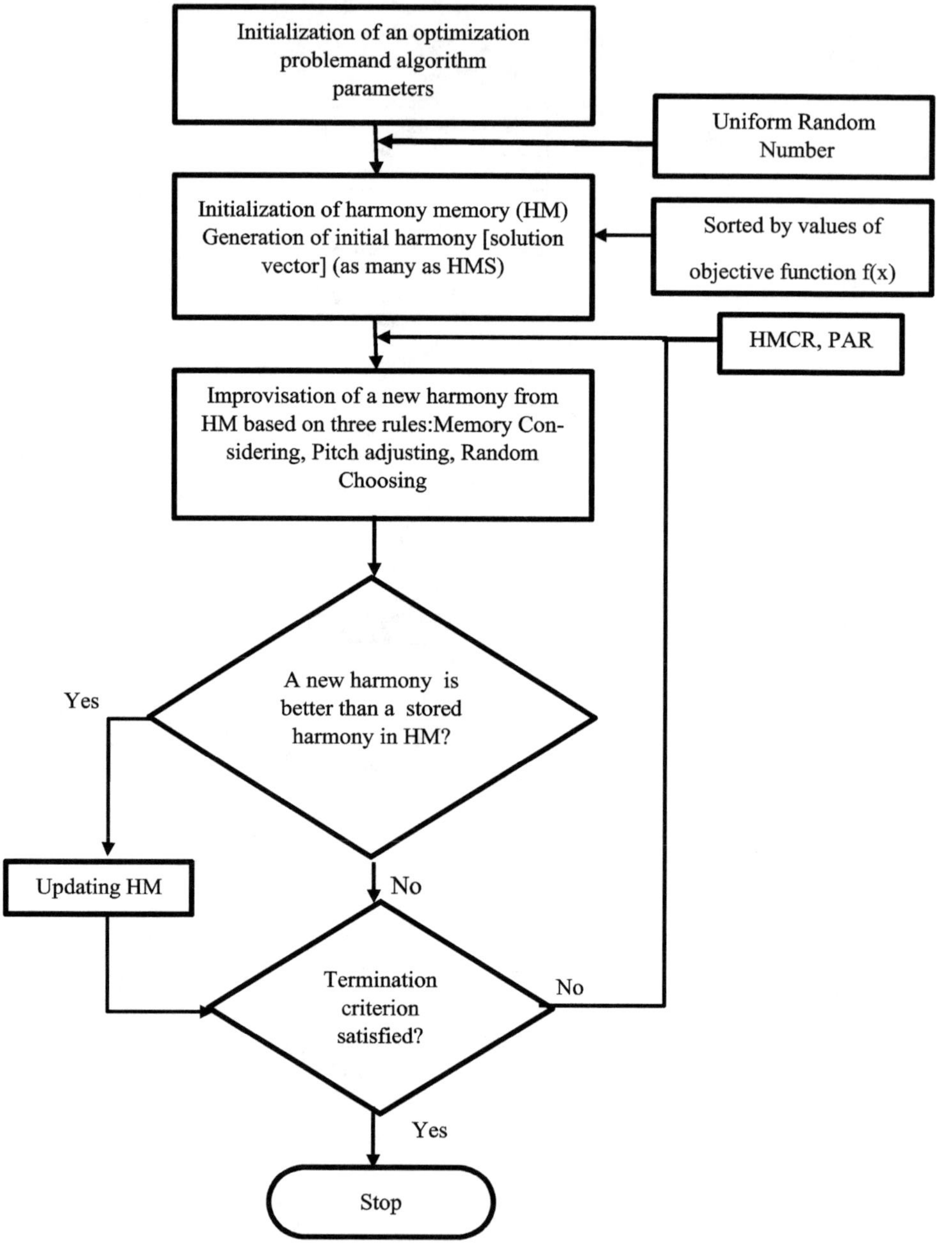

Figure 14.2: Flowchart of the harmony search optimization algorithm.

$$Y_{i,t} \leftarrow move\ (X_{i,t}, F_i^t)$$
$$Z_{i,t} \leftarrow local\ (iteration_{local}, \delta, Y_{i,t})$$
$$X_{i,t+1} \leftarrow select\ (\delta_{t+1}, Y_{i,t}, Z_{i,t})$$

end while

14.2.2 Harmony search optimization algorithm

The harmony search algorithm key concept is that, when musicians compose the harmony, they usually try various possible combinations of the music pitches stored in the memory [24]. The flow diagram of Harmony search optimization algorithm is depicted in Figure 14.2.

The musicians do the music exploration as follows:

(i) Selection of any one pitch from memory.
(ii) Selection of an adjacent pitch from the memory.
(iii) Selection of random pitch from the possible.

The Harmony Search (HS) memory is initialized with solution vectors. The following three conditions are validated:

(i) Selection of any value from HS memory.
(ii) Selecting an adjacent value from the HS memory.
(iii) Selection of a random value from the possible value range.

The steps of harmony search optimization algorithm are summarized as follows:

Step 1: Initialize the HS memory by a number of randomly generated solutions to the optimization problem under consideration.
The harmony memory is depicted below:

$$\mathrm{HM} = \begin{bmatrix} y_1^1 & y_2^1 & y_n^1 \\ y_1^2 & y_2^2 & y_n^2 \\ y_1^{\mathrm{HMS}} & y_2^{\mathrm{HMS}} & y_n^{\mathrm{HMS}} \end{bmatrix},$$

where $[\ y_1^i \ \ y_2^i \ \ y_n^i\ ]$ $(i = 1, 2 \ldots \mathrm{HMS})$ is a solution candidate.

Step 2: The improvement of the solution is done in this step based on HMCR. The HMCR is stated as the probability of selecting a component from the present HM members and 1-HMCr is the probability of randomly selecting the solution. The random solution can be further mutated according to PAR.

Step 3: The harmony memory is updated from the above step. If the fitness of the new solution generated is better than the worst member, it will replace that one.

Step 4: Steps 2 and 3 are repeated until the termination criterion is reached.

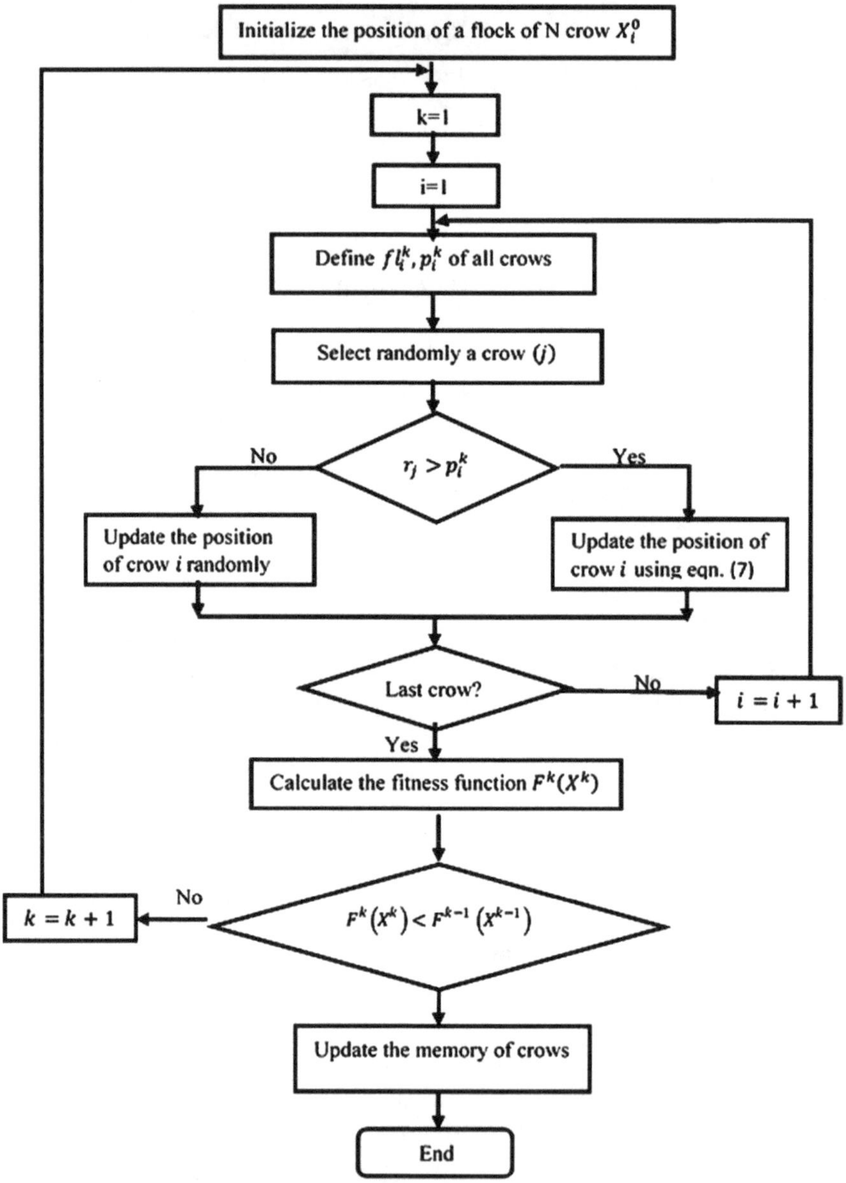

Figure 14.3: Flow diagram of crow search optimization algorithm.

Table 14.1: Parameters of the crow search optimization.

Flight length	Awareness probability	Flock size	Max iteration
2	0.1	20	1000

Table 14.2: Parameters of the HSOA.

HM	HMCR	PAR	DBW	CI
50	0.95	0.5	0.5	1000

Table 14.3: Parameters of the EMO optimization.

Delta	Population	Local iteration	Max iteration
0.025	50	250	1000

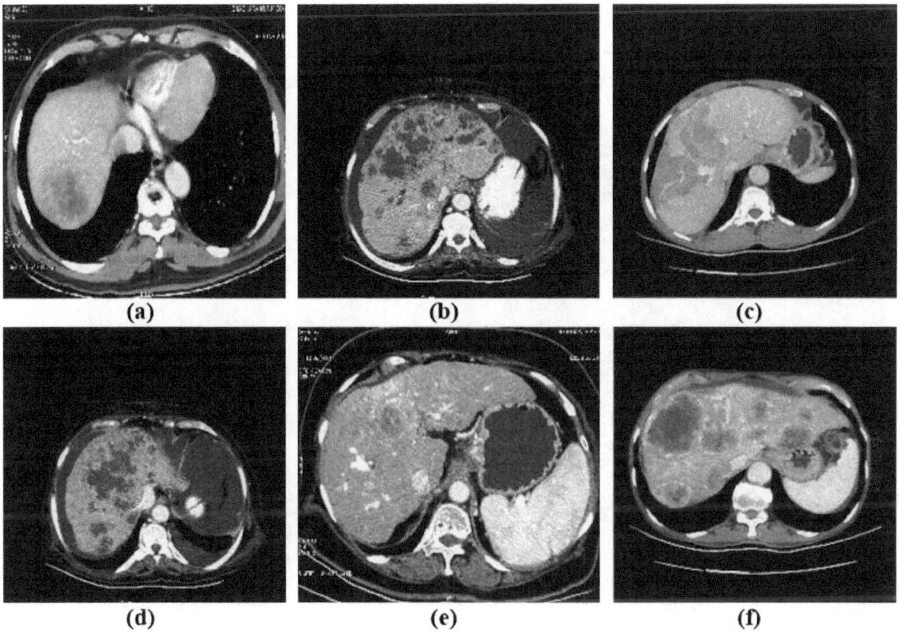

Figure 14.4: Input images for multilevel thresholding optimization segmentation algorithms (ID1–ID6).

14.2.3 Crow search optimization algorithm

The crow search is a metaheuristic bio-inspired optimization that is based on the behavior of crows. It is an intelligent bird with the largest brain size when compared with its body [25].

The crow search optimization is based on the following characteristics:

(i) Crows live in the form of group.
(ii) Crows memorize the positions, where they hide food.
(iii) Crows follow each other to do theft.

Table 14.4: Performance metrics values of multilevel thresholding EMO segmentation algorithm.

Image Details	Level	Threshold values	MSE	PSNR	Energy	Average Cluster variance
ID1	3	93 168	6.167404e+02	20.2298	956036.230490	16.0376
	4	72 135 213	4.091540e+02	22.0119	2339943.063687	18.7702
	5	62 117 166 216	2.333738e+02	24.4503	1442419.748382	11.7697
ID2	3	109 197	8.298072e+02	18.9410	777262.717231	20.4414
	4	83 141 202	4.289262e+02	21.8070	763434.513555	14.4190
	5	1 63 131 202	1.394575e+02	26.6864	1463633.332007	16.0949
ID3	3	1 77	4.737302e+02	21.3755	226971.061124	18.6470
	4	1 65 134	2.471034e+02	24.2020	1298405.483319	15.6448
	5	1 72 134 212	1.109570e+02	27.6793	4839728.543867	16.3265
ID4	3	109 193	8.952728e+02	18.6112	608600.002235	19.4286
	4	87 142 199	5.281969e+02	20.9028	603310.268957	13.3828
	5	1 54 121 193	1.282590e+02	27.0499	1342234.153160	15.1916
ID5	3	138 192	1.536690e+03	16.2649	3570108.226467	11.7728
	4	91 143 195	5.699552e+02	20.5724	5741924.284463	12.0763
	5	85 133 177 215	4.545727e+02	21.5548	5846700.325752	10.5855
ID6	3	142 198	1.704012e+03	15.8161	559647.599963	12.9620
	4	1 86 186	2.823483e+02	23.6230	1845433.946843	20.8698
	5	1 70 138 198	1.223842e+02	27.2536	1206795.079781	14.4535

Table 14.5: Performance metrics values of multilevel thresholding Harmony Search Optimization segmentation algorithm.

Image Details	Level	Threshold values	MSE	PSNR	Energy	Average Cluster variance
ID1	3	68 168	5.340481e+02	20.8550	1690019.723378	23.4546
	4	33 99 183	2.930902e+02	23.4608	1873083.384059	16.9297
	5	33 100 165 220	1.555557e+02	26.2119	2041018.713097	13.7595
ID2	3	48 148	4.308691e+02	21.7874	1316784.212008	24.9883
	4	37 114 197	1.865861e+02	25.4220	1455016.471547	18.6423
	5	27 95 157 200	1.199513e+02	27.3408	1206395.310547	13.8768
ID3	3	65 162	2.041683e+02	25.0309	2962975.310581	18.7993
	4	59 146 197	1.125599e+02	27.6170	3881737.786653	14.7601
	5	48 122 161 207	6.004119e+01	30.3463	3056607.457255	11.3633
ID4	3	46 138	3.594040e+02	22.5750	1132189.150762	21.1802
	4	34 116 201	1.670235e+02	25.9030	1436355.153816	18.3580
	5	30 87 138 203	9.397809e+01	28.4005	1064245.433707	13.3467
ID5	3	58 157	7.037060e+02	19.6569	10404817.287977	26.4822
	4	45 123 196	2.881220e+02	23.5350	10705009.494100	17.2622
	5	32 90 144 198	1.763367e+02	25.6674	8146588.255801	12.5092
ID6	3	59 163	3.099302e+02	23.2182	1164366.104810	20.5381
	4	58 139 198	1.466997e+02	26.4665	1213632.584742	14.5295
	5	43 113 155 204	9.223560e+01	28.4818	1021117.250135	11.9611

Table 14.6: Performance metrics values of multilevel thresholding Crow Search Optimization segmentation algorithm.

Image Details	Level	Threshold values	MSE	PSNR	Energy	Average Cluster variance
ID1	3	68 168	5.340481e+02	20.8550	1690019.723378	23.4546
	4	34 105 179	2.883067e+02	23.5323	1706848.896587	16.4762
	5	32 100 165 220	1.554035e+02	26.2162	2057571.649266	13.8124
ID2	3	48 147	4.307095e+02	21.7890	1286886.935091	24.6691
	4	39 115 197	1.865910e+02	25.4219	1423824.102181	18.4482
	5	32 89 143 210	1.047349e+02	27.9299	1142396.707305	13.8933
ID3	3	65 162	2.041683e+02	25.0309	2962975.310581	18.7993
	4	62 150 211	1.090374e+02	27.7	4317918.838699	15.2830
	5	50 122 165 217	5.799412e+01	30.4970	3252109.329699	11.7121
ID4	3	45 139	3.590525e+02	22.5792	1177663.480840	21.6347
	4	39 113 195	1.635122e+02	25.9953	1330220.038209	17.6775
	5	31 83 134 204	9.341067e+01	28.4268	1049594.396426	13.2824
ID5	3	61 159	7.033483e+02	19.6591	10476126.463699	26.3325
	4	45 123 196	2.881220e+02	23.5350	10705009.494100	17.2622
	5	37 97 147 203	1.736724e+02	25.7335	8272438.030997	12.5512
ID6	3	63 164	3.094294e+02	23.2252	1158590.739710	20.3694
	4	57 141 201	1.462850e+02	26.4788	1259133.520648	14.8017
	5	49 118 160 209	9.138622e+01	28.5220	1031810.165442	11.7778

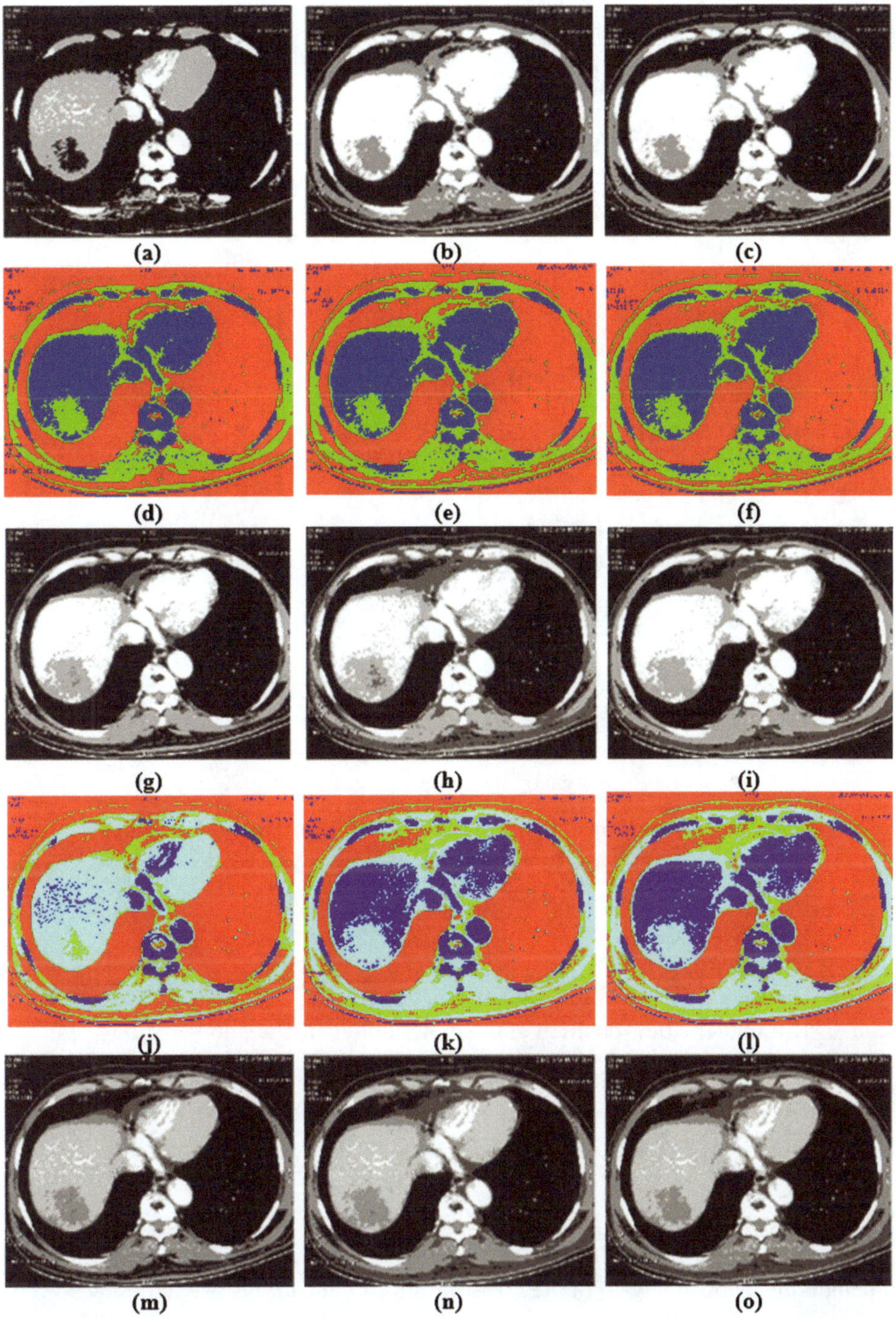

Figure 14.5: ID1 multilevel thresholding crow search optimization segmentation results corresponding to threshold values 3, 4, 5.

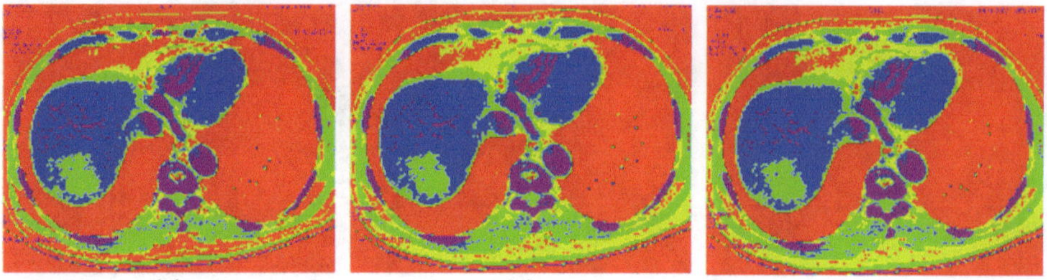
(p) (q) (r)

Figure 14.5: (continued)

The flow diagram of crow search optimization is depicted in Figure 14.3.

The steps in crow search optimization are as follows:

Step 1: Initialization of position of crows in solution space is done randomly.
Step 2: Determine the position of crows.
Step 3: Initialize the memory of crows.
Step 4: Determine the new position of crow.

while $iteration < iteration_{\max}$
for $i = 1 : N$
//*Choose one of the crow randomly* (j)
//*Awareness probability is defined*
if $x_j > A_P^{j,\text{iteration}}$
$y^{j,\text{iteration}+1} = y^{j,\text{iteration}} + x_j \times f_l^{i,\text{iteration}}(m^{j,\text{iteration}} - x^{i,\text{iteration}})$
else
$y^{j,\text{iteration}+1}$ = a random position in solution space
end if
end for
end while

Step 5: Determine the feasibility of new positions.
Step 6: Estimate the new position of the crows update the memory of crows.

14.3 Results and discussion

The algorithms are developed in Matlab 2010a and tested on real-time CT/MR images. The input images are in DICOM format with a size of 512 ∗ 512. The Otsu multilevel thresholding algorithm was coupled with various optimizations like harmony search, electromagnetism and crow search approaches. The output is taken for various threshold values 2, 3, 4, 5. The optimum threshold values in each level are determined by the optimization algorithm.

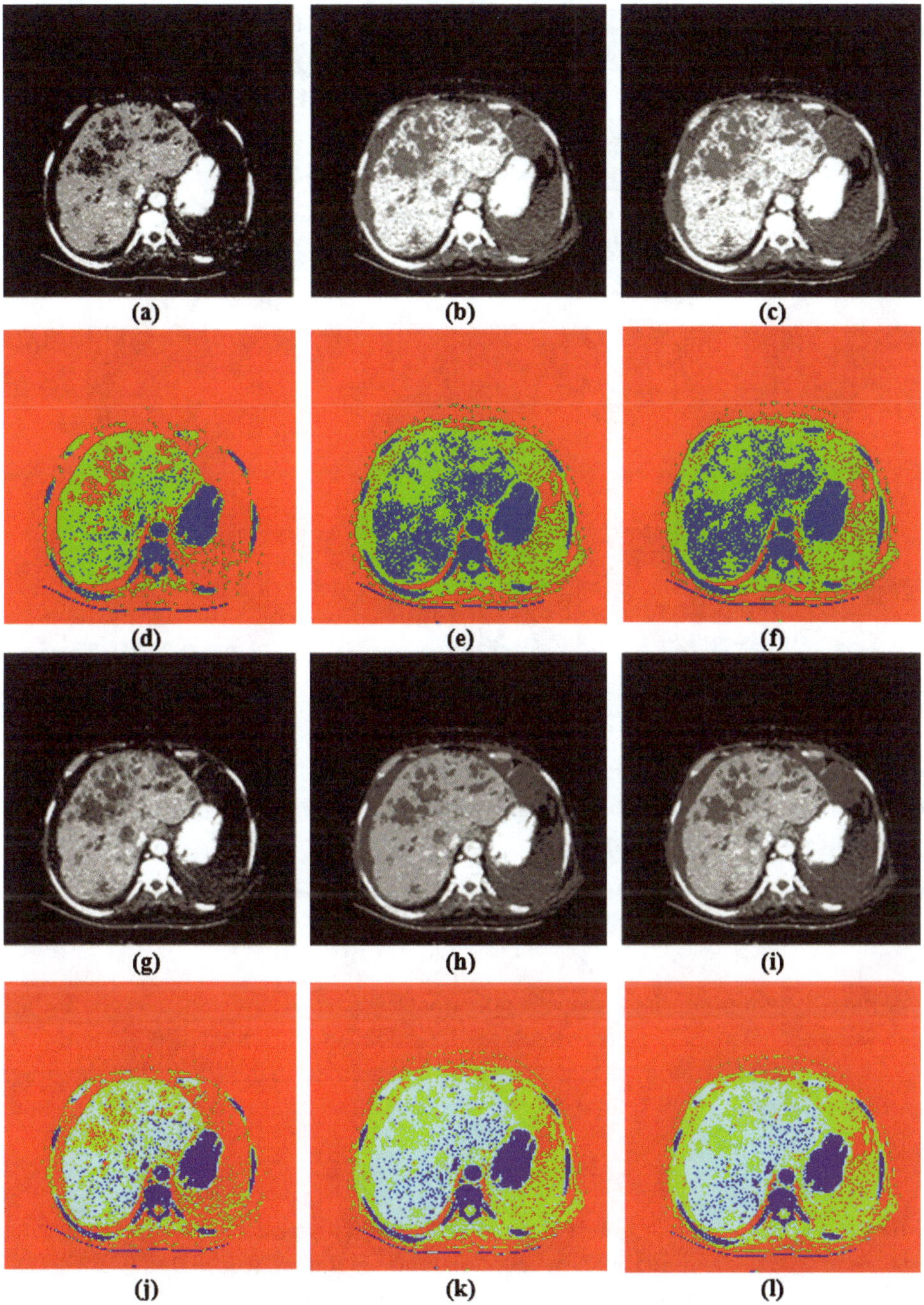

Figure 14.6: ID2 multilevel thresholding crow search optimization segmentation results corresponding to threshold values 3, 4, 5.

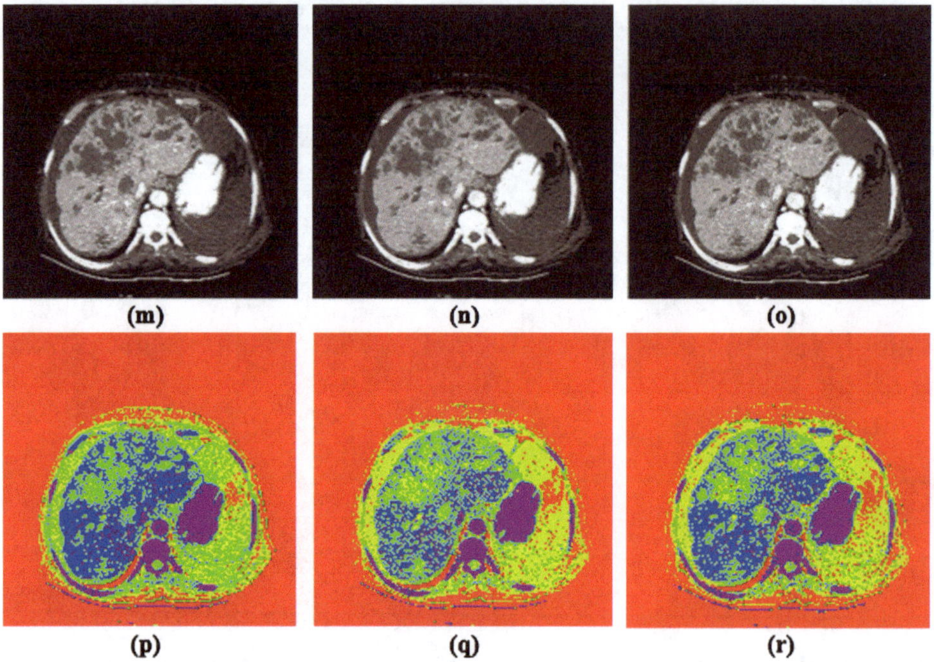

Figure 14.6: (continued)

In the case of crow search optimization employed in multilevel thresholding, each crow depicts "K" different decision elements. The decision elements represent the threshold variables in segmentation:

$$crow = \lfloor y_1^c, y_2^c, \ldots, y_{\text{crow}}^c \rfloor^T$$
$$y_1^c = \lfloor T_1^c, T_2^c, \ldots, T_k^c \rfloor$$

Here, T_1 represents the transpose operator.

The parameters of the crow search optimization is represented in Table 14.1.

In the case of harmony search optimization employed in multilevel thresholding, each harmony depicts "K" different decision elements:

$$\text{Hm} = \lfloor y_1^c, y_2^c, \ldots, y_{\text{HMS}}^c \rfloor^T,$$

where y_1^c the expression is represented above. For RGB images, $c = 1, 2, 3$. For grayscale images, $c = 1$. The boundary of the solution space is limited to $g = 0$ to $g = 255$ that represents intensity levels of the image.

The parameters of the harmony search algorithm are harmony memory (HM), harmony memory consideration rate (HMCR), pitch adjusting rate (PAR), distance bandwidth (DBW) and count of improvisations (CI).

The parameters of HSOA are depicted in Table 14.2.

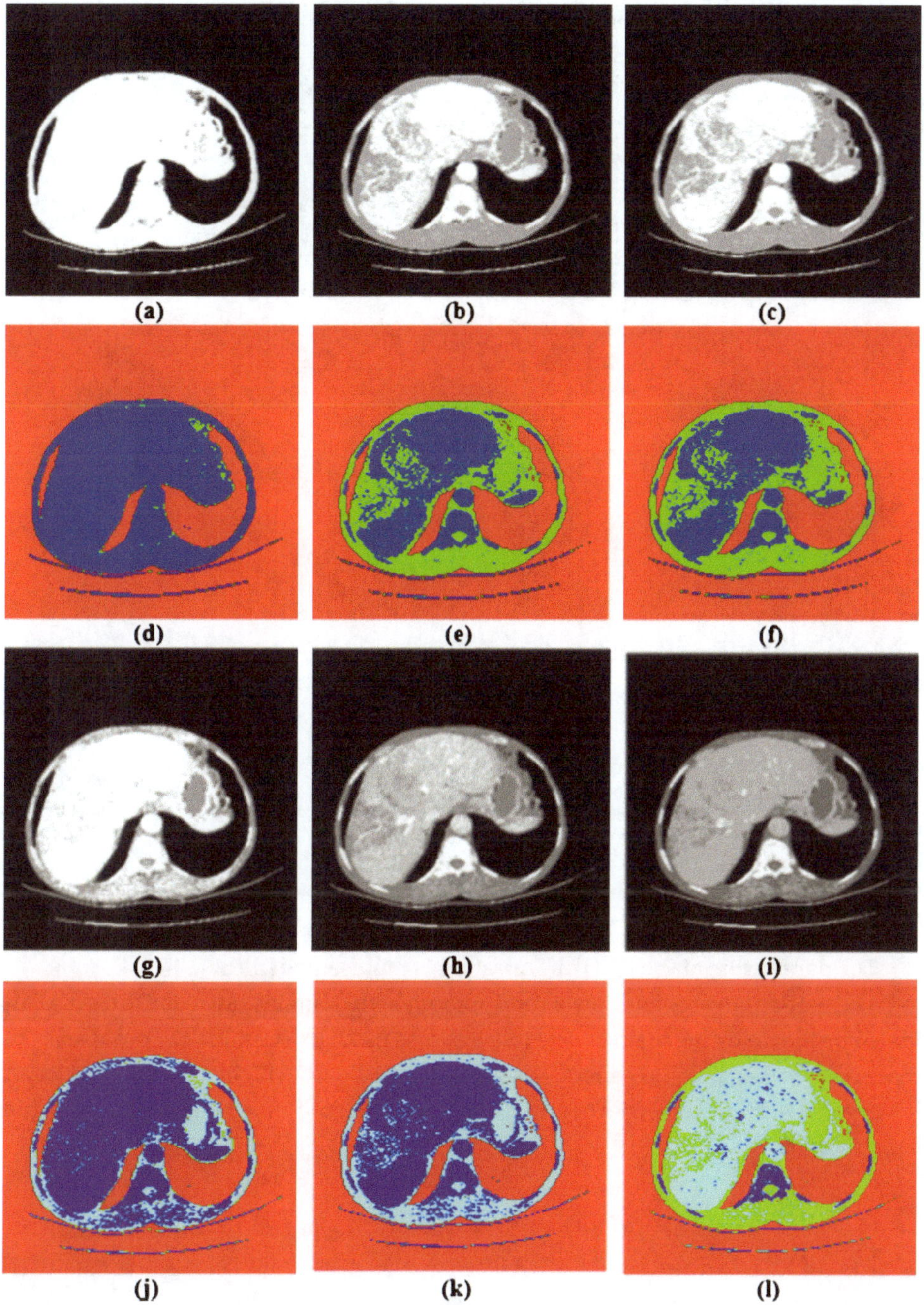

Figure 14.7: ID3 multilevel thresholding crow search optimization segmentation results corresponding to threshold values 3, 4, 5.

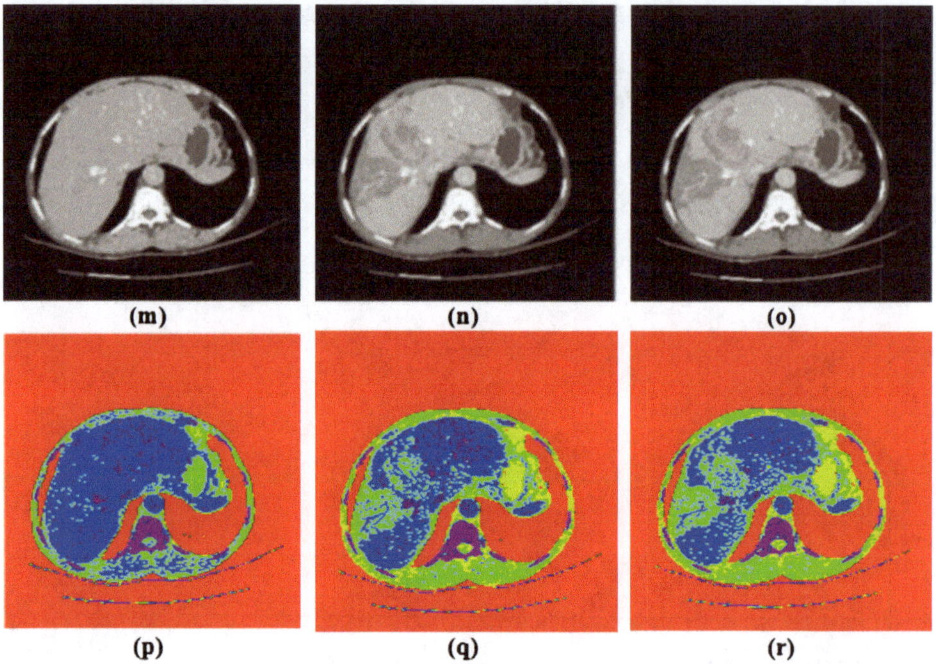

Figure 14.7: (continued)

In the case of EMO employed in multilevel thresholding, each charged particle depicts "K" different decision elements.

$$\mathrm{EMO} = \lfloor y_1^c, y_2^c, \ldots, y_{\mathrm{EMO}}^c \rfloor^T$$

The parameters of the EMO optimization is represented in Table 14.3.

The performance of multilevel thresholding with optimization algorithms are validated by the following metrics. The PSNR can be used to determine the proficiency of the quality of the machine-generated segmentation result. A higher value of PSNR and lower value of MSE represents the robustness of the algorithm. The PSNR and MSE are expressed as

$$\mathrm{MSE} = \frac{1}{N}\sum_{x=1}^{M}\sum_{y=1}^{N}(I(x,y) - \hat{I}(x,y))^2$$

$$\mathrm{PSNR} = 10\log\frac{255^2}{\mathrm{MSE}}$$

The $I(x,y)$ represents the grayscale input image and $\hat{I}(x,y)$ represents the machine-generated segmented image.

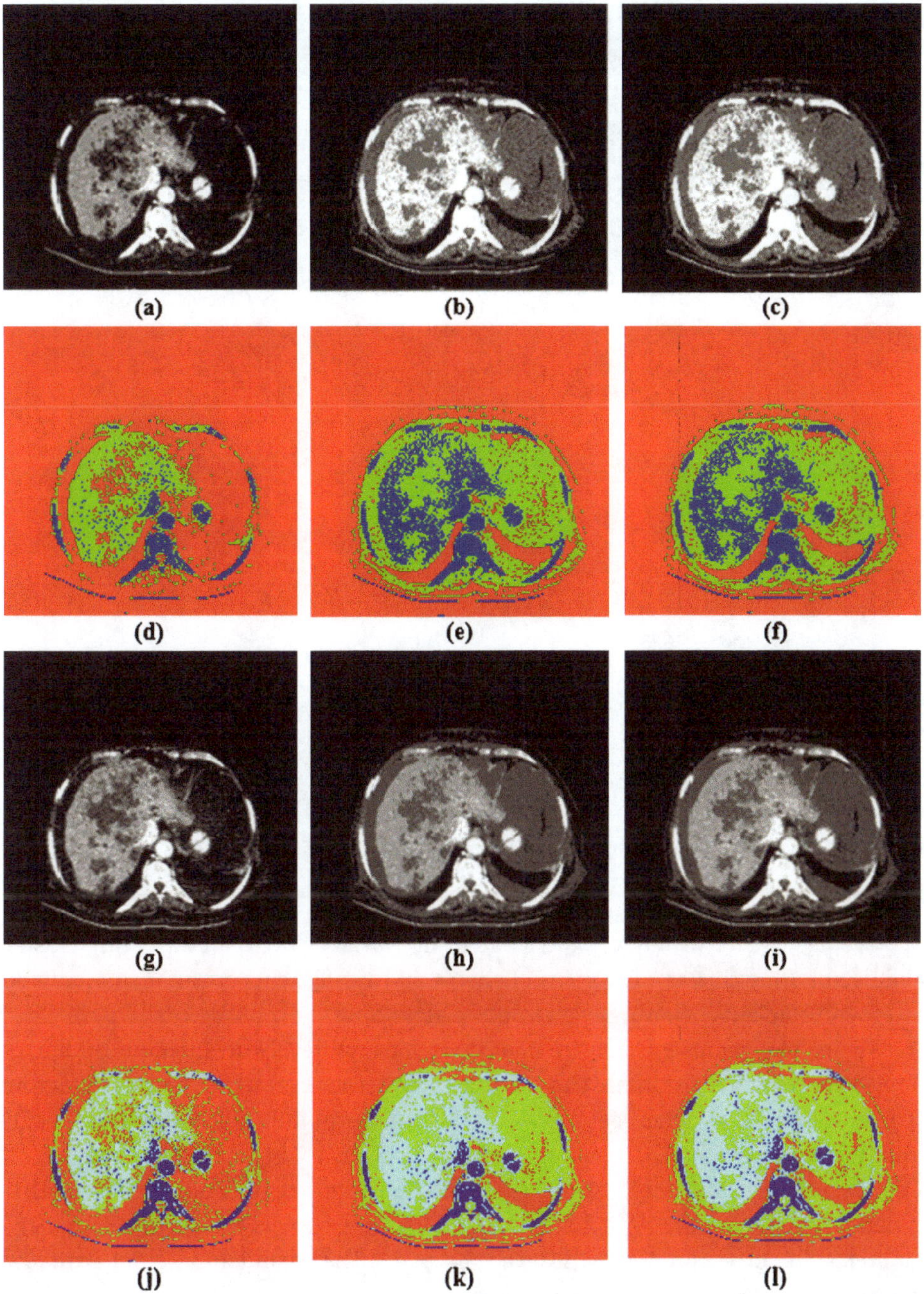

Figure 14.8: ID4 multilevel thresholding crow search optimization segmentation results corresponding to threshold values 3,4,5.

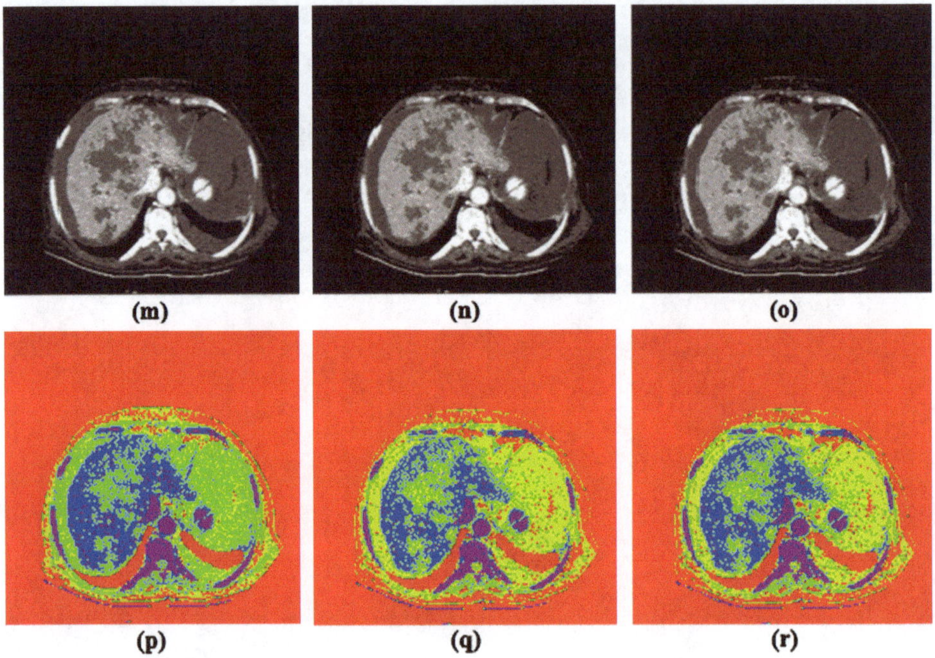

Figure 14.8: (continued)

The expression for energy metric is as follows:

$$\text{Energy} = \sum_{x}^{M} \sum_{y}^{N} \hat{I}(x, y)^2$$

Average cluster variance is the mean of the variance between the centre cluster and neighboring clusters. The value of energy and average cluster variance should be low for an efficient clustering segmentation algorithm.

Figure 14.4 represents the input images for multilevel thresholding optimization segmentation algorithms. The performance metrics evaluation reveals that crow search optimization, when coupled with the multilevel thresholding generates efficient results.

The multilevel thresholding crow search optimization segmentation results of ID1 is depicted in Figure 14.5. The first, third and fifth row depicts the grayscale segmentation results corresponding to the levels 3, 4 and 5. The second, fourth and sixth row depicts pseudo-colored segmentation results.

The multilevel thresholding crow search optimization segmentation results of ID2 is depicted in Figure 14.6. The first, third and fifth row depicts the grayscale segmentation results corresponding to the levels 3, 4 and 5. The second, fourth and sixth row depicts pseudo-colored segmentation results.

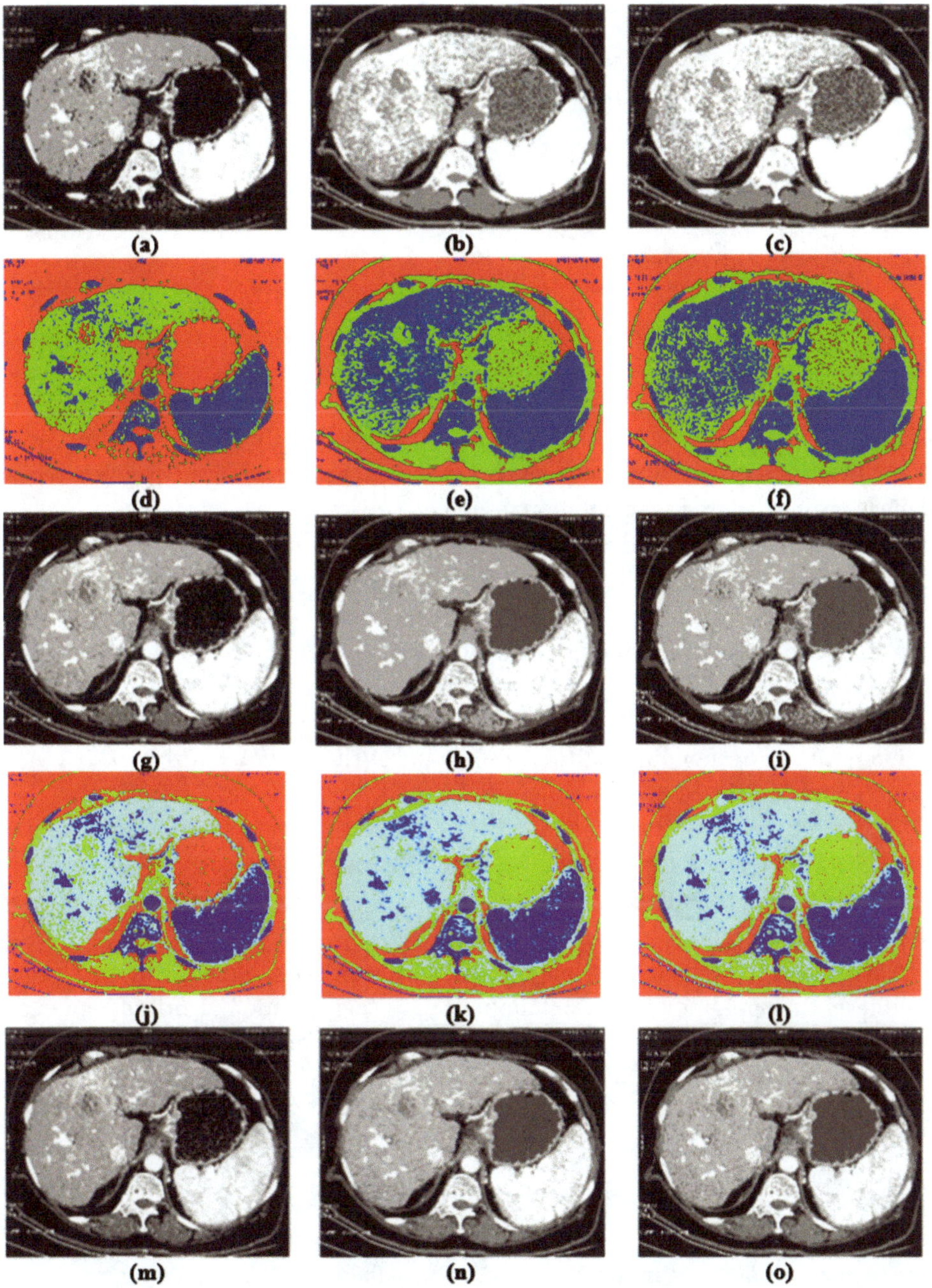

Figure 14.9: ID5 multilevel thresholding crow search optimization segmentation results corresponding to threshold values 3, 4, 5.

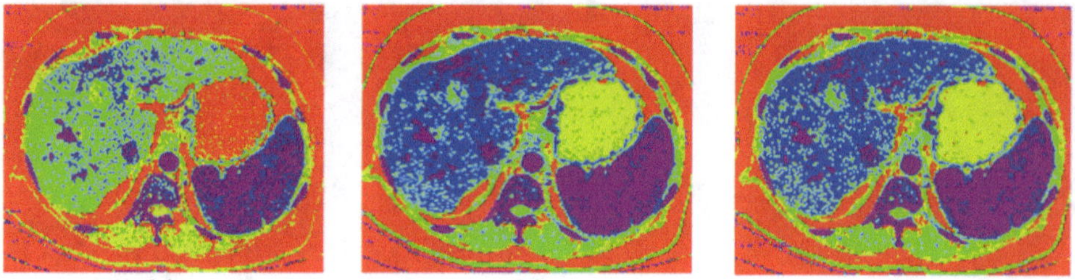

(p) (q) (r)

Figure 14.9: (continued)

The multilevel thresholding crow search optimization segmentation results of ID3 is depicted in Figure 14.7. The first, third and fifth row depicts the grayscale segmentation results corresponding to the levels 3, 4 and 5. The second, fourth and sixth row depicts pseudo colored segmentation results.

The multilevel thresholding crow search optimization segmentation results of ID4 is depicted in Figure 14.8. The first, third and fifth row depicts the grayscale segmentation results corresponding to the levels 3, 4 and 5. The second, fourth and sixth row depicts pseudocolored segmentation results.

The multilevel thresholding crow search optimization segmentation results of ID5 is depicted in Figure 14.9. The first, third and fifth row depicts the grayscale segmentation results corresponding to the levels 3, 4 and 5. The second, fourth and sixth row depicts pseudo-colored segmentation results.

The multilevel thresholding crow search optimization segmentation results of ID6 is depicted in Figure 14.10. The first, third and fifth row depicts the grayscale segmentation results corresponding to the levels 3, 4 and 5. The second, fourth and sixth row depicts pseudo-colored segmentation results.

The performance metrics values in Tables 14.4, 14.5, 14.6 reveal the efficiency of the multilevel thresholding crow search optimization segmentation algorithm. For each input, image outputs are obtained for different levels of threshold and the results are depicted in Figs. 14.5, 14.6, 14.7, 14.8, 14.9 and 14.10.

14.4 Conclusion

The segmentation role is inevitable in the medical field for image-guided therapy and treatment planning. This chapter focuses on the importance of optimization algorithm in multilevel thresholding technique. The optimization algorithm chooses optimum threshold values for the multilevel thresholding algorithm. The crow search optimization, when coupled with the multilevel thresholding generates efficient results when compared with the multilevel thresholding coupled with the harmony search opti-

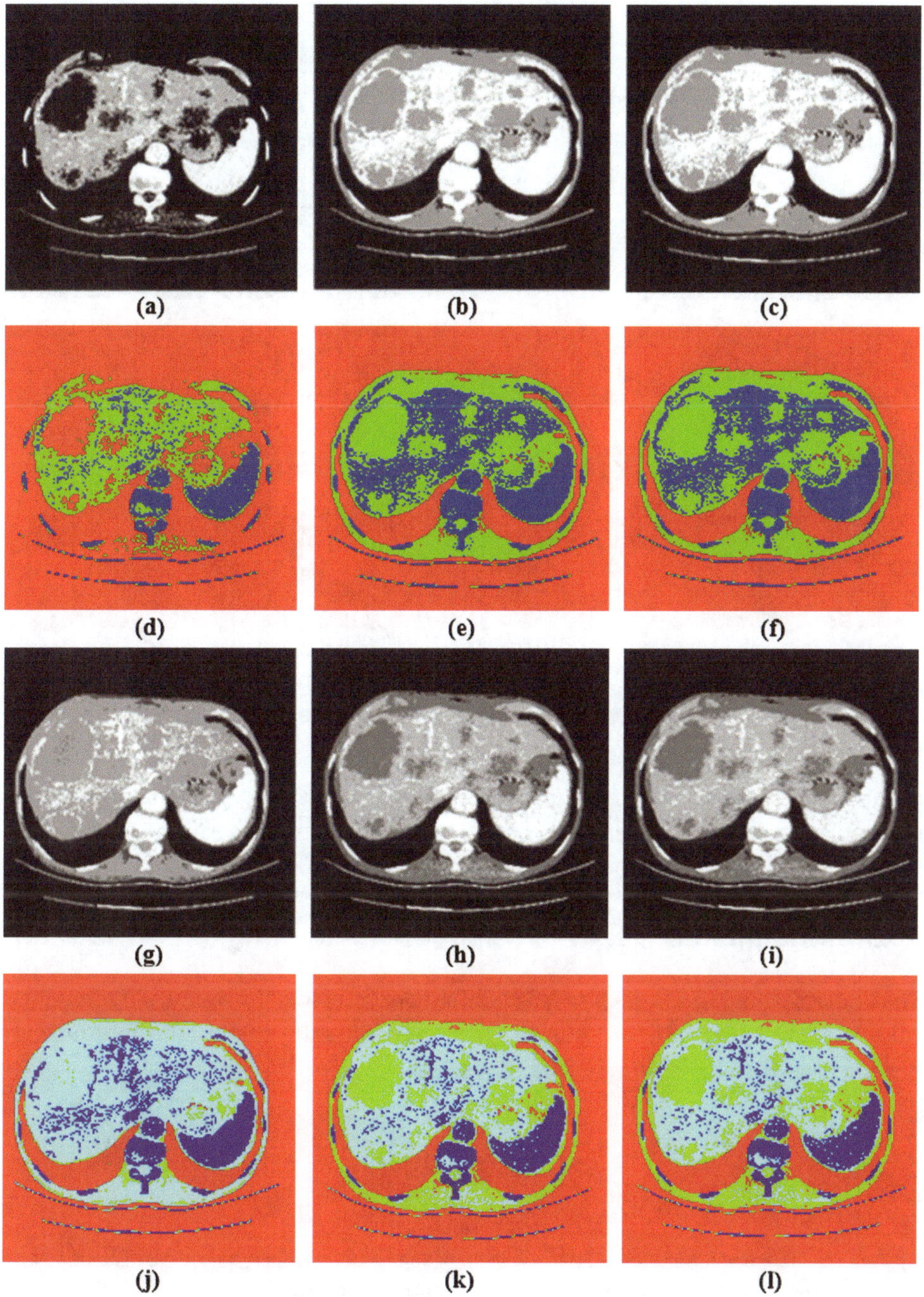

Figure 14.10: ID6 multilevel thresholding crow search optimization segmentation results corresponding to threshold values 3, 4, 5.

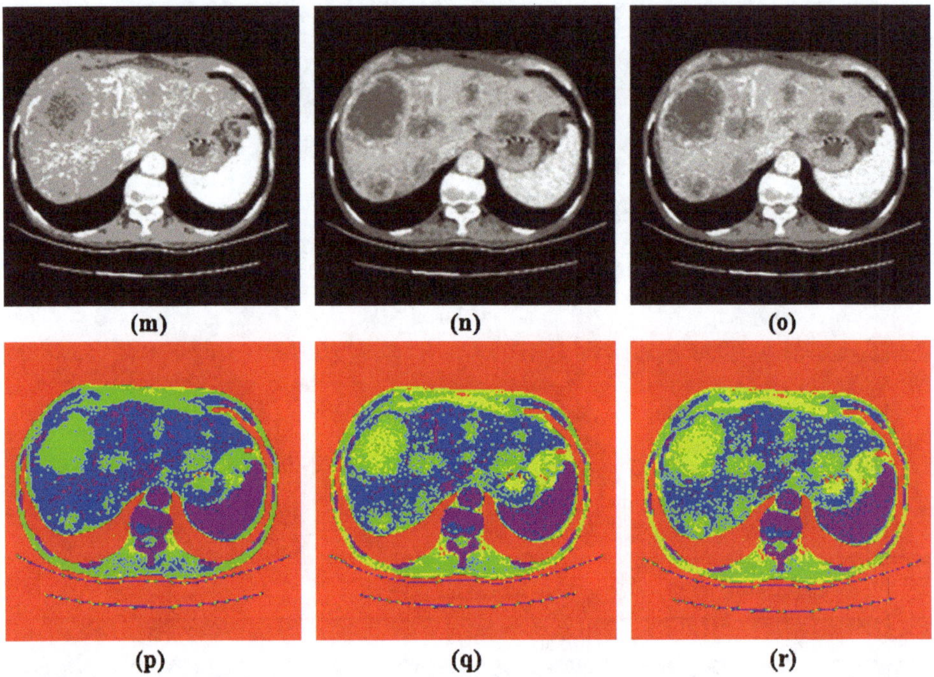

Figure 14.10: (continued)

mization and electromagnetism optimization algorithms. The performance metrics evaluation reveals the efficiency of crow search optimization algorithm.

Acknowledgments

The authors would like to acknowledge the support provided by Nanyang Technological University under NTU Ref: RCA-17/334 for providing the medical images and supporting us in the preparation of the manuscript.

Bibliography

[1] Kumar SN, Fred AL, Varghese PS. An overview of segmentation algorithms for the analysis of anomalies on medical images. J Intell Syst. 2018;29(1):612–25.

[2] Bäck T, Schwefel HP. An overview of evolutionary algorithms for parameter optimization. Evol Comput. 1993;1(1):1–23.

[3] Deb K. Multi-objective optimization using evolutionary algorithms. John Wiley & Sons; 2001.

[4] Glover FW, Kochenberger GA, editors. Handbook of metaheuristics. Science & Business Media; 2006.

[5] Maitra M, Chatterjee A. A hybrid cooperative–comprehensive learning based PSO algorithm for image segmentation using multilevel thresholding. Expert Syst Appl. 2008;34(2):1341–50.
[6] Zahara E, Fan SK, Tsai DM. Optimal multi-thresholding using a hybrid optimization approach. Pattern Recognit Lett. 2005;26(8):1082–95.
[7] Nakib A, Oulhadj H, Siarry P. Image histogram thresholding based on multiobjective optimization. Signal Process. 2007;87(11):2516–34.
[8] Huang DY, Wang CH. Optimal multi-level thresholding using a two-stage Otsu optimization approach. Pattern Recognit Lett. 2009;30(3):275–84.
[9] Horng MH. A multilevel image thresholding using the honey bee mating optimization. Appl Math Comput. 2010;215(9):3302–10.
[10] Akay B. A study on particle swarm optimization and artificial bee colony algorithms for multilevel thresholding. Appl Soft Comput. 2013;13(6):3066–91.
[11] MA EA, Ewees AA, Hassanien AE. Whale optimization algorithm and moth-flame optimization for multilevel thresholding image segmentation. Expert Syst Appl. 2017;83:242–56.
[12] Guo R, Pandit SM. Automatic threshold selection based on histogram modes and a discriminant criterion. Mach Vis Appl. 1998;10(5–6):331–8.
[13] Pal NR, Pal SK. A review on image segmentation techniques. Pattern Recognit. 1993;26(9):1277–94.
[14] Sahoo PK, Soltani S, Wong AKC. A survey of thresholding techniques. Comput Vis Graph Image Process. 1988;41(2):233–60.
[15] Snyder W, Bilbro G, Logenthiran A, Rajala S. Optimal thresholding: a new approach. Pattern Recognit Lett. 1990;11(12):803–9.
[16] Otsu N. A threshold selection method from gray-level histograms. IEEE Trans Syst Man Cybern. 1979;9(1):62–6.
[17] Kapur JN, Sahoo PK, Wong AKC. A new method for gray-level picture thresholding using the entropy of the histogram. Comput Vis Graph Image Process. 1985;29(3):273–85.
[18] Kittler J, Illingworth J. Minimum error thresholding. Pattern Recognit. 1986;19(1):41–7.
[19] Lai C, Tseng D. A hybrid approach using Gaussian smoothing and genetic algorithm for multilevel thresholding. Int J Hybrid Intell Syst. 2004;1:143–52.
[20] Yin P-Y. A fast scheme for optimal thresholding using genetic algorithms. Signal Process. 1999;72(2):85–95.
[21] Akay B. A study on particle swarm optimization and artificial bee colony algorithms for multilevel thresholding. Appl Soft Comput. 2012;13(6):3066–91.
[22] Sathya PD, Kayalvizhi R. Optimal multilevel thresholding using bacterial foraging algorithm. Expert Syst Appl. 2011;38(12):15549–64.
[23] Oliva D, Cuevas E. Electromagnetism-like optimization algorithm: an introduction. In: Advances and applications of optimised algorithms in image processing. Cham: Springer; 2017. p. 23–41.
[24] Oliva D, Cuevas E, Pajares G, Zaldivar D, Perez-Cisneros M. Multilevel thresholding segmentation based on harmony search optimization. J Appl Math. 2013;2013:575414.
[25] Askarzadeh A. A novel metaheuristic method for solving constrained engineering optimization problems: crow search algorithm. Comput Struct. 2016;169:1–12.

Index

accuracy 148, 207
activation functions 199
Adaptive Genetic Algorithm 41
adaptive neuro-fuzzy inference system (ANFIS) 187
ADAptiveLINear Elements (ADALINE) 184
adsorption 184
AHP 64
algebraic operations on TIFN 102
algorithm 184, 186
analogous 183
ant colony optimization 45, 61
Ant Colony System 47
ant programming 53
Ant-Q algorithm 61
applications of genetic algorithm 41
approximations 142
arithmetic operations on trapezoidal fuzzy numbers 97
arithmetic operations on triangular fuzzy numbers 96
artificial intelligence (AI) 184, 198
artificial neural network (ANN) 178, 184, 198
attraction-repulsion 236
attribute reduction 137, 141
AUC 148
average cluster variance 252

balanced and unbalanced TP 94
Basic Probability Assignment 64
basics of fuzzy transportation problem 95
Bayes law 49
Bayesian 49
belief function 63, 64, 66
bilevel thresholding 234
binary coded genetic algorithm 39
binary number encoding 32
binary system 179
bio-inspired optimization 241
black box 188
Boltzmann policy 54
breast cancer 140
building block hypothesis 51

classical logic 179
classical set 182
classification accuracy 142
classification of South Indian dishes 203
combinatorial optimization 45
concept of fuzzy theory 94
confusion matrix 148
convergence 48, 49, 55, 60
convolutional neural network (CNN) 159–162, 164–166, 168, 171, 172, 199
count of improvisations (CI) 248
counter measures 136
credit card fraud 136, 137
crisp 180
criticality 45
cross-flow velocity (CFV) 188
crossover 35
crossover probability 37
crow search optimization 246
cycle crossover 37

DE algorithm 228
deception 53
deceptiveness 45, 50
decision making 64
decision science 64
deep learning 197
defuzzification 65
degree of dependency 138, 147
degree of membership 180
degree of truth 180
Dempster–Shafer theory (DST) of evidence 63
dependency degree 141
dependency function 139
determining criteria weights 77
different schemes of basic operators of genetic algorithm 31
differential evolution algorithm 212
differential evolution (DE) 210, 226
dimensionality reduction 138
discernibility 138
discernibility matrix 139
distance bandwidth (DBW) 248
distance measure 142

efficiency 137
EHR 159–161
electromagnetism 236
electromagnetism like optimization algorithm (EMO) 236

encoding 31
energy 178
entropy 235
evidence theory 66
evolutionary dynamical system 46
expenditure 159–166, 169, 172, 173

feature selection 137, 138, 141
Fisher–Eigen equation 56, 58, 60, 61
fitness 56
FLANN 125
flipping mutation 38
focal element 66, 70
food image classification 197
foulants 178
fouling 178, 191
frame of discernment 66
fraud detection 136
fully fuzzy transportation problem 99
function approximator 188
fuzzy 137
Fuzzy Inference System (FIS) 188
fuzzy information 63
fuzzy information granules 142
fuzzy logic 78
fuzzy number 65
fuzzy rules 180
fuzzy set 65, 180
fuzzy theory 64

genetic algorithm variants 41
gradient descent (GD) 199
greedy quick reduct algorithm 146
group decision making 78

Harmony Search 239
harmony search algorithm 239
height of fuzzy set 182
hexadecimal encoding 32
hill climbing 51
histogram thresholding 232
hydrocarbons 178
hydrophobicity 185
hypertuple 67

I–V characteristics 219, 223
IAE 222, 225
individual absolute error 222
information entropy 142
inputs 183
interchanging mutation 38
intersection operator 193
intuitionistic 137
intuitionistic fuzzy rough set 137
intuitionistic fuzzy transportation problem 100
iTLBO 122

Kapur's method 235
kernel functions 147
kernel logistic regression 147

layers 185
LBWA model 78
Levenberg–Marquardt algorithm 187
linguistic variables 180
local heuristic 48
lower and upper approximations 145
LSTM 159–162, 164–166, 168, 170–173

machine learning algorithms 147
machine learning (ML) 198
MAE 222, 224, 225, 228
MapReduce-based TLBO-FLANN based
prediction model 127
mass 178
mass driven 194
mass function 63
mathematical representation of IFTP 103
Mathew's correlation coefficient 149
Matlab 246
MCDM 64
mean absolute error 222
mean square error (MSE) 125, 250
membership 145
membership function 65, 180
membrane 178
messy genetic algorithm 39
metaheuristic 45
microbial genetic algorithm 40
Min-Max Ant System 55
minimum MSE 127
misclassification 141
MMSE 127
molecular weight cut off (MWCO) 185
momentum 178
Monte Carlo algorithm 46, 55
mTLBO 124
multi-criteria decision making 77

multi-headed 159–162, 165–167, 169–173
multiagent reinforcement learning 53
multilayer perceptron (MLP) 149, 159–162, 164–167, 169, 172, 173
multilevel thresholding 233, 235, 236, 250
multiobjective transportation problem 93
Multiple ADAptiveLINear Elements (MADALINE) 184
multiple point crossover 36
mutation 37
mutation probability 38

nature-inspired optimization algorithm 3
neuro-fuzzy model 179
Newton–Raphson method 222
no-free-lunch theorem 46
nodes 185
noncrisp 180
noncrisp sets 179
nonlinear mapping 188
nonlinear optimization problem 236
nonlinear tensor diffusion filter 233
nonmembership 145

objective function 233, 235
octal encoding 32
off-policy algorithm 55
oily wastewater 178, 191
one-diode model 216, 219
optimally balanced 143
optimization 45, 232
optimization algorithm 45, 236
optimization functions 199
OTLBO 123
Otsu and Kapur 233
Otsu multilevel thresholding algorithm 246
Otsu thresholding 232
Otsu's method 233
Otsu's variance 235

parallel genetic algorithm 40
parameter selection 45
partially matched crossover 37
particle swarm optimization (PSO) 210, 218, 219, 226, 228
pattern mapping 183
permeate flux 188, 191
permutation encoding 32
PES 192
pheromone evaporation 49, 50
pheromone trail 47, 60
pitch adjusting rate (PAR) 248
plausibility function 63, 66
pore plugging 191
positive region 146
prediction performance 135
predictive model 136, 178
preprocessing 137
pressure driven 194
probability distribution 234
probability rule (ACO) 47, 56
probability space 66
project selection 64
proportionate selection 34
PSf 192
PSNR 250
PSO parameters 224
PVDF 192

Q-learning 55

random dynamical systems 45
random forest 149
random search 46
random selection 34
rank based selection 34
ranking technique 97
real coded genetic algorithm 39
Rectified Linear Unit 199
reduct 138
reduct sets 138
reduction algorithms 138
redundancy 137
reinforcement learning 45, 48, 53, 58, 61
reversing mutation 38
Reynold's number 192
ROC 148
rotation forest 149
rough fuzzy transportation problem 104
rough set 103, 137

SARSA 54
segmentation 232, 233
selection 33
self-normalizing 58
semantics 138
sensitivity 148
Sigmoid function 199

significant encoding 33
similarity 141
similarity relation 145
single point crossover 35
single-headed 159–162, 164, 166–168, 172
slope methods 226
SMOTE (Synthetic Minority Oversampling Technique) 143
softmax function 199
solar cell 215, 221
solar cell models 216, 219
solar photovoltaics 215
solar simulator 220, 228
solid transportation problem 100
specificity 148
steady state based genetic algorithm 40
steady state based selection 35
stochastic 178
subset 182
subtractive normalization 56, 57
support of subset 182
susceptible 136

Takagi–Sugeno fuzzy inference model 189
teaching learning based optimization (TLBO) 121
thresholding 232, 233
time horizon 55
time-series 159, 160, 162, 163, 165, 171–173
TLBO-FLANN 125
TLBO-FLANN based prediction model 125
topology 185
tournament selection 34
traditional genetic algorithm 39
training 185, 186
transportation problem 92
traveling salesperson problem 45, 47
tree-like encoding 33
triangular fuzzy number 65
two-diode model 217–219, 224
two-point crossover 35
Type 1 Fuzzy transportation problem 98
Type 2 Fuzzy transportation problem 99

uncertainty 63, 75, 87
uniform crossover 36
union operator 193
universal set 182
unsupervised 139

validation 185, 186
variants of a genetic algorithm 39
variants of fuzzy transportation 98
various approaches to solve variants of fuzzy transportation problems 105
VGG-16 201
viscosity 192
volume concentration factor (VCF) 192

weighted average 57
weighted coefficients 77
working of genetic algorithm 28
wrapper 141
wTLBO 123

De Gruyter Series on the Applications of Mathematics in Engineering and Information Sciences

Volume 2
Sachin Kumar Mangla, Mangey Ram (Eds.)
Supply Chain Sustainability. Modeling and Innovative Research Frameworks
ISBN 978-3-11-062556-1, e-ISBN (PDF) 978-3-11-062859-3,
e-ISBN (EPUB) 978-3-11-062568-4

Volume 1
Mangey Ram, Suraj B. Singh (Eds.)
Soft Computing. Techniques in Engineering Sciences
ISBN 978-3-11-062560-8, e-ISBN (PDF) 978-3-11-062861-6,
e-ISBN (EPUB) 978-3-11-062571-4

www.degruyter.com